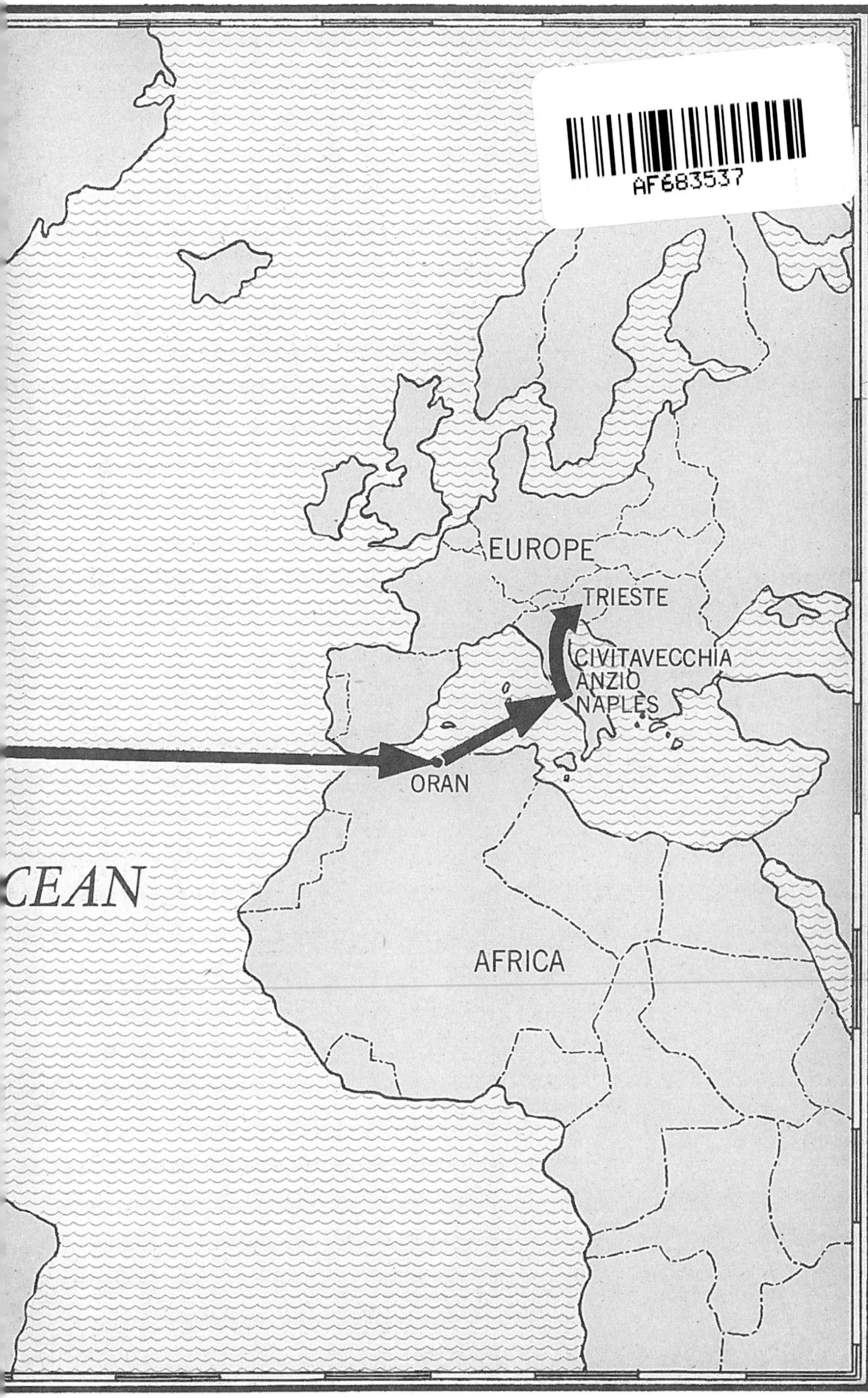
EUROPE
TRIESTE
CIVITAVECCHIA
ANZIO
NAPLES
ORAN
CEAN
AFRICA

AF683537

THE 91st INFANTRY DIVISION

IN WORLD WAR II

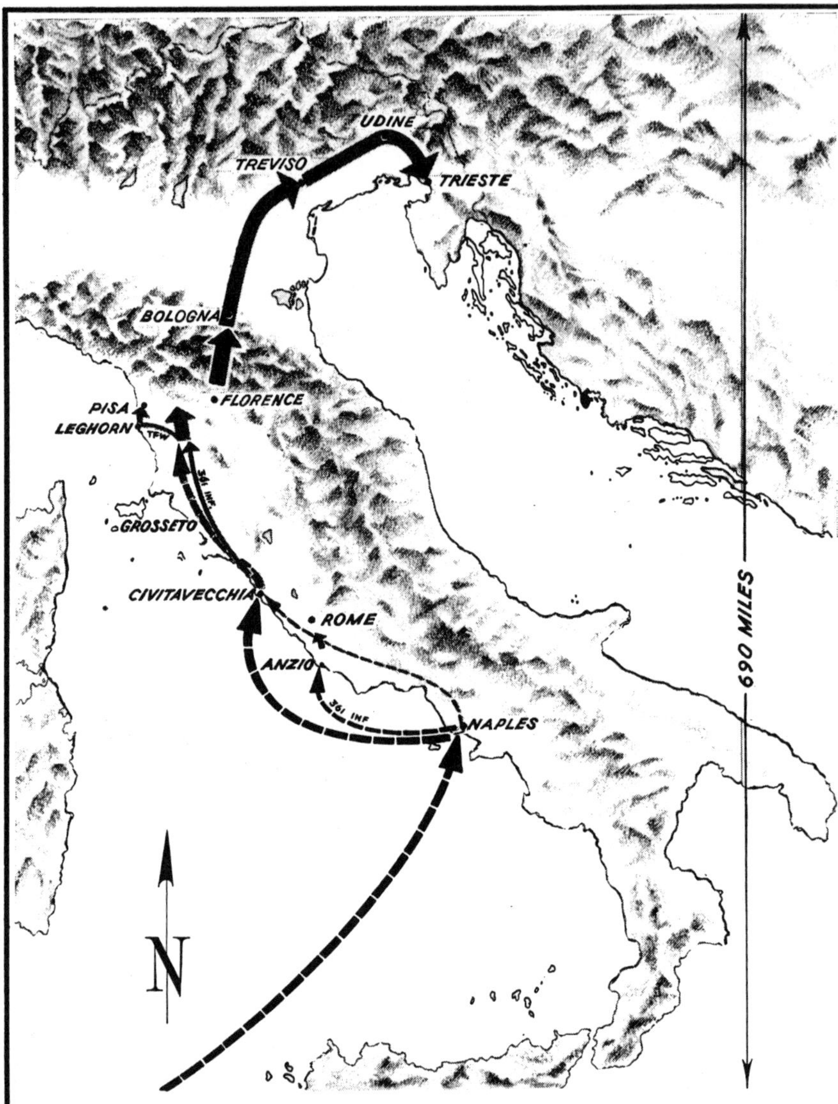

"OPERATION ITALIA"
(Naples to Trieste)

91st INFANTRY DIVISION
WORLD WAR II

The 91st Infantry Division
In World War II

By

MAJOR ROBERT A. ROBBINS

THE BATTERY PRESS
Nashville

Originally published 1947

Reprinted by
The Battery Press, Inc.
P. O. Box 198885
Nashville, Tennessee 37219
Forty-Ninth in The Battery Press Divisional Series
1999
ISBN: 0-89839-297-7
Printed in the United States of America

TO THOSE GALLANT SOLDIERS

OF THE 91ST INFANTRY DIVISION

WHO WERE KILLED IN ACTION

CONTENTS

HEADQUARTERS UNITED STATES FORCES IN AUSTRIA

OFFICE OF THE COMMANDING GENERAL

Vienna, Austria
29 November 1946

To the Officers and Men of the 91st Infantry Division:

It is with deep gratitude that I send you my congratulations and best wishes. You fought a long hard fight through the Italian campaigns, but by your loyalty and devotion to a cause in which you believed, and your unselfish sacrifices, you brought your fight to a glorious and successful conclusion.

This history of the 91st Division is a truly great history—one of which you are a part—one of which you can be justly proud. To many of you it may seem brief and incomplete, but it would take volumes to tell all of the little things which happened to us as individuals on and off the battlefield. However, as you read these pages, they will bring to mind these incidents, and each one of you will read between the lines the part which you individually played in this history.

I know that the names "Gothic Line" and "Livergnano" will strike a resounding note in the memories of all of you Powder River men, as places where the fighting was the bitterest and where we lost so many of our comrades. Their sacred memory will never be forgotten.

Wherever this message reaches you I want you to know that I am proud of you and that I consider it an honor to have had the 91st Infantry Division under my command.

Sincerely,

Mark W. Clark

MARK W. CLARK
General, USA
Commanding

COMMANDING GENERAL'S MESSAGE

This book is intended to record briefly some of the accomplishments of the 91st (Powder River) Division in World War II. As the years pass the details of our experiences will grow dim and these pages will serve to refresh our memories and keep alive the heroic achievements of the Division. This history will also give our friends and families a connected and coherent account of our activities. Our people should know the record of the heroic self-sacrifice and sublime courage that their men displayed along the road we traveled.

The Powder River Division was a harmonious team devoted to a singleness of purpose displayed from the Rogue to the Isonzo. The Division early developed a special pride of service and morale that made every member of it feel that wearing the Fir Tree insignia distinguished him as a member of an organization that carried their mission to completion.

The officers and men of our Division were proud of their record. They were always out in front; they never asked to be relieved; they never said it could not be done; they never complained; they never gave up a foot of ground; they did not criticize others; they were never so tired they could not advance a little farther; they did not boast; they took Monticelli, Futa Pass and Monte Adone.

In a history written so soon after the events it is possible that errors of fact and omission have been committed and that exploits of individuals and units have not been given deserved credit. We must be content with the knowledge that the achievements of the Division were not accomplished by one man or one unit; they were dependent upon all the members of the Division, whose selfless devotion to our cause made possible this glorious history of the 91st (Powder River) Division.

W. G. Livesay

WM. G. LIVESAY,
Major General, U. S. Army

FOREWORD

Shortly after assuming command of the 91st Infantry Division, Major General William G. Livesay directed that arrangements be set up in each unit for the compilation of data looking forward to the eventual publication of a Division history. This was done with the full realization that unless data were assembled, no book would be possible.

After VE-day and during the period when the Division was stationed in the vicinity of Cormons and Gorizia on the Yugoslav border, the time seemed right to determine whether or not the officers and men of the Division desired the publication of a history, and if so, the type of book desired.

Accordingly, a committee of the following officers was formed:

General Livesay
Brigadier General R. E. S. Williamson, Assistant Division Commander
Brigadier General Ralph Hospital, Division Artillery Commander
Colonel Joseph P. Donnovin, Chief of Staff until June 1945
Lieutenant Colonel Ralph N. Woods, Chief of Staff after June 1945
Colonel R. W. Broedlow, Commanding Officer, 361st Infantry
Colonel John W. Cotton, Commanding Officer, 362d Infantry
Colonel W. F. Magill, Commanding Officer, 363d Infantry
Lieutenant Colonel R. F. Scarborough, Division Judge Advocate
Lieutenant Colonel S. J. Taggart, Division Finance Officer

Questionnaires were sent to every unit of the Division to find out if the men wanted a history and if they did, what kind. The response was practically unanimous in favor of a history, and steps were immediately taken to begin the work. Major Richard Gordon, Division Information and Education Officer, initiated the collection of source material. Upon the Division's arrival in the United States, the work was taken over by Major Robert A. Robbins, 361st Infantry, who carried it to completion. The will of the Division, as expressed in response to the questionnaires, was that the history should be in one volume, on good paper, well bound, well illustrated, and not more than five hundred pages long.

The book was to include a short history of the Division in World War I, in addition to the story of its training and combat service in World War II from activation on 15 August 1942 at Camp White,

Oregon, to inactivation on 1 December 1945 at Camp Rucker, Alabama.

It was realized from the beginning that any attempt to record the experiences of a combat division would be subject to most exacting criticism. Because of the nature of the terrain, weather, and enemy action, rarely, if ever, was it possible for one person to observe and record accurately *all* of the details of even a simple operation. Often, within a unit as small as a squad, divergent views would be encountered as to the details of actions occurring within relatively small zones of action.

This is easily understandable when the stress of circumstances and differences in perspective are considered. However, available records, journals, field telephone conversations, radios, oral messages, written messages, and accounts of individuals have been used to the fullest extent. Every effort has been made properly to evaluate this mass of information and to transform it into an accurate, readable record. Unfortunately, the record of operations of some of the units is woefully meager.

Inasmuch as very little of the source material used has been published previously, it has not been necessary to include a bibliography. A list of official documents from which material was taken will be found in the Appendix.

An analysis of plans, tactics, logistics, and technical reports on the participation of the division in the Italian Campaign and their rightness or wrongness does not belong in this narrative. The effort throughout has been to tell a simple story in plain language about the men of the 91st Infantry Division—how they lived and trained and how they fought against what then seemed to be insurmountable obstacles of terrain, weather, and a stubborn and fanatical enemy, and yet came through victorious.

The committee wishes to acknowledge and express deep gratitude to Major Robert A. Robbins, who compiled and wrote this record of some of the accomplishments of the 91st Infantry Division in World War II.

They also wish to acknowledge with gratitude the assistance of Sergeant Richard B. Reynolds, Division Headquarters, who drew the maps and illustrations. Unless otherwise credited, all photographs are by Signal Corps, U.S. Army.

The committee particularly wishes to thank Colonel Joseph I. Greene and his *Infantry Journal* editorial staff for their cooperation and assistance in preparing this book for publication.

91ST DIVISION HISTORY COMMITTEE

PART ONE

THE OLD 91st AND THE NEW

The 91st Infantry Division Monument at Fort Lewis, Washington

CHAPTER 1

WORLD WAR I

THE HISTORY of the 91st Infantry Division begins with World War I, in which it served gallantly and with great distinction.

In 1917, five months after the United States declared war on Germany, the original 91st was activated and organized at Camp Lewis, Washington. Most of the men came from the states of the Northwest, which explains the many distinctively Western traditions and emblems which were part of the heritage of the Division in World War II. It also explains the Division's nickname and battle cry—"Powder River! Let 'er Buck!"—which it adopted after a bunch of Montanans used it in 1918 to answer a sergeant who asked them where they hailed from. The Powder River is a stream in Montana around which much early Western romance is built.

In line with the organizational structure of infantry divisions of the time, the Division consisted of the 181st Infantry Brigade (361st and 362d Infantry Regiments and the 347th Machine-Gun Battalion) and the 182d Brigade (363d and 364th Infantry Regiments and the 348th Machine-Gun Battalion). In addition, it included the 316th Engineer Combat Regiment, the 316th Trains and Military Police, the 316th Field Signal Battalion, and the 166th Field Artillery Brigade (346th, 347th and 348th Field Artillery regiments) the 346th Machine-Gun Battalion, and the 316th Trench Mortar Battery.

Activated under the command of Major General H. A. Greene, the Division immediately, on 5 September 1917, plunged into training for the big task ahead. The men were so eager to commence their training that they began drilling in civilian clothes.

To gather combat experience at first hand so that he could give his men the soundest and most realistic training possible, General Greene spent four months—November 1917 to March 1918—observing the fighting in France. During the few short months between his return and the day of the Division's departure for ports of embarkation, he passed on to his men, through a program of intensified training, the lessons he had learned.

It was just ten months after its activation that the Division made ready to go overseas. On the eve of its departure, General Greene was reassigned and ordered to the Philippine Islands. Brigadier General Frederick S. Foltz, formerly commander of the 182d Infantry Brigade, became Commanding General of the Division. Civilians cheered and feted the soldiers as the troop trains sped across the continent, and

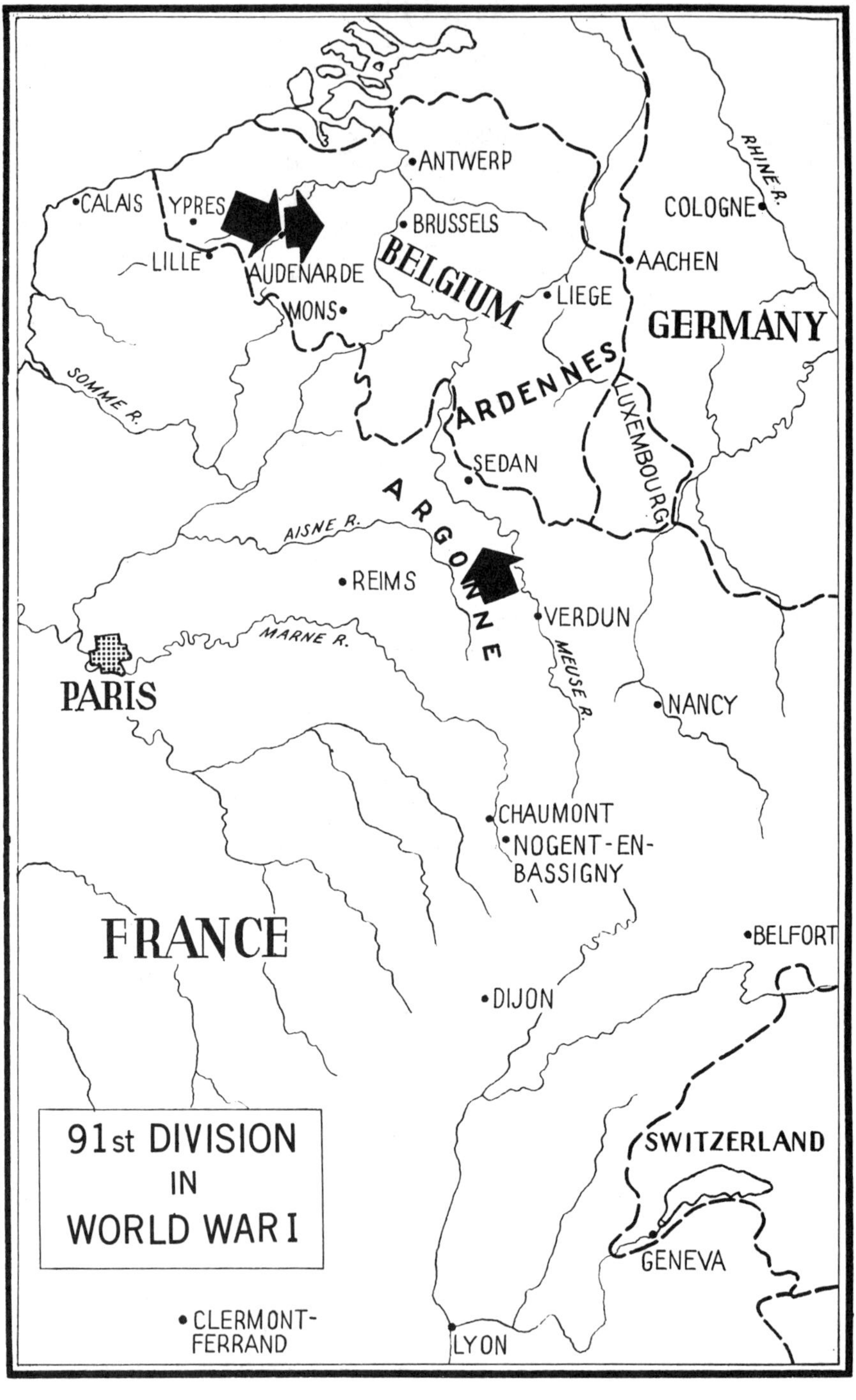
ANTWERP
CALAIS
YPRES
BRUSSELS
LILLE
AUDENARDE
BELGIUM
MONS
LIEGE
AACHEN
COLOGNE
RHINE R.
GERMANY
ARDENNES
LUXEMBOURG
SOMME R.
SEDAN
ARGONNE
AISNE R.
REIMS
VERDUN
MARNE R.
MEUSE R.
PARIS
NANCY
CHAUMONT
NOGENT-EN-
BASSIGNY
FRANCE
BELFORT
DIJON
SWITZERLAND
91st DIVISION
IN
WORLD WAR I
GENEVA
CLERMONT-
FERRAND
LYON

Map 2

after being examined and re-outfitted at the staging area—Camp Merritt, New Jersey—the first elements sailed for France on 6 July 1918.

By August 1918, the infantry brigades were grouped at Montigny-le-Roi, and the artillery brigades at Camp de Souge and Clermont-Ferrand. At these places the men underwent a month of incessant drilling and long hours of marching, until they were declared ready for actual combat. On 29 August Major General William H. Johnston assumed command of the Division, and on 7 September it was assigned to the reserve of the American First Army during the St. Mihiel Offensive. Its headquarters were at Sorcy.

When the success of the St. Mihiel Offensive was assured, the 91st entered the Meuse–Argonne sector prepared to attack. Nearly every other division in the Meuse–Argonne operation had had previous combat experience; the 91st had not. Yet it gave a notable account of itself. On 25 September, General John J. Pershing, Commander in Chief of the American Expeditionary Forces, visited General Johnston to express his confidence in the 91st before it went into battle. The next day the Division proved that General Pershing's confidence had not been misplaced by breaking through two German lines and penetrating a third, advancing eight kilometers. The enemy was driven from the strongpoints of Véry, Épinonville, Gesnes, Éclisfontaine, and Tronsol Farm.

Despite the fact that this offensive was the Division's first experience in combat, it captured more artillery, machine guns, and prisoners and advanced a greater distance under fire than many divisions with previous combat experience.

These achievements brought high tribute from Major General George H. Cameron, Commanding General of V Corps, who said, in an order to General Johnston:

> At a time when the divisions on its flanks were faltering and even falling back, the 91st pushed ahead and steadfastly clung to every yard gained. In its initial performance, your Division has established itself firmly in the list of the Commander in Chief's reliable units. Please extend to your officers and men my appreciation of their splendid behavior and my hearty congratulations on the brilliant record they have made.

On 4 October, the Division was relieved by the 32d Division and assembled near Contrisson.

The 181st Brigade's period of rest and rehabilitation was of short duration, however, for on 6 October it was detached from the Division, and its elements were assigned to the 1st and 32d Divisions. From the

7 to 12 October, the 361st and 362d Infantry Regiments, plus the 347th Machine Gun Battalion, fought for a second time in the Meuse–Argonne Offensive. The 181st Brigade was finally relieved on the night of 11 October by the 64th Brigade of the 32d Division.

One of the great honors given the Division came on 16 October, when, along with the 37th Division, it was made part of the armies in Flanders. These armies, under Albert, King of the Belgians, were about to launch the final crushing drive against the enemy in Belgium. The 91st attacked in the early morning mists of 31 October. From then until the Armistice (1100 hours on 11 November), the Division drove the enemy steadily back. Although the German troops had been ordered to hold the heights between the Lys and the Escaut (Scheldt) rivers to the death, the 91st smashed them the first day and by evening of 1 November was on the outskirts of Audenarde. The next day the town was secured, and the Division pushed on to capture, in rapid succession, Welden, Petegem, and Kasteelwijk.

In recognition of the courage and fighting ability the 91st Division had shown, the following order was issued by Major General DeGoutte, who had resumed command of the French Sixth Army:

> In addressing the divisions of the United States Army who covered themselves with glory in the Château-Thierry Offensive, I said that orders given by the commanding officers were always accomplished irrespective of the difficulties arising thereby or the sacrifices to be made.
>
> I have found the same spirit of duty and discipline freely given in the 37th and 91st Divisions, United States Army, which brings about valiant soldiers and victorious armies.
>
> On the heights between the Lys and the Escaut, the enemy was to hold "to the death." The American troops belonging to these divisions, acting with the French divisions of the Flanders Army Group, smashed them on October 31, 1918, and after hard fighting threw them back upon the Escaut.
>
> Then in an operation of extraordinary daring, the American units crossed the Escaut under the enemy fire and maintained themselves on the opposite bank, notwithstanding enemy counterattacks.
>
> Glory to such troops and to such commanders. They have bravely contributed to the liberation of a part of Belgian territory and to final victory.
>
> The great nation to which they belong can be proud of them.
>
> The Commanding General of the Army:
>
> DEGOUTTE

After the Armistice, elements of the 166th Field Artillery Brigade moved into Germany and occupied the village of Wittlich until February 1919. The infantry of the Division, after parading triumphantly with the men of the 37th Infantry Division before the cheering crowds of Brussels, patroled the Franco-Belgian border west of Poperinghe, from Beveren to Warande, for a short time. On 2 January the first contingent of men sailed for home, and echelons sailed thereafter as transportation became available. Division Headquarters, last of the Division to leave France, sailed on 6 April. Final demobilization of the Division was completed at camps in California, Washington, and Wyoming by 14 May 1919.

During the years between World War I and World War II, the memory of the heroic accomplishments and rich traditions of the 91st was perpetuated by the 91st Infantry Division Association, organized at La Ferté Bernard, France, before the Division embarked for its journey home. Strongest in the major cities along the West Coast, active chapters of the Association have—by holding annual reunions, picnics, patriotic celebrations, and performing civic functions—kept in touch with each other and served conspicuously the communities in which they live.

CHAPTER 2

REACTIVATION AND TRAINING

THE 91st Infantry Division was reactivated in World War II at Camp White, Oregon, on 15 August 1942. On 8 July, five weeks prior to the reactivation date, Major General Charles H. Gerhardt arrived at Camp White to supervise the organization and special training of the officer and enlisted cadre of the new division. By 15 July most of the officers had reported for duty, and by 19 July the enlisted cadre had arrived. The officers assigned to the Division had been selected by the War Department from all parts of the Army; the enlisted cadre had come for the most part from the 1st Cavalry Division, stationed at Fort Bliss, Texas.

On Sunday, 19 July, General Gerhardt called the officers of the Division together for the first time. The occasion served as an orientation period, with an address by the General, who said:

> The 91st Infantry Division is to be a combat division. It is a Division that will see active service. . . . In an emergency the Division may be employed even prior to the completion of the training program. Therefore, to meet this possibility, the training programs must be arranged so as to provide a progressive training in essential subjects in order to secure the maximum amount of combat training before such a movement.
>
> The goal of an effective battle team is to *march, shoot,* and *obey.* If we cannot at present march, shoot and obey, we can at least *march, salute,* and *obey.* Discipline and esprit, both individual and organizational, spring from efficient training. Personnel confident of their stamina and ability to march and shoot are eager to *obey.*

At the elaborate reactivation ceremony at Camp White on 15 August, 529 officers and 1,279 enlisted men listened to the roll call of the dead of World War I and took part in the presentation of colors. In addition to the military personnel, thousands of civilians from the surrounding towns of Medford, Ashland and Grants Pass witnessed the colorful ceremony. Thus, the history of the 91st Infantry Division in World War II began.

Immediately following the reactivation ceremony, all efforts of the Division were concentrated on building up its strength, and then on training the men for the battle trials that lay ahead. Under the broiling-hot sun of the treeless Agate Desert, officers and enlisted men worked side by side in the strenuous physical-conditioning and training schedule.

At the end of the first week in September, the Division commenced its famous "91-mile march." Undertaken to instruct the cadre in marches and bivouacs and to test the physical condition of the officers and enlisted men, the march was conducted with full field equipment over the rough roads and winding trails of the Cascade Mountains, which surround Camp White. The full distance was covered in twenty-eight and three-quarters hours of actual marching time. The 91-mile march is still the subject of spirited conversation when members of the original cadre gather and reminisce.

The training of the original cadre continued at a grueling pace in preparation for the coming of 12,000 recruits, who arrived at Camp White during the months of October and November from all parts of the United States. With few exceptions they had had little or no military training. And so, on 15 November, instruction in the fundamentals began. The training period, based on the War Department's Mobilization Training Program, covered thirty-nine weeks and was divided into three parts. The first period, basic training, in which the soldiers learned how to *march, shoot,* and *obey* lasted until 15 February 1943.

Slogging along through mud and rain at all hours of the day and night, the men quickly developed a physical stamina and "can-do" attitude which would enable them to uphold and extend the finest traditions of the Division in World War I. The men and officers probably would not have minded so much the wretched weather of the 1942-43 fall and winter in southern Oregon if they had known that it was only a sample of what they were to experience two years later in the rugged Italian Apennines. Firing on the ranges was hampered by fog and constant rains, but in the fall of 1944 many of those same infantrymen were firing at the enemy through the even heavier mists and low hanging clouds which blanketed the German Gothic Line on the Italian front.

Following basic training came training in platoon and company formations, which occupied the next thirteen weeks, 15 February to 13 May. Slowly but surely, the pattern of teamwork began to develop within the units. While the riflemen were learning the fine art of small-unit infantry tactics around the Roxey Ann and Dry Creek areas, the artillerymen were busily engaged in selecting gun positions and man-handling their big howitzers in and out of action in the Beagle area. From 20 March to 3 May, the artillery participated in the Army Ground Forces firing tests at the Field Artillery Training Center, Yakima, Washington.

The engineers improved roads, built bridges, developed demolition techniques, laid and cleared minefields; the men of the medical department got plenty of experience in the care of blistered feet and "achin' " backs; wire crews, repair and maintenance teams and message center personnel of the Signal Company worked long and late on communications; on foot and by motor, members of the Reconnaissance Troop covered the network of roads surrounding Camp White; the Quartermaster Company bent its efforts toward smoothing the flow of supplies; crews of the 791st Ordnance Company were fast learning the importance of their job on the Division team; and in Headquarters and Special Troops coördination of effort was constantly being improved upon.

The final 13-week training period was devoted to tactics of battalions and regimental combat-team exercises. Potato Hill and both Upper and Lower Tablerock were the scenes of many engagements; Iron Mountain fell a number of times before the onslaught of the Powder River battle teams. Extensive training in battalion and regimental problems was also carried on in the Grants Pass area. Climaxing this training, the entire Division moved out on the morning of 21 June to participate in the maneuvers (designated D Series) held north and northeast of Camp White.

At the conclusion of D Series on 10 July 1943, the Division returned to Camp White and undertook a period of intensive training to correct the errors and weaknesses revealed during the maneuvers. In addition, units were instructed in the assault of fortified areas and attacks on enemy villages. This phase of the training was directed by Major General William G. Livesay, who assumed command of the Division on 14 July 1943. Under his supervision the Division prepared for the IV Corps maneuvers at the Bend Manuever Area in east-central Oregon.

The Division closed at the IV Corps Maneuver Area on 3 September 1943. Two other infantry divisions, the 104th and the 96th, together with IV Corps troops and attachments, took part in these maneuvers. The maneuver area extended over approximately 10,000 square miles, ranging from the hot, dusty, trackless wastelands of the central Oregon desert to the treacherous lava formations and mountains of Deschutes National Park.

The maneuvers were broken down into eight separate problems, in which the three divisions took part in both offensive and defensive operations over desert and mountain terrain. The last two problems emphasized the attack and defense of river lines.

The extremes of heat and cold, the excessive dust, snow and rain, and the difficult terrain taxed the ingenuity and physical endurance of every officer and man in the Division. Participation in the maneuvers accomplished much toward preparing the 91st for the big test coming.

The troops emerged a physically hardened, confident, and well trained battle team almost ready to take its place in combat.

From the Bend Maneuver Area the Division moved to Camp Adair, Oregon, closing into camp on 4 November 1943. While at Camp Adair it underwent vigorous training to maintain its high level of mental and physical fitness and to correct weaknesses revealed by the maneuvers. In anticipation of an early alert for movement overseas, every effort was made to fill in deficiencies. The alert was not long in coming. The next problem would be a "wet run."

CHAPTER 3

MOVEMENT TO COMBAT

AFTER eighteen months of thorough, realistic and patient training, the day toward which all of the 91st's efforts had been directed became a reality on 20 January 1944.

The dramatic news came in the form of a secret teletype message from III Corps to General Livesay:

ASSUME RESPONSIBILITIES AND EXECUTE FUNCTIONS OF SENIOR TACTICAL COMMANDER . . . READINESS DATE OF 1 MARCH 1944 HAS BEEN ASSIGNED THE 91ST INFANTRY DIVISION.

The statement somehow intimated, despite its terse phraseology, that a tremendous undertaking lay ahead of the Division. We had forty days in which to accomplish the incalculably difficult operation of re-equipping more than fourteen thousand men and preparing them and thousands of tons of equipment for movement to another continent. Within the hour of that first notice, the total energy of an excellently coordinated and throughly groomed division was applied to the painstaking task of carrying out the order.

Events moved rapidly during the next few days, in a swirl of intensive planning of the entire movement. Simultaneously with the first orders from III Corps, Army Service Forces assigned top priority to the Division for those items of equipment still short.

The message from Major General John Millikin, Commanding General, III Corps, directed the Division to "conduct immediate inspections, intensify training, complete firing, and expedite shortage lists at once" and to "submit a personnel status report on or before 28 January 1944, showing the exact status of training."

Instruction in boxing and crating, servicing of vehicles and weapons for overseas shipment, personal affairs, and malaria control was immediately begun. The Division cleared personnel ineligible for overseas service and received replacements; equipment was procured and issued to units; furloughs were granted to men eligible for them; security was maintained; the physical condition of the men was tested—improved when possible—by the Division Surgeon; and immunizations were given.

On 13 February 1944, Colonel W. Fulton Magill, Jr., was assigned command of the 363d Infantry, replacing Lieutenant Colonel Harry A. Austin, who had been placed in temporary command of the regiment following the reassignment of Colonel Paul Murray to Camp Roberts, California, in January.

Colonel Joseph P. Donnovin, Chief of Staff

On 8 March 1944, forty-eight days after the readiness order, General Livesay reported to III Corps that all arrangements short of the last-minute details were complete and that the Division was ready to move. Four days later, the War Department set the final readiness dates: 12 March for the advance detachment; 14 March for the impedimenta; and 20 March for personnel and accompanying equipment.

Thus ended the first phase of the Division's movement overseas, and thus began the second phase—perhaps even more crucial. As a whole the job went smoothly and with little confusion. It was an achievement of which all men of the 91st may be proud. Now, ready and confident, the Division stood poised for action.

Hampton Roads Port of Embarkation

On 11 March, General Livesay called a conference of his staff and unit commanders and outlined the final plans for movement. Orders

Brig.Gen. Ralph Hospital, Division Artillery Commander

were issued the following day, and the advance detachment and the staging-area detail left Camp Adair on 14 March. The movement was under way.

Also on 14 March, the Division received orders from III Corps to move half its personnel and accompanying equipment to arrive at Camp Patrick Henry, Virginia, on 24 and 25 March. This echelon cleared Camp Adair on 21 March. General Livesay and Colonel Joseph P. Donnovin, Chief of Staff, followed by air. Brigadier General Ralph Hospital, Division Artillery commander, was left in charge of the rear echelon. The last of the Division left Camp Adair on 29 March and closed at Camp Patrick Henry on 3 April.

At the staging area the Division was under the command of the station commander, who assigned each unit to a specific area. Processing—including issue of equipment, individual physical examinations, orientation on shipboard procedure, and debarkation training—was conducted under the supervision of the staging area commander. The Division's

Brig.Gen. Raymond E. S. Williamson, Assistant Division Commander

principal problem was to supervise the preparation of passenger lists and make certain that it left the port with full T/O strength. This involved last-minute replacements for hospitalized men and a few AWOL cases who had to be transferred out of the Division before it sailed.

The movement overseas was planned by the port authorities. There were four echelons: (1) the advance party; (2) the first half of the Division; (3) the remainder of the Division, less (4) the 2d Battalion of the 363d infantry, which became the fourth echelon because of lack of shipping space.

At 1100 on 30 March, the advance party—four officers and six enlisted men under command of Brigadier General Raymond E. S. Williamson, Assistant Division Commander—boarded the ship for overseas. Destined for Naples, Italy, the detachment landed at Casablanca on 9 April and then flew to Algiers and Naples. It arrived on 11 April 1944.

Members of 91st MP Platoon aboard Liberty Ship en route to Africa

General Livesay, accompanied by Colonel Donnovin and Captain Frederick F. Lash, Aide-de-Camp, left by air from Washington on 5 April and, after conferences in Algiers, arrived in Naples on 10 April. Due to a change in plans for the 91st, they flew back to Algiers on 16 April and thence to Oran, the Allied base in Algeria, to await the arrival of the Division.

The second echelon left the port of embarkation on 3 April. It also started for Naples, but while it was at sea its destination was changed to Oran. On 14 April the advance detachment, already at Naples, had to return to Africa, where it reestablished the Division's advance headquarters at No. 10, Rue Galliéni, Oran.

The first half of the Division arrived in convoy off the shores of North Africa on 18 April. The troops debarked at Mers-el-Kebir, Port of Oran, and were moved by truck to the Mediterranean Base Section Staging Area near Fleurus. Two days later they moved to bivouac areas in the vicinity of Port-aux-Poules, where headquarters were established. The third echelon began its move across the Atlantic from the United States on 12 April and arrived at Mers-el-Kebir on 30 April. These elements were debarked and moved by truck directly to their bivouac areas.

The 2d Battalion of the 363d Infantry embarked from Hampton Roads on 21 April and arrived in North Africa nineteen days later, on 10 May. This officially closed the Division in North Africa, and General Livesay so notified the Commanding General of the North African Theater of Operations. The 7,500-mile operation had taken exactly fifty-four days.

Training in North Africa

The long voyage across the Atlantic, and the landing in North Africa were strange, exciting experiences for the doughboys of the 91st. Life aboard the slow-moving Liberty ships, with all its inconveniences and its lack of entertainment facilities, had been anything but dull. The Battle of the Atlantic was still in progress, and the merchant seamen added tension to the already tense atmosphere by telling lurid stories of the air and submarine attacks they had experienced. However, the men relaxed at amateur shows, wrestling bouts, and boxing matches, which were staged with huge success on the large, canvas-covered hatchways. Between periods of planned entertainment, the troops busied themselves repairing and cleaning their equipment, playing cards, and watching the progress of ships in the convoy.

Excitement ran high as the troopships neared North Africa and the first sight of land in over two weeks. The huge convoys narrowed down to a column of fours to pass through the Strait of Gibraltar. Much humorous speculation was carried on as to whether the large Prudential Insurance Company sign would be seen on the "Rock." It wasn't there.

From a distance, Oran, sprawling lazily over the rolling North Africa countryside and bathed in the dazzling sunlight which blazed down out of a clear, blue Mediterranean sky, appeared as clean and white as Monday morning's wash. However, closer observation of the city and surrounding villages disclosed, with startling impact, the poverty and accumulated filth in which the native population existed. The pungent odors which emanated from the highly congested towns bore a strong resemblance to those surrounding a hog-pen burrow in the middle of August.

The white-domed Moslem shrines, rising in alabaster brilliance out of the most desolate terrain, were "off limits" to Christian infidels, but they provided an unusual background for many GI snapshots. The veiled Arab women and the colorful French Colonials added to the atmosphere of intrigue and adventure for newcomers to the Dark Continent. Men of the Division laughed at the common sight of old Arabs riding along the dusty roads on their decrepit-looking jackasses while their barefooted wives trudged meekly along behind. It was all like a gigantic three-ring circus.

The predatory habits of the natives, however, did not strike the troops as funny. Articles of clothing and equipment began disappearing as if into thin air. Sentries guarding the bivouac areas had a tough job spotting the elusive bushrangers. One regimental intelligence officer's face was mighty red when he woke up one morning and discovered that his shoes and trousers had been lifted neatly from under his sack during the night.

The serious business of training did not permit extended sightseeing tours outside of the regular training areas, unfortunately, but time passed swiftly as the Division's dry-run days drew to a close. After being assigned to the Seventh Army, under the command of Major General Alexander M. Patch, the Division launched an intensive amphibious training program planned to simulate every possible phase of battle. This training, scheduled to last approximately six weeks, was under the supervision of a special Invasion Training Center with headquarters at Port-aux-Poules. The 65th Infantry Regiment (Puerto Rican), minus one battalion, was attached for training to the Division from 18 April to 27 May. The 361st Regimental Combat Team—composed of the 361st Infantry Regiment, the 916th Field Artillery

In North Africa, the men received training in removal of mines and booby traps

Battalion, Company A of the 316th Engineer Battalion, and Company A of the 316th Medical Battalion—began its training on 3 May.

The program was divided into basic phases: (1) individual and small-unit training, and (2) battalion and—in one case—regimental landings. The men learned organization of boat teams, wire breaching, debarkation, demolition, the use of rockets and flame throwers, and a dozen other aspects of invasion technique. It was at this time, too, that they found out the meaning of the warning shouts, "Bang-a-lore!" and "Fire in the hole!" During the final period, units in battalion strength practiced day and night landings on the beaches of Arzew Bay. After landing, the troops had to push through a heavily fortified zone—complete with barbed-wire entanglements, pillboxes, and tanks—climb mountainous terrain, land artillery from the sea, and fight under the gunfire of naval support. It was the Division's toughest single period of training.

The 361st Regimental Combat Team, commanded by Colonel Rudolph W. Broedlow, completed its course on 15 May and left the Division for detached service with the Fifth Army in Italy.

The 362d RCT—made up of the 362d Infantry Regiment and the 346th Field Artillery Battalion, Company B of the 316th Engineer Battalion, and Company B of the 316th Medical Battalion—began its training on 11 May. Its course was the same as the 361st's except that

Knocked-out American tanks were grim reminders of the job that lay ahead

it executed a practice regimental landing, while the 361st had operated only in battalion strength. The 362d's amphibious training ended on 19 May, shortly before that of the 363d RCT—the 363d Infantry Regiment, plus the 347th Field Artillery Battalion, Company C of the 316th Engineer Battalion, and Company C of the 316th Medical Battalion—began, and by the end of May all of the combat teams had completed the course.

Operation Tarheel—Prelude to Combat

On 1 June the Division began preparing for a mock invasion of the Arzew beaches as the finale to its training and the prelude to its movement to the Italian front. On 3 June it moved from Port-aux-Poules to Mediterranean Base Section Staging Area No. 2 near Fleurus, where final plans for the Arzew landing, or Operation Tarheel, as it was called, were developed. At the same time, the more comprehensive preparations required by the impending movement to Italy were carried on.

D-day for the all-out assault against "enemy" positions on the coast of Arzew Bay was to be 11 June. It was assumed that the area from Oran to Mostaganem was held by elements of "German Infantry Division W," with "Y Grenadier Regiment" at Oran, "X Grenadiers" at St. Cloud, and one battalion of "Z Grenadier Regiment" at Mostaganem. The entire stretch of beach was strongly fortified with pillboxes, barbed-wire entanglements, antitank traps, and strongpoints. There was also a force of "enemy" tanks, prepared to repel an attack, reported in the vicinity. The Division's mission was to seize the port of Arzew and the airport at the town of Renan, and its final objective was the high ground west of Renan and south of Kléber. The plan called for a coordinated attack by the 362d and the 363d RCTs, with engineer, medical, AAA, air, and naval units in direct support.

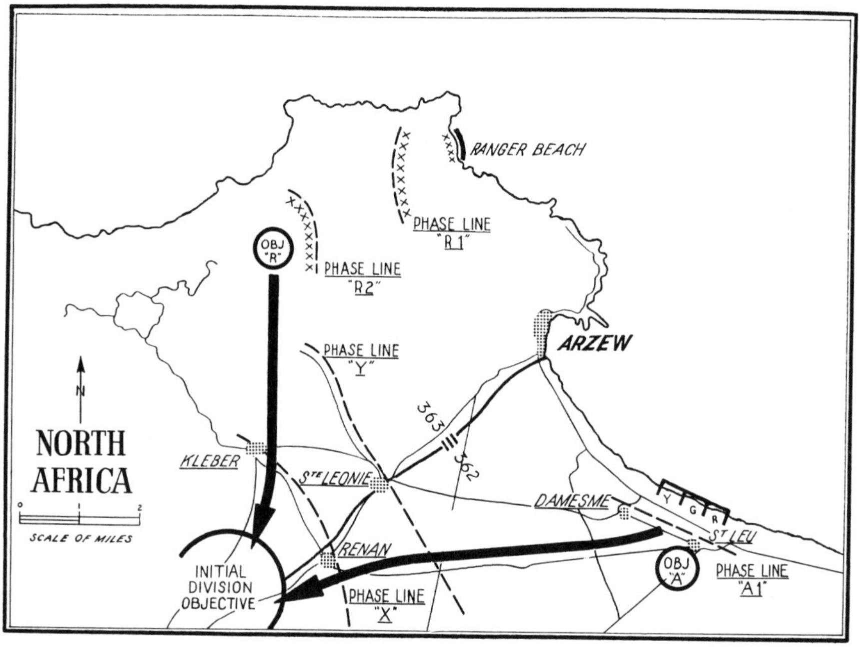

Map 3

Two hundred and fifteen landing ships were massed for the "invasion"—LCVPs, LCMs, DUKWs, LSTs, AKAs, and APAs. The Division command post was located on the USS *Biscayne,* and an alternate CP was established on HMS *Derbyshire.* Troops began embarking on 9 June. The following day was spent in briefing the commanders and men on the specific plans for the operation. Then, late that same night, the ships moved slowly out of the harbor of Oran and steamed ten miles out to sea opposite Cap Carbon and the shore southeast of Arzew.

The troops were loaded into the small landing craft, while a heavy sea rolled against the ships. Plans called for loading the boats from both sides of the parent ships, but because of the rough sea it was necessary to disembark from only one side.

Live ammunition was used throughout the maneuver, starting forty-five minutes before the 0400 H-hour, when Navy destroyers laid down a heavy barrage on the beaches. Under this protective covering the first assault teams moved toward the "hostile" shore. The 363d RCT, delayed in reaching the beach when its landing craft were forced off course by the wind, landed at 0440 on narrow "Ranger" beach, just south of Cap Carbon. It came in in fourteen waves, and when it hit the beach, it struck hard, supported by all available naval fire. Within twelve minutes, the first wave had breached the initial wire entangle-

Training in use of bangalore torpedoes to breach barbed-wire obstacles paid big dividends a few months later

ments at the beach's edge and was moving south to its next phase line. By the end of the first hour, the team had knocked out all "enemy" pillboxes with flame throwers and demolition charges.

Meanwhile, at 0515, the 362d RCT, with all three battalions taking part in the initial attack, moved against the "enemy's" defenses on the shore east of Arzew, cut the road leading to the city, and struck a thousand yards inland to its first objective. By 1000 the entire team was ashore, and the town had been captured. Reverting to the approach-march formation, with the 3d Battalion in the lead, the team advanced on the Division objective. By 1400 it had joined forces with the 363d on its right flank and seized the high ground south of Kléber.

This problem, one of the most extensive ever undertaken by the Division, was observed by high-ranking officials of the Navy, Allied Forces Headquarters, Seventh Army, the French Forces of Liberation, and the Brazilian Expeditionary Forces. The split-second timing had been disrupted by rough weather which interfered with the Navy's debarkation schedule, but the operation, viewed as a whole, was declared a success, and much of the credit went to Colonel Lee C. Bissell, senior instructor of amphibious training in the Mediterranean Theater.

ITALY

The troops of the 91st embarked with grim anticipation on the last leg of their journey toward combat. Already the Allied forces had smashed the vaunted Hitler and Gustav Lines south of Rome, had liberated the Eternal City, and were relentlessly driving the remnants of Marshal Kesselring's German armies northward up the Italian

Peninsula. In France, the huge Allied invasion forces had struck the Normandy coast and were locked in a mighty life-and-death struggle to exploit their initial successes on the beachheads.

The 91st's move to Italy began on 15 June, when it left Staging Area No. 2 and went aboard ships in the harbor of Oran. The following day, minus its rear echelon headquarters and the 361st RCT, already in Italy, it steamed out of Oran for Naples and the smaller port of Bagnoli.

The men of the Division gazed wistfully at the legendary Isle of Capri as the troopships slipped silently by and into the Bay of Naples. Then they had their first view of Mt. Vesuvius, rising dramatically above the hazy background of the drab Italian landscape. It was a thrilling, never-to-be-forgotten sight. In the port itself, the men saw the concrete causeways, built over the hulls of sunken ships, which put the Port of Naples back into operation after its almost complete destruction by demolition and naval and aerial bombardment.

Thus, on 15 June 1944, ended a long journey—and thus began a new chapter in the lives of the Powder River doughboys. The appallingly squalid conditions in the Neapolitan harbor area depressed the men and at the same time reminded them by contrast of the kind of life they were on their way to fight for. The close proximity of war fired imaginations. The thoughts of most were on the future—and its uncertainty.

The Division quickly assembled in staging areas in the vicinity of Bagnoli, and preparations for combat began. On 20 June the 91st was assigned to the Fifth Army. Five days later part of the Division staff and more than two hundred officers and noncommissioned officers were attached to IV Corps to receive front-line experience with the troops fighting southeast of Leghorn (Livorno).

On 27 June General Livesay received a telegram from General Mark W. Clark, Commanding General of the Fifth Army, ordering the Division to move on approximately 30 June to the vicinity of Civitavecchia, north of Rome. One combat team was to be ready for immediate movement north from Civitavecchia on Fifth Army orders. General Livesay alerted the 363d RCT for this assignment.

The over-all plan for the Division's move called for its organic transportation to move by land, the dismounted troops by water. The 363d left Bagnoli on 1 July, followed the next day by part of Headquarters and Headquarters Company, Special Troops, the 316th Medical Battalion, and the 346th and 348th Field Artillery Battalions. The 362d RCT moved out on 4 July.

First a university, then a Luftwaffe headquarters, Bagnoli then became a staging area for the 91st

Powder River doughboys bargain for souvenirs with Bagnoli's street vendors

The Division command post was opened at 0800, 2 July, at Montalto di Castro, four miles east of Civitavecchia. Next day the 363d Infantry, with combat team attachments, was attached to the 34th Infantry Division for a short period to gain combat experience. At the same time the 91st's Reconnaissance Troop was assigned to IV Corps, which in turn assigned it to work with the 804th Tank Destroyer Battalion as right-flank security for the 34th Division.

Events followed in rapid succession as the rest of the Division made final preparations to go into the line. The 361st RCT, which had been fighting with the Fifth Army since 1 June, reverted to Division control on 4 July. On 7 July the 348th Field Artillery Battalion was attached to the 34th Division as general support, reinforcing the fires of the 347th and 151st Field Artillery Battalions. The Reconnaissance Troop rejoined the Division on the night of 10 July. The field artillery units attached to the 34th Division reverted to 91st Division control on 11 July. The 363d Infantry, however, continued in the line until 12 July, when it automatically came back under Division control as the 91st attacked.

General Mark W. Clark addresses men of Combat Team 361 the day before entry into combat

PART TWO

COMBAT

CHAPTER 4

361st REGIMENTAL COMBAT TEAM

EVENTS moved rapidly for the 361st Regimental Combat Team following the receipt of secret orders sending it to Naples, Italy, and thence to an active sector of the Fifth Army front. Personnel of the team were assembled on the evening of 21 May 1944, prior to their departure from North Africa, to hear a final address by General Livesay. The General told the men that the reputation of the entire Division would be based on the record they made in their initial engagements, since the 361st would be the first of the Division to see action. He assured them of his confidence in their success and wished them good luck.

Loading for the trip to Naples commenced on 20 May, with the advance detail, vehicles, and heavy impedimenta sailing on the 22d. The men boarded His Majesty's Transport *Samaria* at 1500, 23 May. The ship weighed anchor in the port of Mers-el-Kebir, Algeria, at 1330, 24 May, and arrived in the harbor of Naples at 0940, 27 May. The troops went directly to the Peninsular Base Section Staging Area near Bagnoli, Italy.

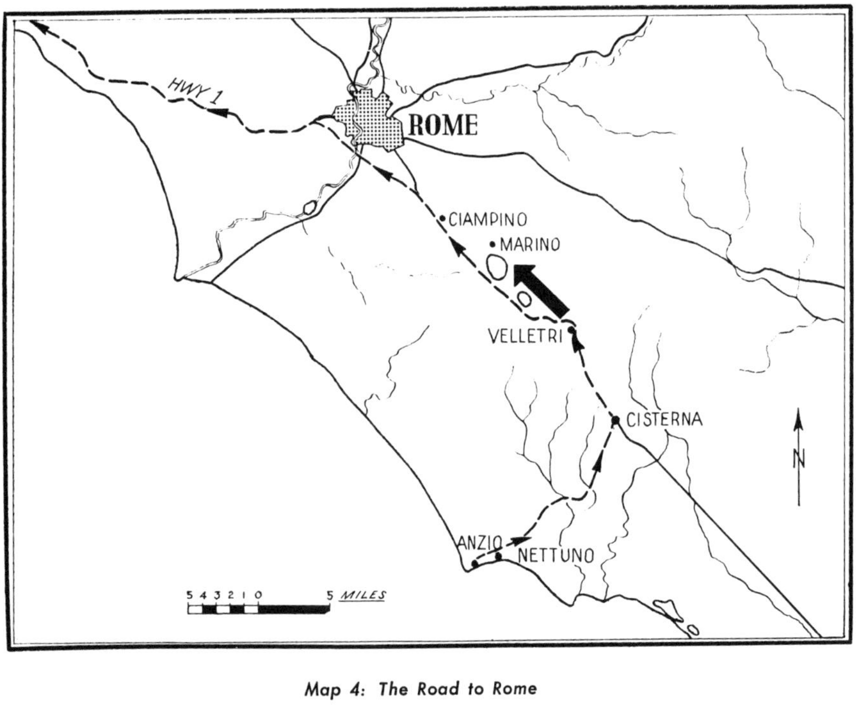

Map 4: The Road to Rome

Col. Rudolph W. Broedlow, CO, 361st Infantry

In the meantime, Colonel Rudolph W. Broedlow, regimental commander, and Major David F. Hawkinson, regimental S-3, had returned to Bagnoli after visiting the front line, and on 30 May, Colonel Broedlow gave his men an informal orientation talk. The next day, the troops marched with full field packs from Bagnoli to the adjoining naval base at Nisidia, where they boarded LSTs for Anzio.

The convoy of LSTs transporting the combat team closed in Anzio harbor at 0930, 1 June. The effect of Anzio was electric. An atmosphere of death and destruction hung heavily over the ruins of the once-beautiful resort city. Ambulances in endless procession drove up to the bomb-scarred wharves and deposited their loads of litter-borne wounded in waiting barges for quick transfer to the hospital ships lying at anchor in the bay. The constant rumbling of heavy-artillery fire told of the mighty struggle being waged in the Alban Hills south of Rome. The nearness of battle was everywhere.

The 361st was rushed from the harbor to its initial bivouac near Tre Concelli, where the regimental command post was established by 1530. At 1030 the next morning, all members of the combat team were assembled to hear an address by General Clark, who welcomed the

Men and equipment of Combat Team 361 unload at Anzio

361st into the Fifth Army, expressing confidence in its ability to carry out any mission which might be assigned to it.

Later an order from Fifth Army attached the 361st to VI Corps and assigned it to the 36th Infantry Division. Its position was on the ridge north of Velletri, and that night the troops rode in trucks to the railroad yards in the center of the town where they dismounted. Columns formed quickly and moved out in complete silence over the road winding north out of Velletri, which only a few hours earlier had been the scene of a fierce battle. It was the first time the men had smelled the sickening odor of death.

The regimental command post was established on a hillside north of Velletri, and by midnight the entire regiment was in position behind the 141st Infantry Regiment of the 36th Division.

At 0200, 3 June, the regiment made contact with the 141st Infantry, and plans were coordinated. The regimental commander returned from the 36th Division command post at 0740 and issued orders for the 2d Battalion, with the Cannon Company and one platoon of the Antitank Company attached, to move out at 0830 along Via del Laghi, following the 141st Infantry, and seize and hold Marino and Mt. Crescenzo. The 1st Battalion, on the east, was to assist the 2d Battalion. The 3d Battalion would be in reserve.

The 2d Battalion moved out promptly at 0830 and received its baptism in artillery fire and its first battle casualties at 0930 on Via del Laghi. At 1100 a field message from 36th Division Headquarters assigned the 361st a new mission: relief of the 141st Infantry. A few moments later, Colonel Broedlow was ordered to use one battalion to clear out pockets of enemy resistance by-passed in the rapid advance of the 141st.

The 2d Battalion was committed, with Company E on the right and

Company F on the left. Company G, in reserve, followed Company F. Company E moved forward on a broad front, advancing to the east flank of the 141st Infantry, where it became engaged in a heavy fire fight with the enemy forces which were holding up the advance of the right of the 141st. After routing the Germans with small-arms and mortar fire and overrunning their positions, it advanced to its objective. Company F engaged an enemy force in machine-gun positions cleverly concealed in the underbrush of a thickly wooded area west of Via del Laghi. Although suffering heavy casualties, the company succeeded in clearing the enemy positions and pushed on north to secure its objective. Company G, committed around the left flank of Company F, assisted the advance with small-arms and mortar fire. The battalion consolidated and held these positions for the night.

At 0420, 4 June, Colonel Broedlow ordered an attack in accordance with instructions received the previous afternoon from the 36th Division. The order was rescinded a half-hour later after receipt of new instructions from the 36th Division.[1]

---

[1]After the war Colonel Broedlow issued the following clarification of the orders he received from the Commanding General, 36th Infantry Division:

"On the early morning of 3 June I issued the necessary orders for the movement of the combat team to a battle assembly area just south of the front-line battalions of the 141st Infantry, 36th Division. Major General Fred Walker, the Commanding General, had advised me that while his troops were very tired and losses were heavy from continuous fighting, he was of the opinion that his present front-line troops could crack the German line. He said he would use the 361st RCT only to relieve the 141st by passing through it . . . this in case the German line failed to crack on 3 June or that night . . . or to exploit the breakthrough in the event the German cracked on 3 June. By noon it was apparent that the Germans were holding and casualties were mounting. General Walker warned me to be ready to move forward on several occasions. He held back, however, stating that he did not care to commit the 361st unless it was absolutely necessary. The turn of battle on the night of 3-4 June proved that General Walker's decision to hold the 361st in division reserve was sound. A complete regimental attack during daylight hours through the 141st Infantry was not warranted.

"During the morning hours I noted many men of the 36th Division being killed or wounded by snipers firing from bypassed machine-gun positions. On several occasions General Walker and I were forced to ditches by heavy fire from small arms and machine guns. In the early afternoon I asked General Walker to permit me to commit one battalion to clean out the German snipers and bypassed German troops which were doing considerable damage to troops and vehicles along the highway between the battalions and the advance CP of the 141st Infantry. As you well know, it turned out to be more than a sniper fight, once the 2nd Battalion was committed. Enemy machine-gun nests and small combat groups in the woods gave the 2nd Battalion a stiff fight. Companies E, F, G and H pushed up to the front lines of the 141st Infantry, and I feel that the 2nd Battalion, with the support of the 916th FA Battalion, not only cleared up a bad situation, but bolstered the morale of the 141st Infantry, which had fought so hard and suffered so much.

"We know what the front-line GIs said: 'Fresh troops! Oh, boy!' We will also remember that the Krauts pitched all their ammunition at us that night as they disengaged and withdrew what they could to the north of Rome and across the Tiber River. I cannot help but believe that the attack of the 2nd Battalion on the afternoon of 3 June had some bearing on the hasty German withdrawal that night. They could not help but know that American reinforcements were at hand and that their withdrawal was mandatory if they expected to save anything in the front of the 36th Division sector."

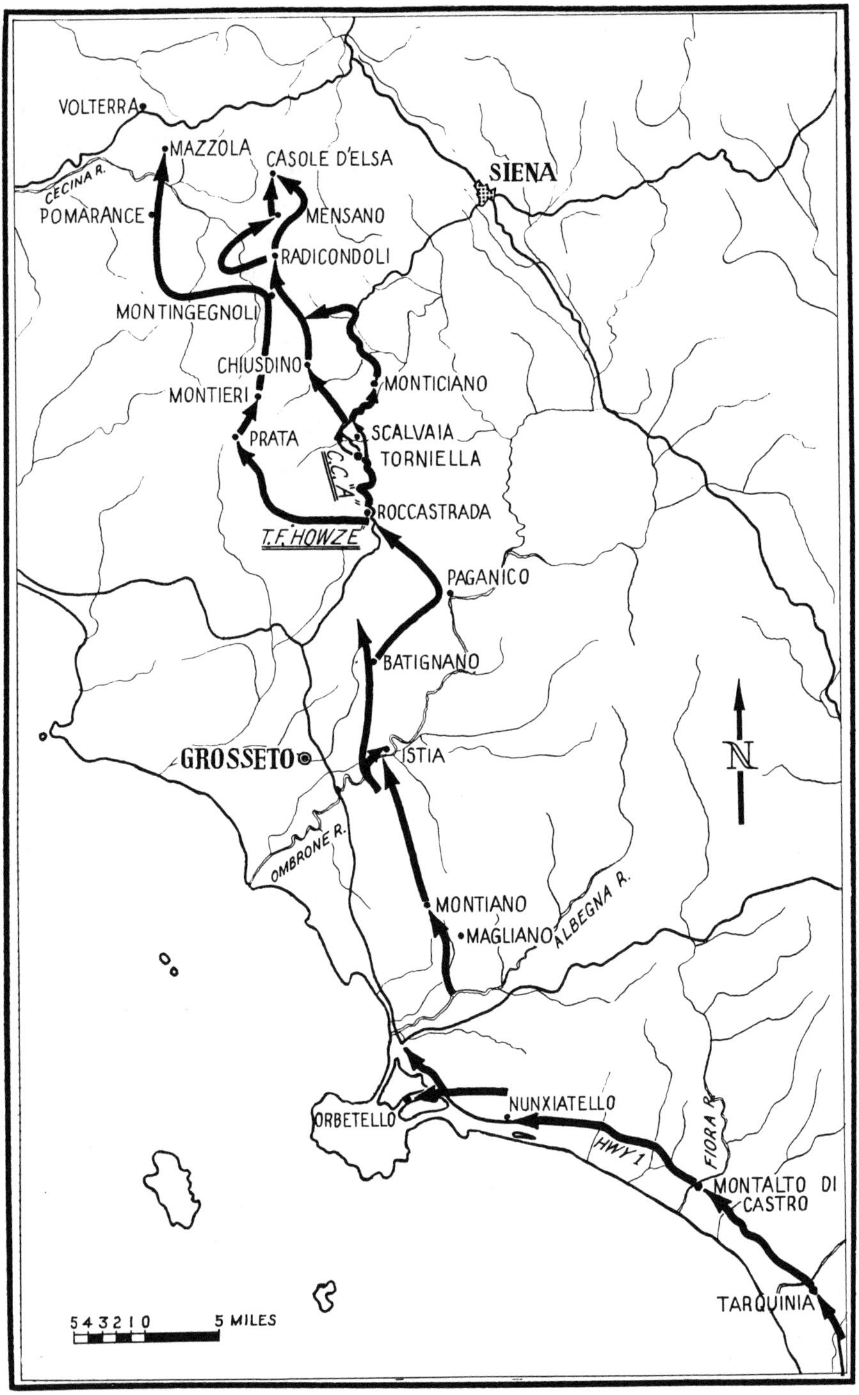

Map 5: Tarquinia to Mazzola

At 0715, the 361st RCT (less Batteries A and B and Headquarters and Service of the 916th Field Artillery Battalion) reverted to VI Corps reserve, but remained in position, prepared for rapid movement. The 916th FA Battalion was to see further action with the 36th Division for several days prior to its return to control of the 361st.

Pursuit North of Rome

After the fighting in the Alban Hills, the 361st was attached to the 34th Infantry Division and moved rapidly along Highway No. 1 north of Rome. During the night of 8-9 June, it relieved the 133d Infantry of the 34th Division and reverted to control of the 36th Division. The team was ordered to attack and capture Tarquinia and then continue north along Highway No. 1 toward Montalto. Two companies of the 753d Tank Battalion were attached to the 361st for the mission.

The attack jumped off at 0530, 9 June, with the 1st Battalion on the left, along Highway No. 1. The 2d Battalion, on the right, approached Tarquinia from the southeast. The 3d Battalion, in reserve, followed the 2d Battalion.

The attack went smoothly, and, after taking the town and eliminating the sniper fire that still persisted, the regiment pushed on north. The advance was delayed slightly by a blown-out bridge across the Marta River, but the foot troops soon got across the river, and the vehicles followed when a crossing was found.

Continuing rapidly against slight small-arms resistance, the forward elements of the regiment pushed into Montalto di Castro at 1520. It was here that Lieutenant Colonel (then Major) Bertram N. Sheff, Regimental S-2, captured an armored scout car single-handed, killing the entire crew. Lieutenant Colonel Sheff was leading the Intelligence and Reconnaissance Platoon.

At 0530, 10 June, the 361st, led by the 3d Battalion, moved out along Highway No. 1. The advance was slowed at 1300 by German 88mm artillery and automatic-weapons fire, but deploying astride the highway, the 3d Battalion slowly pushed ahead to Nunxiatello. The advance was made against extremely accurate fire from artillery, self-propelled guns, and automatic weapons.

During the heavy fighting, a platoon of Company I, advancing across an open field to the right of Highway No. 1 to clear a route for the armor, was pinned down by machine-gun, automatic-pistol, rifle, and intermittent artillery fire. A German Mark IV tank rumbled toward the platoon, firing its machine guns and cannon. In the face of this additional heavy fire, Private First Class Joseph D. Loe, an automatic

"Colonel Sheff captured a scout car single-handed, killing the entire crew."

rifleman, crawled fifty yards out ahead of his platoon and opened fire on the tank. He poured such a volume of fire at it that it had to stop and close its ports. Then the tank crew spotted Private First Class Loe and opened up on him with a machine gun. Bullets pounded into his pack and cut his ammunition belt from his waist, but he refused to give ground. He kept on firing his BAR until another machine-gun burst from the tank struck and mortally wounded him. Still he refused to budge and pumped lead at the tank until he lost consciousness. Thanks to Private First Class Loe, the platoon had time to reorganize and send a bazooka team up to force the tank to withdraw. Private First Class Loe's display of courage and determination will always be an inspiration to those who saw him that day.

In the meantime, the 2d Battalion, moving in close support of the 3d, had established itself on high ground on the flanks. The 361st held these positions until relieved by the 141st Infantry at 2000.

Because of a German counterattack against the 141st, the 361st moved up to positions in close support early on the morning of 11 June. Late that same night, it was ordered to relieve the 141st again and be prepared to attack early the next morning.

All three battalions, less Company K, attacked at 0600, 12 June, across the rugged terrain east of Orbetello. Company K, supported

"Loe kept on firing his BAR at the tank until he lost consciousness . . ."

by tanks, exerted pressure from positions on Highway No. 1 near Nunxiatello.

Once the high ground dominating Orbetello on the east had been secured, the 361st was able to direct well observed artillery fire on enemy positions and troop movements. It inflicted heavy casualties on the Germans, and they left behind quantities of equipment and material and many prisoners. Steady pressure on the right flank enabled Company K, with its supporting armor, to push swiftly along Highway No. 1 through the town of Orbetello to the south bank of the Albegna River.

At 1730, the 361st was relieved by the 143d Infantry of the 36th Division.

After a brief rest, the 361st (less the 3d Battalion, which was held in 36th Division reserve) swung to the right of Highway No. 1 on 14 June. In a series of rapid moves, it by-passed the town of Magliano, seized Montiano, and at 1900 on 15 June, reached the south bank of the Ombrone River. The regiment could not cross the river because of blown bridges, so the 2d Battalion took up positions among the hills on the south bank, overlooking Ponte d' Istia, and the 1st Battalion deployed along the river to the west. These positions were held during the night.

On 16 June the 3d Battalion, released from 36th Division reserve, found a partially destroyed power plant and dam two and a half miles downstream from Ponte d' Istia. It infiltrated in single file across the dam and had established itself on the north bank by 1100, 16 June. Then, under cover of dense underbrush, it moved upstream about a mile and a half and launched a surprise attack against the German garrison holding Ponte d' Istia.

In an effort to beat off the attack, the enemy used artillery and mortar fire, as well as fire from the many machine guns he had carefully placed in the houses and orchards in and around Ponte d' Istia. Companies K and L attacked hills 61 and 66 to the north in a flanking move designed to relieve the strong resistance at Istia.

During the attack on Hill 66, the right assault platoon of Company L ran into enemy machine-gun fire from a nearby building. Private First Class Thomas Gibson, a member of the platoon, ignoring the fact that any movement brought instant, deadly fire from the enemy, rushed across the open ground, fired on the enemy position, and forced one German machine-gun crew to withdraw. Then another machine gun opened up on him, but Private First Class Gibson stood his ground and returned the fire, enabling his squad to advance and enter the building. Private First Class Gibson then killed six Germans and wounded several others as they tried to flee. Refusing to take cover, he braved direct fire from another enemy position and knocked out a machine gun by killing its four-man crew. Later, when the Germans tried to counterattack under cover of mortar and automatic-weapons fire, Private First Class Gibson stopped them cold with his return fire. Next the enemy sent out two decoy prisoners, and when four other Germans jumped up and fired on the squad, Private First Class Gibson killed all six of them. The Germans then concentrated mortar fire on his position and totally blinded him. Private First Class Gibson's heroism saved the lives of his comrades, accounted for at least 16 dead Germans and made it possible for his company to accomplish its mission.

Thus, despite severe casualties, the 3d Battalion continued to move forward. By 2030, Companies K and L had taken Hills 61 and 66, and while they were reorganizing, Company I completed clearing Istia of all resistance.

Company I's battle for Istia, the intensity with which it was fought and the courage displayed by the men of the company in closing with and destroying the enemy, will make it live forever in the annals of the 91st Division.

Company I attacked Istia with two platoons abreast and the support

"Gibson killed six Germans and wounded several others as they tried to flee."

platoon committed to the left in a flanking attack. The town, situated on a low hill, afforded the enemy excellent observation of all approaches, and as the left assault platoon, led by Lieutenant (then Technical Sergeant) Roy H. Dullum, advanced through an olive grove, it was hit by a mortar concentration which caused several casualties. Lieutenant Dullum coolly reorganized his men and led them forward against heavy small-arms fire. As it moved across an open field, the platoon was stopped by three enemy machine guns. Lieutenant Dullum instructed an automatic-rifle team to cover him and then moved forward to where he could fire into the machine-gun nests. He killed the entire crew of one emplacement, wounded two of the Germans in another, and captured the crew of the third. This resulted in the capture of 13 Germans.

Leading his men on into the left edge of the town, Lieutenant Dullum cleared the nearest building with hand and rifle grenades and killed two of the enemy as they were attempting to escape. A gun down the street was pouring fire at his platoon then, so Lieutenant Dullum, alone, crept toward it from building to building, clearing each as he went. He killed three more Germans on his way to the enemy machine gun. When he got close enough to the gun, his rifle fire so

"Lieutenant Dullum's rifle fire so harassed the crew that his platoon was able to drive them back."

harassed the crew that his platoon was able to drive them back. On the other edge of the town, Lieutenant Dullum killed seven more Germans as they retreated.

Lieutenant George G. Wilson led the right assault platoon through heavy enemy fire to the main street of Istia, where his platoon ran into machine-gun fire from German positions at the far end of the street and on the right flank. The buildings between the platoon and the enemy were occupied by snipers. Lieutenant Wilson sent Staff Sergeant Harry W. Douglass with a squad to neutralize the machine gun on the right, while he led another squad down the street toward the other enemy weapon. Wilson's squad was scattered by a concentration of artillery fire, but he went on with only four men, moving from house to house toward the end of the street. Fired at by a sniper from an upper-story window, he stepped into the open and threw a hand grenade through the window.

Later Lieutenant Wilson observed enemy reinforcements moving forward to attack his platoon and killed two of them from a position in an open window. An artillery shell crashed into the building and exploded in the next room, but Lieutenant Wilson continued to fire,

"Lieutenant Wilson continued to fire, killing three more Germans."

killing three more Germans. Then he moved on to kill another sniper who had pinned down one of his squad leaders. His next victims were two Germans who were preparing to fire on his men with a light machine gun.

Finally, nearing his objective, Lieutenant Wilson crawled along a low hedge line toward the machine gun which had been firing down the street. He threw a hand grenade which killed one of the crew, and as he rose to throw another, a sniper's bullet killed him. His intrepidity, skill, and devotion to duty inspired his men and aided materially in clearing the town.

In the meantime, Sergeant Douglass' squad had knocked out the machine gun at the right edge of the town by flanking it and attacking it from the rear. Almost immediately the squad drew fire from a machine pistol in a nearby group of buildings. Sergeant Douglass approached one of the buildings alone, entered it, and with his rifle and hand grenades, forced the enemy to evacuate. As Sergeant Douglass' men advanced toward the newly won building, a machine gun opened fire on them, and at the same time a heavy concentration of 88mm artillery fire began falling in the area. Undaunted, Sergeant Douglass

"Douglass threw a grenade and wiped out the nest."

jumped from roof to roof toward the machine gun. A shell struck the roof of a building as Sergeant Douglass crept across it, but he kept on. As he entered another building, enemy riflemen in the upstairs windows of the house across the street opened fire and threw potato-masher grenades at him. Sergeant Douglass advanced to the next building, fired on the snipers and killed three of them.

Continuing his perilous journey Sergeant Douglass saw a German squad coming up the street, fired on it and completely disorganized it. Finally, unable to go farther across the roofs, he continued his advance along the street and reached the machine gun which had been holding up his squad—the gun Lieutenant Wilson had tried to silence. He took cover behind the hedge Wilson had used, threw a grenade, and wiped out the nest. His courage and skill not only enabled his squad to secure the right flank of Company I but also eliminated the last enemy stronghold in Istia.

Combat infantrymen were not the only heroes of the battle for Istia. Attached to Lieutenant Wilson's platoon as an aid man was Technician Fifth Grade Rosario V. Lerma, who was with the support squad when its leader was shot down in the main street of the town. Several attempts were made by members of the squad to rescue their leader, but

"Lerma calmly walked through the fire until he reached safety."

each attempt drew a hail of fire from the machine gun at the end of the street.

Technician Fifth Grade Lerma worked his way up to the building from which the squad leader had started to cross the street. Disregarding the advice of the others, he stepped boldly into the street in plain sight of the enemy machine gunner. He reached the wounded man and decided that the only route for evacuation was straight down the street the German machine gun was covering. Picking up the squad leader, he calmly walked through the fire until he reached safety.

During the rest of the day Technician Fifth Grade Lerma cared for five other casualties under artillery and small-arms fire. His courage and devotion to duty were an inspiration to his fellow soldiers.

During the bitter engagement at Istia, heavy casualties were inflicted on the German forces, and approximately 80 prisoners were taken. The 2d Battalion, from its position overlooking the town, could see the entire action and rendered close support with extremely accurate artillery and mortar fire.

On 17 June the 1st Battalion, 361st Infantry, captured Poggio di Moscona. The 2d Battalion captured hills 192, 135, and 177 to the east of Moscona and then was relieved by the 143d Infantry, 36th

Division. On 18 June the 361st was again on the move with orders to capture Batignano and a series of hills several kilometers north of it. The attack jumped off at 0600 with the 3d Battalion leading. This attack was launched in one of the worst rain storms encountered by the combat team during the entire campaign. Thunder and lightning blotted out the fire of the enemy and of the 916th Field Artillery Battalion. Thanks to this storm and the inaccuracy of the German artillery, the 3d Battalion was able to move forward rapidly. Within two hours the enemy was forced to abandon three 155mm cannons, many horses, and other valuable equipment. By 1000 the weather had cleared and German tanks several miles in rear of the former artillery positions caused some casualities and slowed down the advance. Although progress was slow, it remained steady, throughout the day, and by night the three battalions consolidated positions several kilometers north of Batignano. Supply was most difficult due to the unusual darkness on the nights of 18 and 19 June. Many individual soldiers became lost, and vehicles could not move on the roads without a guide, which slowed down the supply operation.

On 19 June the 361st was ordered to reassemble just south of Batignano and await further orders. While in the assembly area, the regiment was incorporated in special task force units of the 1st Armored Division. The 1st and 3d battalions were attached to Combat Command A, consisting of task forces A, B, and C. The 2d Battalion remained for a while in 36th Division reserve and later was attached to Task Force Howze. The 916th FA Battalion and Company A, 316th Engineer Battalion were detached from the 361st and attached to the 36th Division until early in July.

During the following two weeks the men of the 361st saw much action. The main axis of advance of the command was northward along the Batignano–Roccostrada–Monticiano highway. The mountainous terrain was ill suited to armor, and the enemy exploited this disadvantage to the utmost by clever use of mines and extensive demolitions and by lashing back fiercely with strong rear-guard actions.

One of the most bitterly contested of these engagements occurred in the vicinity of Torniella on 24 and 25 June. Highway No. 73, the main axis of advance for Combat Command A above Roccostrada, crosses the Farma River one kilometer northwest of Torniella. When the Germans were forced to withdraw across the river, they blew the bridge at this point. The approaches were too steep and the water too deep to permit Combat Command A's heavy armor to ford the stream, so the infantry waded the river and secured the high ground to the east and north to allow the engineers to work in comparative safety while they repaired the bridge.

Since the tanks were unable to advance, the infantry attached to Combat Command A reverted to the control of Colonel Broedlow, who immediately began the difficult task of assembling the widely deployed troops so that the advance could be coordinated. Company A, 1st Battalion, had already pushed a bridgehead across the Farma River and was engaged with the enemy along Highway No. 73 to the north. Company C had crossed the river behind Company A and was swinging toward Scalvaia, the enemy strongpoint on the high ground to the northeast. Company B, south of the river, followed in close support of Company A. Throughout the night of 24-25 June the 3d Battalion assembled in the vicinity of Torniella.

Company C continued its advance toward Scalvaia on the morning of 25 June. At about 0900, the enemy, in hidden positions on the flanks, allowed the forward elements of Company C to pass through and then attacked the weapons platoon in the rear with small-arms and machine-gun fire. As the forward elements of Company C attempted to pull back in an effort to eliminate this threat to their rear, they became engaged in a heavy fire fight, losing a number of men as prisoners and suffering numerous casualties.

When Company C withdrew, seven wounded men were left between the American and enemy forces. They were in positions that afforded very little cover or concealment, but Private First Class Bruce K. Turner, company aid man, crept through intense machine-pistol, machine-gun, rifle, and intermittent artillery fire to administer first aid to them. He made his way from one to another, covering approximately two hundred yards, until he had tended the wounds of all. As he reached the last man, the platoon leader, Turner came face to face with a German, who ordered him to put up his hands. Turner pointed to his medical brassard and continued to dress the officer's wounds at the point of a gun. When a burst of machine-gun fire sent the enemy soldier to cover, Turner moved the wounded officer to a ditch, which afforded some protection.

He remained with his patient the rest of the day, under machine-gun, mortar and rifle fire coming from both sides. After dark he returned to the platoon and led litter squads forward to evacuate the wounded. By his courage, complete fearlessness, and keen sense of responsibility, he saved the lives of all of them.

In the meantime, Company B had moved from its position southwest of Company A to the aid of Company C, but as it advanced eastward,

"Turner moved the wounded officer to a ditch . . ."

it encountered stiff enemy resistance and was unable to continue. Company A, already engaged with the enemy, was unable to support it. At 1525 Company B moved back to the west flank of Company A, and together the two companies advanced toward Scalvaia.

Because of the small-arms, mortar, and machine-gun fire coming from enemy positions on the hill on which Scalvaia is located, Colonel Broedlow, at 1500, ordered a heavy concentration of artillery fire on the town as troops from the 1st Battalion on the west flank and the 3d Battalion on the east flank advanced toward it. Three hours later, Company I, spearheading the 3d Battalion's attack, neared the crest of the hill before the town and by 1830 was engaged in a fire fight with the Germans in Scalvaia. By 1930 the 3d had occupied the town and called for additional artillery fire on the retreating Germans to the north. With this threat removed, the engineers were able to complete a bridge across the Farma River, and next day the advance continued under the control of Combat Command A.

Another bitterly contested enemy strongpoint was Casole d' Elsa, where the fighting raged for four days, 1 to 4 July, before the town was finally captured.

While the 361st's 1st and 3d Battalions were fighting with Combat

Command A, the 2d Battalion, attached to Task Force Howze, was involved in the same type of fighting in its sector. After capturing Prata, Montieri, and Montingegnoli, Task Force Howze continued its attack northwestward, crossed the Cecina River north of Pomerance, and, under the frowning heights of the ancient city of Volterra, attacked and seized Mazzola on 2 July.

Although this thrust denied the use of the important lateral Highway No. 68 to the Germans, Task Force Howze's position proved extremely vulnerable. Any attempt by the armor to advance in support of the extended infantry positions was clearly observed by the enemy, who still held the high ridge to the north, and brought a large volume of extremely accurate artillery fire into the area. The position was held, however, until relief came during the night of 3-4 July.

On 4 July, the 361st RCT (less the 2d and 3d Battalions, which remained near the front in IV Corps reserve until 6 July) moved to a new assembly area at Lago della Accesa, near Massa Marittima, and reverted to 91st Division control. After a full month of almost continuous fighting on the Fifth Army front, the 361st had established itself as an aggressive battle team and had indicated what could be expected of the 91st Division as a whole.

Troops of the 363d moving up to their line of departure in their first day of combat

CHAPTER 5

363D REGIMENTAL COMBAT TEAM

ON 4 JULY 1944, the 363d Regimental Combat Team, commanded by Colonel W. Fulton Magill, Jr., went into action with IV Corps on the Fifth Army front near Riparbella. It was attached to Task Force Ramey under the 34th Infantry Division.

In its first day of battle, the 363d, with the 3d Battalion in the lead, advanced 2,000 yards north until it was slowed by artillery and small-arms fire, but by nightfall the 3d Battalion had secured Hill 553.

Continuing forward the next morning, the 3d met heavy enemy resistance 500 yards north of Hill 553 and held there until 1700, when the regimental commander received orders to move it to Mt. Vitalba (Hill 675) in the zone of the 168th Infantry, 34th Division. Moving laterally across the front, Company I reached the rear slope of Mt. Vitalba at 2115.

In the meantime, Company A, with elements of the 776th Tank Destroyer Battalion, had been given the mission of screening the right flank of Task Force Ramey. It moved forward, on 5 July, to positions approximately three miles north of Casaglia, where it ran into determined resistance. Extremely heavy enemy artillery concentrations sent the tank destroyers to cover in the rear. With the threat of the armor removed, the enemy greatly increased his small-arms and machine-gun fire, causing heavy casualties, and the forward elements of Company A withdrew 1,000 yards, reorganized and prepared to renew their effort.

During the four days that followed, 6 to 9 July, the troops of the 363d RCT experienced the severest action in the campaign in taking Mt. Vaso (Hill 634), a large, rocky, bush-covered hill which was a key terrain feature, dominating the area. The rocky soil prevented the men from digging foxholes deep enough to afford adequate protection against small-arms and artillery fire. The hill's approaches were open and exceedingly rugged, making troop movement a hazardous operation and the supply of troops very difficult. Moreover, the hill gave the enemy excellent observation of the entire regimental zone, and he used it to the limit.

At 0730, 6 July, the 3d Battalion attacked Mt. Vitalba, and after slight opposition from snipers, Company L reached the top. By noon the entire battalion was well established on the hill. Company L pushed forward to a second Hill 553, the intermediate objective between Mt. Vitalba and Mt. Vaso. At 1400, Company L reported it was on Hill 553, overlooking the Chianni–Castellina highway, and was reorganiz-

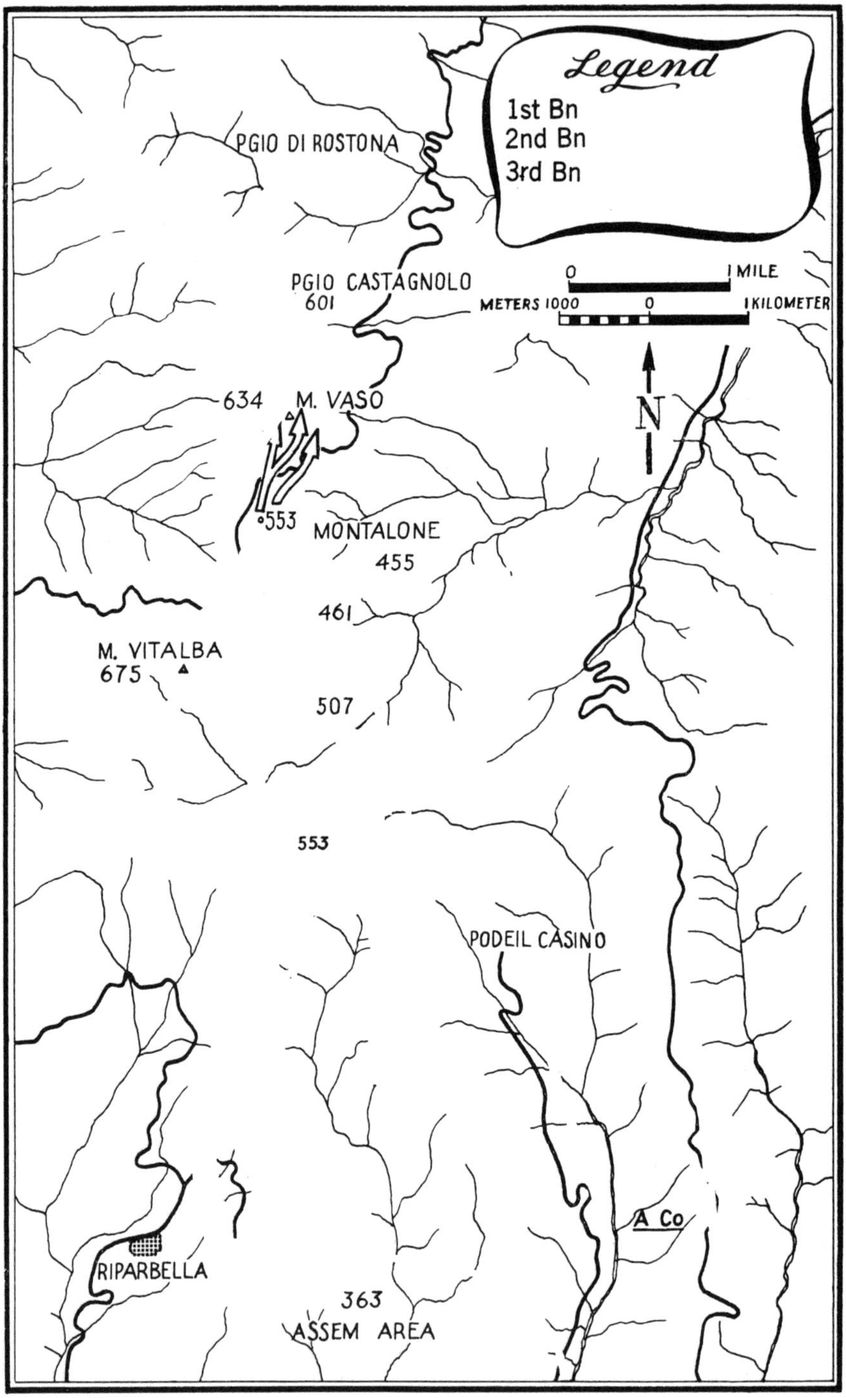
Legend
1st Bn
2nd Bn
3rd Bn
0
1 MILE
METERS 1000
0
1 KILOMETER
N
PGIO DI ROSTONA
PGIO CASTAGNOLO
601
634
M. VASO
553
MONTALONE
455
461
M. VITALBA
675
507
553
PODEIL CASINO
A Co
RIPARBELLA
363
ASSEM AREA

Map 6

Col. W. Fulton Magill, CO, 363d Infantry

ing. Company I, in the meantime, was occupying the left side of the hill, tying in with Company L.

While occupying Hill 553, Company L repelled a counterattack with BAR and light machine-gun fire. At 1700, orders were received to continue on and take Mt. Vaso. As Companies I and L pushed forward, a strong enemy counterattack started from the top of Mt. Vaso. It was repulsed by artillery and 60mm mortar fire on suspected strongpoints, plus grenades and small-arms fire. By 2030, Company L and Company I were both well up on the hill and busily engaged in reorganizing and digging in on its rocky surface.

Throughout the night of 6 July and the following day, a see-saw battle for Mt. Vaso raged furiously. The Germans, determined to wrest this vital position from the 3d Battalion, launched counterattack after counterattack.

When the first counterattack was over, Staff Sergeant Alexander M. Greig of Company L collected eight or nine men to protect the right flank of the company position. Then the enemy charged up the rocky slope again. Several members of the hastily formed squad began to waver, and it looked as if the line might break.

"Greig charged toward the advancing Germans, firing from the hip. . ."

Greig jumped out of his firing position, shouted to his men not to let the enemy bluff them, spun around, and charged toward the advancing Germans, firing from the hip.

His men were so inspired by Greig's courage and disregard for his own safety that four of them leapt up and followed him. The rest of the squad took new positions of vantage and reopened fire. There was no longer any question of the line breaking.

Greig and all but two of the men who had followed him were killed, but around them lay a score of enemy dead. The counterattack was repulsed with heavy losses to the enemy, and the battalion's right flank was secured.

Meanwhile, on the left flank, Staff Sergeant (then Technician Fifth Grade) William A. Montooth, an extra cook and barber in Company L who had recently replaced a rocket-launcher man, had volunteered to act as company observer.

When the Germans counterattacked in the morning, Montooth undertook to protect his company's left flank from his observation post. He stood up, exposing himself to enemy small-arms fire, and fired his M1 until he had killed twelve Germans and delayed the enemy advance. When his rifle ammunition was exhausted, he borrowed a carbine and killed two more Germans. During the second counterattack of the day,

"Montooth took a rifle from a wounded man and killed at least five additional enemy soldiers."

he took a rifle and ammunition from a wounded man and killed at least five additional enemy soldiers. The counterattack was broken up.

Finally, after suffering extremely heavy casualties, its ammunition supply dangerously low and its troops near exhaustion, the 3d Battalion was ordered to withdraw to Hill 553 and Mt. Vitalba. By 2400 it was in its new position.

While Company L was withdrawing from Mt. Vaso, Sergeant Montooth again distinguished himself. With shells falling all around him, he remained in his position, firing every bit of ammunition he could lay hands on in an effort to cover the withdrawal. He picked up the rifles of wounded men and fired them until no ammunition was left. Then he helped to carry wounded over the top of the hill to the reverse slope. Finally, when he was too exhausted to go on himself—he had been fighting for three days and nights with little food and less water—he reported to a battalion officer the location of other wounded so that medical assistance could be gotten to them.

The 2d Battalion, which had been protecting the right flank of the combat team, advanced on 6 July and by dark was established on Hill 553 with its elements echeloned to the right rear of the 3d Battalion.

From an outpost of Company F, Staff Sergeant Robert J. Hutson saw an enemy machine gun fire on a team of litter bearers who were trying to reach several seriously wounded men. Followed by a com-

"Hutson charged forward with his BAR blazing at his hip."

rade, he crawled to the top of a hill about twenty-five yards from the machine gun and discovered that the weapon was protected by at least a squad of German riflemen and two snipers. Undaunted, he flung a hand grenade into the position, sprang to his feet, and charged forward with his BAR blazing at his hip. After cutting down two Germans as they tried to flee, Hutson fought it out with the rest until nine of them lay dead around the gun position. The litter bearers were able to continue their work unmolested.

On 7 July the 1st Battalion occupied Hills 507 and 450 to the right rear of the 2d Battalion, and on 8 July it advanced between the 3d and 2d Battalions to seize and hold Mt. Vaso. Throughout the day and night it pushed across the rough terrain against scattered resistance, and by 1145, 9 July, it had secured the bitterly contested objective.

At 0500, 10 July, the 363d RCT moved north toward the high ground west of the town of Chianni and there reverted to 91st Division control at 0300, 12 July.

In nine days of continuous fighting across extremely difficult terrain, the 363d, in its initial action, had spearheaded the Fifth Army advance. It had cut the vital Chianni–Castellina highway, which had been a focal point of enemy resistance, and it had stormed and captured the key enemy stronghold of Mt. Vaso. Two combat teams of the 91st Division had met the enemy and established their superiority to him.

CHAPTER 6

THE ARNO RIVER CAMPAIGN

FOR a full month following the announcement by the Fifth Army, on 5 June 1944, of the occupation of Rome, the Wehrmacht was driven steadily north—to a point 150 miles above the Eternal City. However, with the fall, on 3 July, of Cecina, a small town at the mouth of the Cecina River on the Ligurian coast, it became apparent that the enemy intended to defend to the utmost the approaches to Leghorn (Livorno). To the north of Leghorn, third largest port in Italy and a prime military objective, flows the Arno River, which was the first natural defense barrier before the enemy's last bulwark south of the Po Valley, the Gothic Line. Small wonder that the German high command massed troops and artillery on the Fifth Army front at this point.

It was evident that a direct frontal assault on Leghorn would be extremely difficult and costly, so a flanking movement was decided on. We would attack north to the Arno, inland from the coastal road, thereby isolating Leghorn and at the same time providing a more favorable direction of attack against the heavily defended port city.

While the mountainous terrain involved in this plan was far from ideal for military operations, the plan offered a greater chance of success than battering head-on at the defenses of Leghorn from the relatively open coastal plains to the south. We hoped with one main effort to accomplish two great objectives of the current campaign: the capture of Leghorn and the control of the Arno River.

The Fifth Army began massing its forces during the first weeks of July. The veteran 34th Infantry Division, with many attached units, was hammering at the outer defenses of Leghorn, while the 88th Infantry Division, on the right flank, was striking for the high ground south of the Arno. The 91st Infantry Division was assigned the IV Corps' central sector, between the 34th and the 88th Divisions.

The 91st received orders from Corps on 9 July to move to forward assembly areas behind the front lines in preparation for a coordinated attack to secure the south bank of the Arno River, northeast of Leghorn. For the first time the full weight of the Powder River battle team would be thrown against the enemy, beginning a brilliant combat record which was to end with the complete and overwhelming defeat of the German forces in Italy.

Twenty-two kilometers of rough, mountainous terrain lay between the advance positions of the Division and its objective, the dominating ridge south of the Arno, which hinged on the city of Ponsacco.

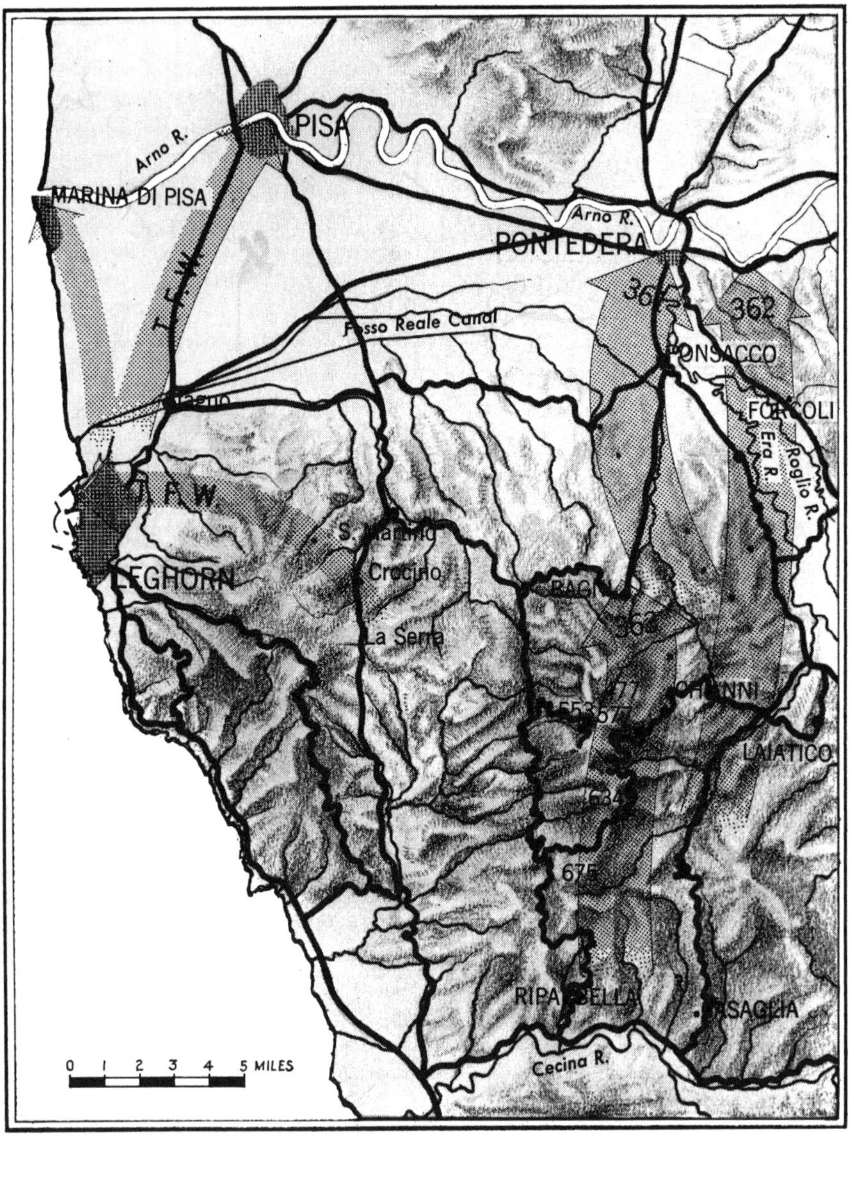

Map 7: The Capture of Leghorn and Pisa

Because the enemy had the advantage of prepared positions in a country that was ideal for defense, heavy opposition was expected. According to our information, the 2d Battalion of the 67th Panzergrenadier Regiment was facing the 363d Infantry. The enemy was also known to be massing a small force of tanks, to be mining all the routes of advance, and to be taking every advantage of the mountainous terrain.

When the 363d RCT reverted to Division control, it was holding a four-mile front south of Chianni and west of the little town of Laiatico on the left flank of the Division sector. The 362d Infantry, which had not been committed to action prior to this time, was deployed on the right front north of the Cateste Hills, with the Sterza River and the Casaglia–Capannoli highway as its main axis of advance. The 91st Reconnaissance Troop was given the mission of screening the right flank of the Division and maintaining contact with the 88th Division on the right. The 361st Infantry was held in Division reserve. Supporting the regiments, in addition to the regular organic Division units, were the 107th and 889th AAA Battalions; the 776th Tank Destroyer Battalion, minus Company A; the 19th Engineers; and the 2d Battalion (minus one company) of the 1st Armored Regiment, 1st Armored Division.

PHASE I

The first objective of the Division was the town of Chianni and the lateral road running to the east of it. Late at night on Tuesday, 11 July, the 362d Infantry, under the command of Colonel John W. Cotton, moved to forward assembly positions. The movement was delayed slightly by four blown bridges and several mined roads along the route. All four of the bridges had been undamaged when the road was reconnoitered earlier. The 316th Engineer Battalion quickly repaired or bypassed these obstacles. Later, mines again delayed the forward elements of the 362d short of the line of departure, forcing foot troops to leave the road and march across country.

At 0300 on the morning of 12 July, after a 45-minute artillery preparation, the 91st launched its first coordinated attack, in conjunction with the 34th and 88th Divisions. By 0320 the 1st Battalion of the 363d Infantry had smashed across the line of departure, but the 362d found its line of departure occupied by the enemy and had to fight to gain it. Antipersonnel mines added to its difficulties. All along the line, stiff enemy resistance impeded the advance; both regiments received heavy small-arms, mortar, and artillery fire. The Ger-

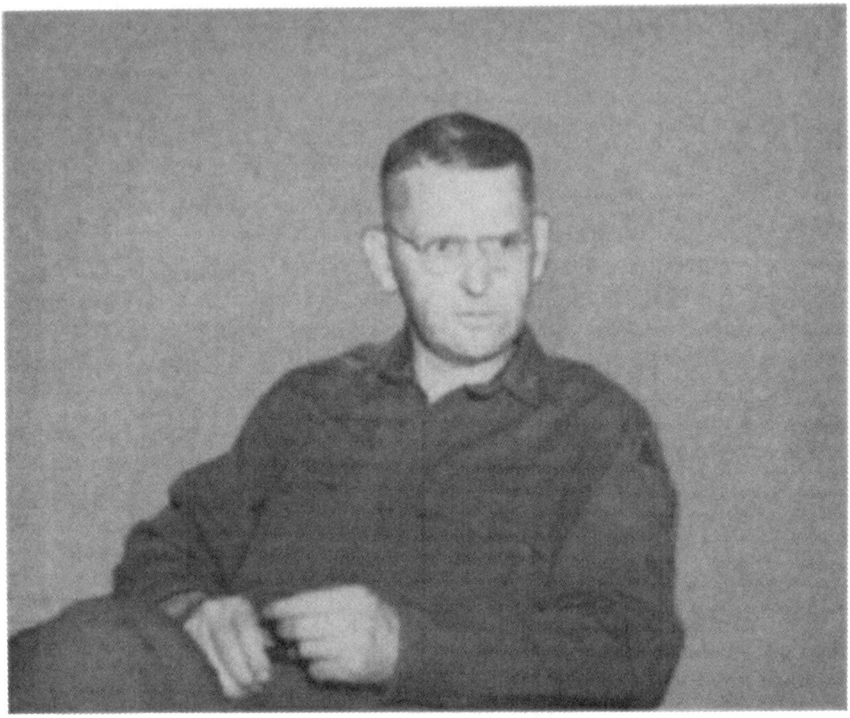

Col. John W. Cotton, CO, 362d Infantry

mans were determined to make the attackers pay dearly for every inch of ground, but progress during the Division's first day of combat was most gratifying.

On the left, the 363d Infantry had reached the vicinity of Hill 577, west of Chianni, shortly after noon. At this point it encountered extremely accurate fire by enemy riflemen on Hill 577 and heavy artillery fire from the vicinity of nearby Hill 401.

In order to overcome this opposition, the 1st Battalion of the 363d seized Hill 533 in a surprise flanking attack and forced the enemy to withdraw. This gave the battalion positions from which to assault and capture Hill 577. By 2330, 12 July, the 363d had captured the important high ground dominating the approaches to Chianni.

Throughout the remainder of the night of 12-13 July, the 363d's dog-tired infantry reorganized and tried to snatch a few minutes of rest. After ten days and nights of the bitterest kind of fighting, they were no longer battle rookies. They knew what it meant to keep moving forward—to attack again and again although their aching muscles almost refused to respond to orders from their tortured brains. They had learned quickly to respect the sickening crackle of small-arms fire, the paralyzing crunch of exploding mortar shells, and the soul-splitting scream of the deadly German artillery. They had known, in a few short days that seemed like eternity, what only men who walk "through the valley of the shadow" will ever know. Sherman knew what he was talking about—war and hell are synonymous.

On the morning of 13 July the 363d's 2d Battalion passed through the 1st Battalion and advanced without opposition. By 1320 it had occupied Hill 477, two kilometers west of Chianni, and was organizing for an attack on Hill 401 to the north. The mission of the 3d Battalion, 363d Infantry, was to seize Chianni, but before the action started, Italian patriots reported that the enemy had withdrawn from the town. At their request, artillery fire was withheld. Investigation proved that the enemy had pulled back to Bagni, and it was decided that the 3d Battalion should bypass Chianni, sending patrols in to occupy the town. The battalion's main force proceeded north along the ridge west of the town of Rivalto.

While the 363d was fighting in the hills southwest of Chianni, the 362d was striking for the important lateral road leading to the town. Shortly before H-hour, 12 July, the regiment moved out of its assembly area in a column of companies. The winding road descending into the Sterza Valley was heavily mined, and the bridges over the tributaries of the Sterza River had been blown. But Company B of the 316th Engineers, working ahead of the infantry, cleared the route of mines and bypassed or repaired the bridges almost as fast as the 362d's vanguard could advance.

At 0415, the 362d's Company C made contact with the first enemy outpost, two miles southeast of Chianni. The company was met by automatic-weapons fire from enemy positions on commanding ground. The forward elements continued to advance, however, until they reached the crest of a hill held by the enemy, where they were stopped cold by a deadly grazing fire that fanned out from several German machine-gun positions. The leading platoon of Company C, caught in an exposed position, faced almost certain annihilation.

In an attempt to rescue the beleaguered platoon by neutralizing the German fire, Sergeant Roy W. Harmon led his squad forward along a draw to the right of the trapped unit. His objectives were three key enemy positions which were pouring murderous fire at his helpless comrades. When Sergeant Harmon's squad got within range, it fired tracer bullets in an attempt to set fire to three haystacks near the enemy positions. Seeing that this was ineffective, Sergeant Harmon ordered his squad to hold its positions.

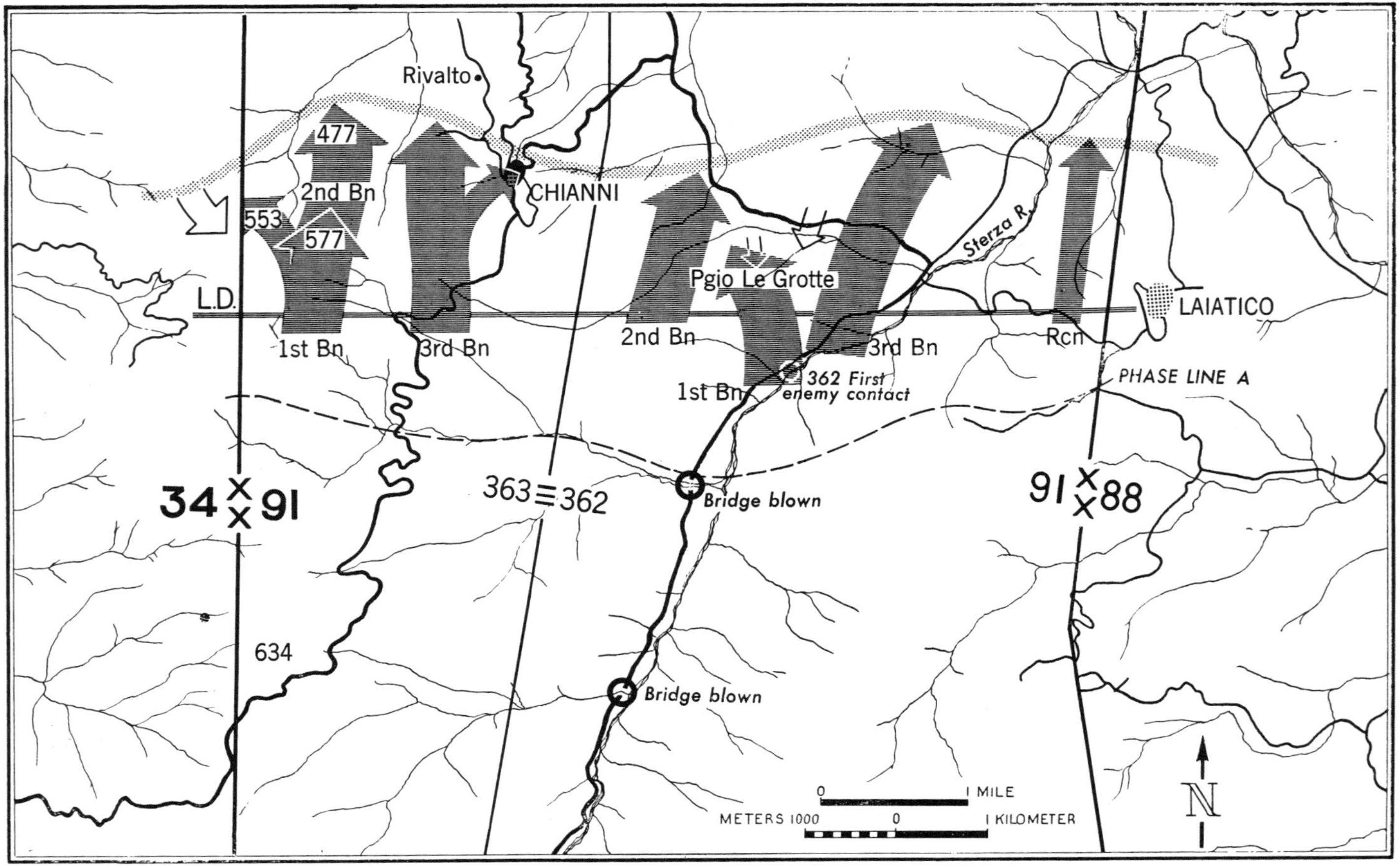
Rivalto
477
2nd Bn
553
577
CHIANNI
L.D.
1st Bn
3rd Bn
2nd Bn
Pgio Le Grotte
Sterza R.
3rd Bn
Rcn
LAIATICO
362 First enemy contact
1st Bn
PHASE LINE A
34 XX 91
363 ≡ 362
Bridge blown
91 XX 88
634
Bridge blown
0 1 MILE
METERS 1000 0 1 KILOMETER
N

Map 8

Two Company B engineers work on one of several by-passes south of Chianni

"I'm going to see what I can do about this," he said quietly and moved out alone.

He skillfully crawled to within twenty yards of the first position, while the enemy machine gunners tried frantically to cut him down. Tossing a phosphorus grenade, he set fire to a haystack directly in front of the position, and when the two machine gunners fled, killed them both.

He started for the next enemy position, nearly the length of a football field away. Heavy machine-gun bullets ripped up the ground all around him, and finally, just short of the embrasure, he was wounded. Refusing to give up, he dragged himself to within twenty yards of the strongpoint and threw another phosphorus grenade, setting the position afire and killing two more Germans.

Then he started for the third position. He dashed thirty yards and then crawled straight into the fire of the enemy guns. He was wounded again, hesitated for a moment, almost fell, then kept going. By this time a trail of blood marked his advance.

When he had reached a point about twenty-five yards from the enemy position, Sergeant Harmon got up on his knees to throw a grenade. Enemy machine-gun and rifle fire from the position knocked him down. He got up again and, with a final heroic effort, threw his last phosphorus grenade. For a brief second the enemy poured fire into him. He crumpled to the ground, dead and literally riddled by bullets. His grenade fired the third position and destroyed it.

"With a final heroic effort, Harmon threw his last phosphorus grenade."

Sergeant Harmon's extraordinary heroism, gallantry, and self-sacrifice saved the threatened platoon and made it possible for his company to advance.

While Sergeant Harmon was carrying on his one-man war, Company B was committed around the left flank of Company C in an attempt to push forward and secure the enemy's main positions. By the time Company B had fought its way up next to Company C, Sergeant Harmon had eliminated most of the enemy resistance holding up the advance, and the two companies pushed forward toward the Chianni road, a half mile to the north.

In their delaying tactics, the Germans were employing scattered pockets of resistance rather than continuous lines of entrenchments. At 0845, the 362d's advance was again checked by stiffening enemy resistance. Company B, battling on the left flank of the 1st Battalion, suffered a grievous loss when its commander, Captain Hugh S. French, was killed by an enemy sniper. Company A was immediately committed on the right flank of Company C in an attempt to catch the Germans off balance. The maneuver was unsuccessful, however, and the advance was temporarily halted.

At 1020, the 362d's 2d Battalion attacked to seize the high ground on the left of the 1st Battalion. The thrust was met by twelve enemy tanks, but when Division Artillery placed concentrated fire on them, destroying one, the remaining eleven withdrew. In the meantime, the 1st Battalion held tenaciously to its positions until 1420 when the enemy unleashed a determined counterattack. Concentrated fire of the entire Division Artillery stopped the attack with very little loss of ground.

Despite fanatical enemy resistance, the 362d renewed the attack at 1640. The 2d Battalion pushed ahead eight hundred yards, but the 1st Battalion was held in its positions on the low ground just north of Poggio Le Grotte. The 2d Battalion attacked again at 2010, advancing to within a half-mile east of Chianni, where it encountered direct cannon fire by enemy tanks and self-propelled guns. Company I was moved to the northeast to relieve pressure on the 2d Battalion's left flank.

While the 2d Battalion was scoring steady gains in the left half of the 362d Infantry's sector, the Germans struck back in the most vicious counterattack yet encountered along the Division front. The full weight of the enemy counterthrust was directed at the 1st Battalion. To meet it, Colonel Cotton committed the 3d Battalion through the right-flank positions of the 1st Battalion, its mission being to destroy

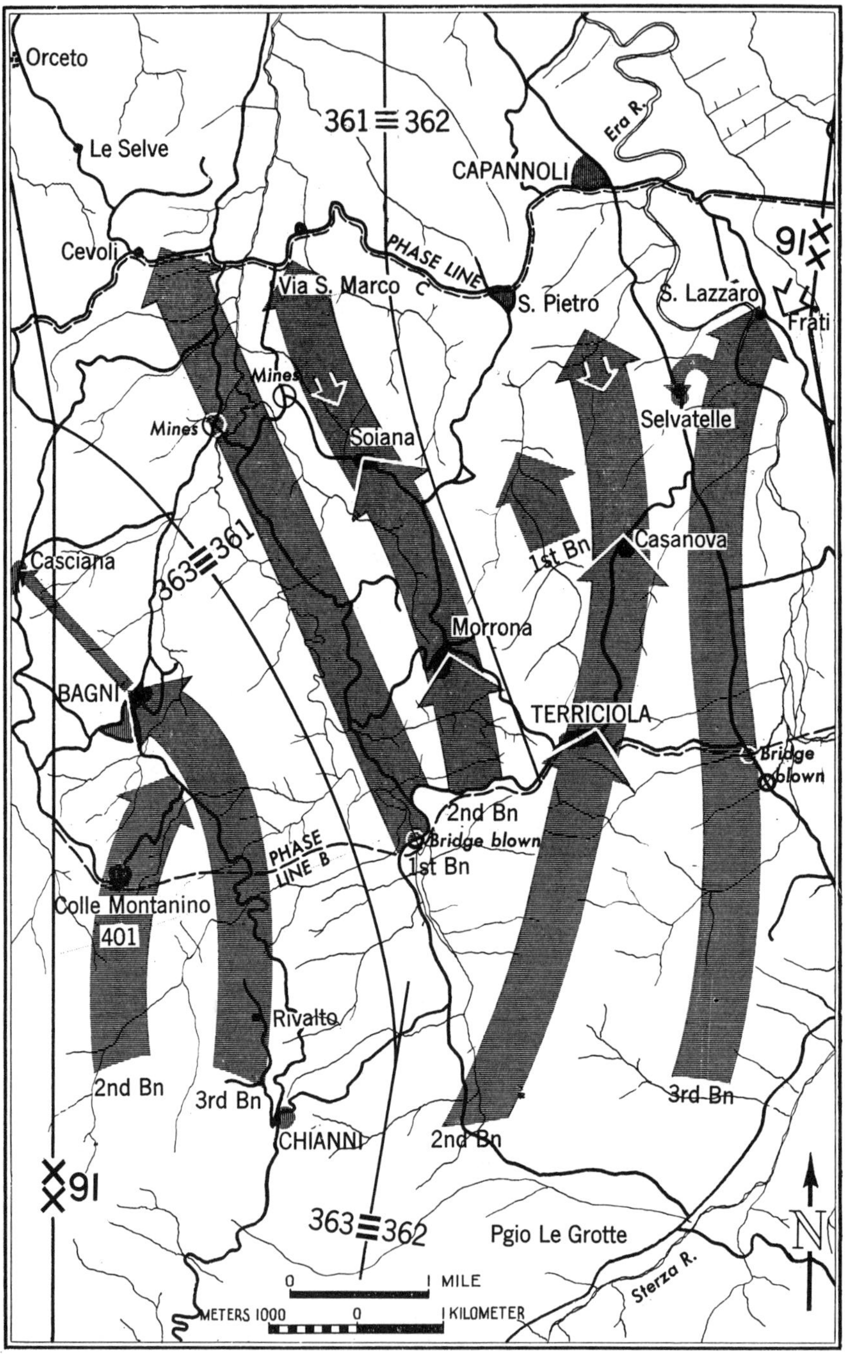

Orceto
Le Selve
361 ≡ 362
CAPANNOLI
Era R.
91
Cevoli
PHASE LINE C
Via S. Marco
S. Pietro
S. Lazzaro
Frati
Mines
Mines
Soiana
Selvatelle
1st Bn
Casanova
Casciana
363≡361
Morrona
BAGNI
TERRICIOLA
Bridge blown
2nd Bn
PHASE LINE B
Bridge blown
1st Bn
Colle Montanino
401
Rivalto
2nd Bn
3rd Bn
3rd Bn
CHIANNI
2nd Bn
91
363≡362
Pgio Le Grotte
Sterza R.
N
0
1 MILE
METERS 1000
0
1 KILOMETER

Map 9

the enemy in its zone and seize control of the German-held Chianni road. The maneuver succeeded; by midnight the 362d had the road.

July 12 1944 was a momentous day in the history of the Powder River Division. To the men of the 362d Infantry it meant what 3 June and 3 July 1944 meant to the men of the 361st and 363d Regimental Combat Teams: their first day of action against the enemy. Little has been said about the shock of the first day of combat, because words cannot describe it. They say that anticipation is greater than realization, but you would have a hard time convincing a combat veteran that his most terrifying expectations came anywhere near the horror of that first day in action. The sight of violent death struck men hard with the realization of the brutality of warfare.

Phase II

During the night of 12-13 July, the enemy withdrew northward, and when the 91st attacked again at 0400, it met only scattered resistance. Thus, with Chianni and the lateral road running east from it firmly in the Division's possession, the first phase of the current battle was completed, and the second phase begun.

The second phase was characterized by fluid fighting, centered principally about the towns of Bagni, Soiana, and Terriciola. Opposing the Division across its front, from left to right, were elements of the German 1059th Regiment, the 1027th Infantry Regiment, and the 67th and 9th Panzergrenadier regiments, supported by the 93d Artillery Regiment. Although two counterattacks of considerable force were launched against the Division during this period, the advance was never seriously threatened, indicating the increasing familiarity of the troops with battle and the terrain.

After the 362d Infantry crossed the lateral road running east of Chianni on the morning of 13 July, it advanced steadily northward. The 3d Battalion, on the right, met only slight delaying action, but the 2d, on the left, ran into heavy artillery fire from Terriciola to the north and fire from self-propelled guns in the vicinity of Chianni to the southwest. This fire was so heavy that the battalion was pinned down for several hours. Division Artillery, directed by one of its air OPs, destroyed a self-propelled gun near Chianni, and Corps Artillery shelled Terriciola. Under this protective fire, the 2d Battalion advanced slowly and by 1755 was abreast of the 3d Battalion, six hundred yards north of the Chianni road.

The 2d was held up again when its command and observation posts were brought under extremely accurate self-propelled-gun fire from a point southeast of Chianni, but the Cannon Company of the 362d Infantry soon destroyed one of the dug-in 88mm guns and forced the remaining two to withdraw.

To the west of Chianni, the 363d Infantry advanced slowly northward. While the 3d Battalion patrolled Chianni, the 1st Battalion dug in to protect the left flank of the regiment against possible counterattack. The 2d Battalion occupied Hill 477.

On 14 July the 363d continued to advance against scattered enemy resistance through mountainous country. By 2000 the 3d Battalion had seized Bagni and was reorganizing there, while the 2d Battalion had reached a point north of Colle Montanino, just south of Bagni. The 1st Battalion remained west of Rivalto, protecting the left flank of the regiment. Patrols were sent out both to the left and right. One of the latter was so zealous that it reached Terriciola, where it assisted in the capture of the town by the 362d Infantry.

With the harassing artillery and self-propelled-gun fire neutralized, the 362d moved forward at 0400, 14 July to seize the high ground between Terriciola and Casanova. The regiment advanced with the 3d Battalion on the right and the 2d Battalion on the left. The rugged terrain limited tank support to the roads, with the result that the tanks had to be used as indirect artillery support.

At 1800, 14 July, Division Field Order No. 3 was issued, in which the 361st Infantry, which had been in reserve, was ordered to move northwest from Poggio Ginistra, past the 363d Infantry at Bagni, to the high ground overlooking Ponsacco. It was deployed with the 1st Battalion on the left, the 2d Battalion on the right, and the 3d Battalion in regimental reserve.

At 0400, 15 July, the 361st began its attack. Meeting little enemy resistance, the battalions pushed rapidly forward through Morrona. Although sporadic artillery fire was encountered during the morning, the first enemy action of importance was met by the 2d Battalion in the vicinity of Soiana.

The 3d Platoon of Company E had the mission of seizing Soiana, which was known to be occupied by the enemy. The plan was to envelop the town by sending one squad to the left to block the road and deny the enemy that escape route. Simultaneously, the right squad would enter Soiana from the west.

Shortly after the attack was launched, the support squad, led by Technical Sergeant Clyde Lloyd, second in command of the 3d Platoon, struck the center of the town from the south. As Sergeant Lloyd,

"Sergeant Lloyd shot and killed the driver and seriously wounded one of the riders."

in the lead, entered the town square, he came face to face with a German soldier approaching from the west. Shooting from the hip, Sergeant Lloyd dropped the enemy soldier in his tracks before he could reach for his machine pistol.

At the same instant a motorcycle came bearing down on Lloyd from the east. The sergeant wheeled and shot the driver, killing him instantly. As motorcycle and driver came skidding to his feet, Sergeant Lloyd heard another motorcycle approaching from the west. With their escape route cut off, the three men in the cycle and sidecar were evidently attempting to reach the center of town to warn the main enemy force of the attack. Sergeant Lloyd shot and killed the driver of the motorcycle and seriously wounded one of the riders. Before he could reload his rifle, the motorcycle careened off the road and hurled the other occupant over the wall into the ravine beyond.

Then Sergeant Lloyd noticed a number of enemy soldiers attempting to escape across country to the north from a large building directly in front of him. Rushing forward, he killed one enemy soldier as the man ran out of a door and dropped another who was attempting to escape by jumping from a window. He captured the one German remaining in the building. Later, Sergeant Lloyd remarked, "I was so scared, all I could do was aim my gun and shoot Krauts. They were everywhere."

In the meantime, the right squad, pushing rapidly forward from the west end of Soiana, had occupied the high ridge running along the north edge of the town. The enemy, fleeing in wild disorder before Sergeant Lloyd's blitz in the center of the town, were fired on by this squad as they attempted to reach the cover of a draw and the village of San Pietro to the northeast.

Thus, in an action lasting only a few minutes, the town of Soiana, held by a force of approximately 35 enemy soldiers, was captured. So completely were the Germans taken by surprise that they abandoned many articles of individual equipment, weapons, plates of half-eaten food, and a large radio, still turned on.

With Soiana secured, Company E reorganized and prepared to press forward in pursuit of the enemy. The advance elements had cleared the town by little more than a few hundred yards when all hell broke loose. The 2d Battalion was halted by an extremely heavy pounding of enemy artillery fire, which, combined with the lack of cover or concealment, forced it to move to positions on the reverse slope south of the road and in the town of Soiana. The fire extended across the 361st's entire front and reached back to the rear installations as well.

The situation of the regiment was further complicated by the development of a gap between the 1st and 2d Battalions. When the enemy counterattacked from the northeast through this gap, Company A was ordered to close it and repel the counterattack, which it did. At this point, General Livesay ordered the regiment to hold, outpost its position, and resume the attack the next morning.

When the attack started again at 0700, 16 July, practically no resistance, outside of harassing artillery fire, was encountered, and the divisional Phase Line C, near Via San Marco, was reached by the regiment at 0920.

In its sector on the right flank of the Division, the 362d Infantry had also met and repulsed a severe counterattack on 15 October. Just south of Casanova the 362d's 1st and 2d battalions were subjected to moderate to heavy artillery, mortar, and machine-gun fire from Selvatelle. Mortars were used to neutralize the hostile automatic-weapons and small-arms fire, while the supporting fire of the Cannon Company and the 346th Field Artillery Battalion was directed against the German artillery and mortars.

After an intense preparation on Selvatelle, the town was by-passed by the 362d, with the 2d Battalion moving around it on the west and the 3d Battalion skirting it to the east. Forward movement was greatly impeded, however, by continued heavy artillery and mortar fire. By 1730, 15 July, this fire became so intense that both attacking battalions of the 362d were forced to seek covered positions just north of Selvatelle.

Colonel Cotton ordered the 3d Battalion to move to the right in a flanking movement. The 2d Platoon of Company K spearheaded the battalion's advance, contacting the enemy below the military crest of a hill near San Lazzaro. A fierce fire fight developed in which the 2d Platoon, commanded by Second Lieutenant Thomas G. Tarwick, pushed the German force onto the flat terrain on top of the hill. As the fight became more intense, the enemy threw in reinforcements. In order to overcome this increased enemy resistance, the 3d Platoon of Company K was committed around the right flank of the 2d Platoon, but was immediately checked by hostile fire. Both platoons fought furiously to hold their ground against the German force entrenched on the top of the hill. The enemy was equally determined to hold his positions and continued to pour down a murderous volume of small-arms fire.

Seizing his advantage, the enemy then sent two tanks into action, one on the left flank and one on the right. Staff Sergeant Henry H. F. Wendt, of the 2d Platoon, Company K, crawled forward in the face of machine-gun and tank fire to a partially defiladed position two hundred yards in front of the company. He fired a bazooka at the right tank, silencing its machine gun and forcing it to withdraw. Then he crawled laterally across the front and scored a direct hit on the cannon of the second tank, putting it out of action and forcing the tank to withdraw. Wendt's courageous feat forced the Germans from their positions.

While the 3d Battalion was fighting for the enemy-held hill near San Lazzaro, the 2d Battalion, attacking in the left sector of the 362d Infantry, was also battling its way northward against increasingly stubborn enemy resistance. The leading elements of the 2d Battalion had reached a point approximately nine hundred yards to the northwest of Selvatelle when, at 2100, the whole battalion was subjected to an intense shelling from both artillery and mortars. The Germans immediately followed this barrage with a counterattack. Company F, covered by Company E, was forced to withdraw to the high ground eight hundred yards to its rear. By 2200, however, the front lines were reorganized and dug in.

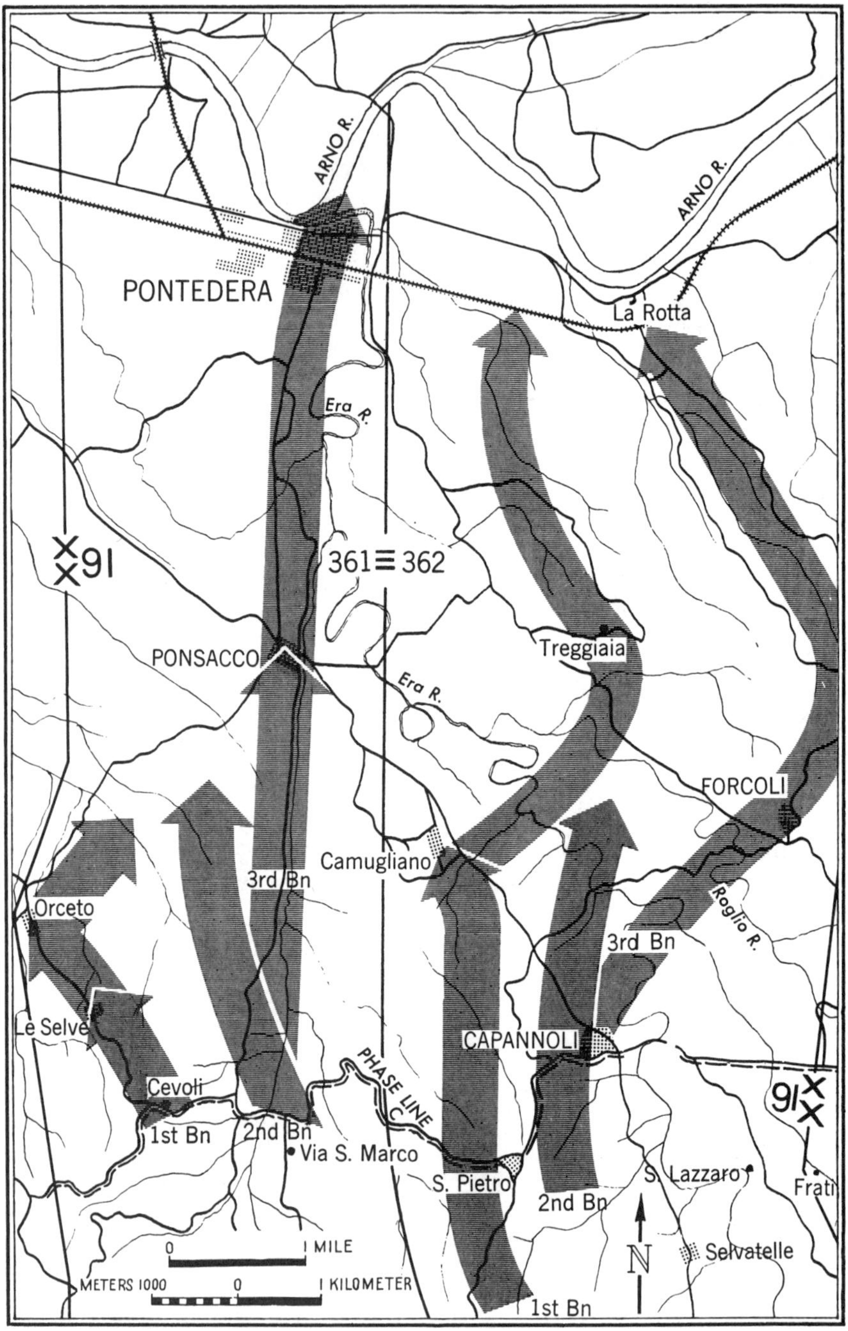
ARNO R.
ARNO R.
PONTEDERA
La Rotta
Era R.
91
361
362
PONSACCO
Treggiaia
Era R.
FORCOLI
Camugliano
3rd Bn
Orceto
Roglio R.
3rd Bn
Le Selve
CAPANNOLI
PHASE LINE C
Cevoli
91
1st Bn
2nd Bn
Via S. Marco
S. Pietro
S. Lazzaro
Frati
2nd Bn
0
1 MILE
N
Selvatelle
METERS 1000
0
1 KILOMETER
1st Bn

Map 10

At 2245, 15 July, Colonel Cotton issued a new attack order for the next morning. The 2d Battalion was to continue its attack to the north; the 3d Battalion was to hold in its present positions and revert to regimental reserve; and the 1st Battalion was to attack in a northwesterly direction. On 15 July the 363d Infantry, which now constituted Division reserve near Bagni, sent out extensive combat patrols as protection for the left flank of the Division, and elements of the regiment assisted the 133d Infantry, 34th Division, in capturing Casciana.

Phase III

By day's end on 15 July the second phase of the 91st's initial battle had come to a close. Steady progress had been made all along the Division front despite increased enemy resistance. Two severe counterattacks, one at Soiana and one just north of Selvatelle, had been repelled, and the situation had developed so successfully that a breakthrough to the Arno River was imminent. During the next three days —16 through 18 July—the operations of the third phase saw the achievement of the Division's mission: the occupation of the south bank of the Arno River.

In its positions on Phase Line C, a line extending along the lateral road running east from Caponnoli, the 361st Infantry reorganized and prepared to push on. At 1030 on 16 July the 1st and 2d Battalions moved forward in a coordinated attack. The enemy, fighting tenaciously to retain control of the last high ground south of the Arno River, opposed the regiment's advance with small-arms and automatic-weapons fire as well as artillery concentrations.

Despite mounting resistance, the 361st pressed its attack steadily northward. The 1st Battalion, fighting in the left sector of the regiment, observed the enemy forming for a counterattack in the vicinity of Le Selve, and shortly before 1600 a tremendous volume of fire was poured into the assembly area by the 916th, 346th, and 348th Field Artillery battalions, the Cannon Company, and the 81mm mortars of Company D, breaking up the attack before it could be launched and causing heavy enemy losses of men and vehicles. As a result of this overwhelming fire power, the enemy withdrew to Orceto, where they were shelled again. Orceto, however, continued to remain a focal point of enemy resistance.

Following the German withdrawal, both attacking battalions of the 361st pushed forward aggressively. Late in the afternoon, during an attack against strongly fortified enemy positions on the ridge southwest of Ponsacco, the 2d Battalion was subjected to murderous artillery and

"Lieutenant Anderson reached a point fifteen yards from the hostile weapon and hurled two grenades at it."

mortar fire. Unable to advance over the open terrain which lay between it and its objective because of the temporary superiority and deadly accuracy of the German fire power, the battalion committed the reserve company, Company F, around the left flank in an effort to catch the enemy off balance.

Advancing with great stealth and skill, Company F succeeded in launching a surprise attack against the west flank of the enemy-held positions. When the company's advance was checked by intense fire from an enemy machine gun on a small knoll, Lieutenant John W. Anderson, leader of the assault platoon, deployed his men to form a base of fire and then crawled two hundred yards across open terrain toward the enemy strongpoint.

He reached a point fifteen yards from the hostile weapon and hurled two hand grenades at it. Before the grenades exploded, the enemy machine gunner caught Lieutenant Anderson in the side with a burst of bullets, knocking him down. Momentarily stunned, he rose on his elbows in time to see one of the enemy attempting to escape from the position. He shot and killed the fleeing soldier. Then, in spite of the gaping wound in his side, he charged the enemy position, but the two grenades had done their work: the other two Germans lay dead beside their smashed gun.

Almost immediately, a second machine gun farther along the ridge

Two Mark IV tanks captured by the 361st Infantry at Ponsacco

opened up on Lieutenant Anderson. Weakened by loss of blood, he nevertheless unhesitatingly crawled forward to destroy his new adversary. As he struggled across an open spot, a sniper's bullet pierced his helmet and killed him. Company F, inspired by Lieutenant Anderson's courage, renewed its flanking attack and enabled the remainder of the 2d Battalion to advance and seize its objective.

At 2200, 16 July, Colonel Broedlow issued a field order for the continuation of the attack the next morning. The 361st's 3d Battalion, with Companies D and E of the 1st Armored Regiment, 1st Armored Division, attached and the 916th Field Artillery Battalion in support, was to pass through the right-flank positions of the 2d Battalion, attack north along the main north-south road leading into Ponsacco, and drive the enemy north of the Arno. The tanks were to precede the infantry. The 2d Battalion was to continue its attack northward echeloned to the left rear of the 3d Battalion, protecting its advance. The 1st Battalion was to capture Le Selve and Orceto, reorganize, and be prepared to follow the forward elements of the regiment to the river.

At 0500, 17 July, the attack jumped off. In addition to heavy artillery and automatic-weapons fire, the enemy employed tanks to oppose it. It was estimated that twenty-five enemy tanks—Mark IIs, IVs, and Vs—were operating in the regimental zone. Against this opposition, the 3d Battalion pushed on, and by 1030, the front lines stretched from south of Le Selve across the Ponsacco–Pontedera road to a point southwest of Camugliano. All morning the 1st Battalion encountered determined German resistance from Orceto, where enemy tanks and infantry constantly threatened to launch a serious counterattack. At 1045 the 361st's Cannon Company fired for thirty minutes to forestall a tank attack, but at 1200 a new and more serious threat developed. The 1st Battalion was supported by the 689th Field Artillery Battalion, which fired twenty-five rounds of 240mm shells into Orceto and its vicinity, thoroughly reducing the threat from there.

By 1130, 17 July, Companies I and K of the 3d Battalion had reached a point one-half mile south of Ponsacco. Taking advantage of the relatively flat terrain, the tanks of the 2d Battalion, 1st Armored Regiment, enveloped Ponsacco with a swiftly executed pincer movement and commenced shelling the enemy installations located in the town. Coincidentally with the tank attack, two platoons of Companies I and K pushed into Ponsacco from the south. Hopelessly surrounded, with all avenues of escape cut off, the German garrison surrendered without further resistance.

The 3d Battalion rapidly occupied Ponsacco, while the 2d Battalion consolidated its positions along the high ground a mile and a half to the southwest.

After only a brief rest, the regiment took up the pursuit of the enemy again. The 1st Battalion attacked at 0500 on 18 July and found Orceto an easy objective. By 0715 the battalion had moved forward through Orceto and reached the regimental objective.

At the same time, the 3d Battalion, together with the 2d Battalion of the 1st Armored Regiment (minus one company) and one company of the 776th Tank Destroyer Battalion, drove rapidly north. Less than three hours after the attack opened, the 1st Platoon of Company K entered Pontedera. An hour later, having disposed of enemy machine-gun and sniper fire, the company pushed to the south bank of the Arno, becoming, at 0900 on 16 July, the first element of the Fifth Army to reach the river.

Company K—with one section of heavy machine guns, a section of 81mm mortars from Company M, and the 3d Battalion AT Platoon—was ordered to hold Pontedera, while the remainder of the 361st Infantry quickly consolidated and organized positions along the high ground south of the Arno.

Although the Germans had managed to get most of their artillery across the Arno, many tanks and armored vehicles were found abandoned south of the river. On the morning of 19 July tankers of the

Riflemen pick their way through the ruins of Pontedera

1st Armored Division startled the troops by driving to the rear several undamaged German tanks draped with white sheets and preceded by an officer in a jeep who shouted to all units along the road to hold their fire.

Coincidentally with the brilliant drive of the 361st up the Ponsacco–Pontedera road, the 362d Infantry, attacking in the right sector of the Division, moved steadily northward in its zone of action. The 362d attacked at 0700, 16 July, with the 1st Battalion on the left and the 2d Battalion on the right. The 3d Battalion held in its position in the vicinity of San Lazzaro and became regimental reserve.

Pushing forward relentlessly against heavy small-arms, mortar, and artillery fire, the two assault battalions by nightfall, 16 July, had reached positions north of San Pietro, where they dug in and reorganized.

Hardly had the regiment set up its all-around defense for the night and settled down for a few hours of much-needed rest, when new orders were received. Each battalion was to make a night march of approximately four kilometers through enemy-held country and, at dawn, be prepared to attack from these new positions for a final drive to the Arno River. With superhuman effort, the 362d's infantrymen, advancing steadily throughout the night, reached their new line of departure shortly after daybreak.

By 0700, 17 July, the 2d Battalion had reached positions west of Forcoli, while the 1st Battalion, attacking in the regiment's sector, had entered the southern outskirts of Camugliano. At this stage the enemy loosed a terrific barrage of 88mm artillery and mortar fire. Both assault battalions deployed quickly to engage the enemy, but the tremendous volume of high-explosive shells landing in the area made all forward movement extremely difficult and costly. Although the 346th Field Artillery attempted to silence the opposition, limited observation prevented successful accomplishment of the mission.

Throughout the daylight hours of 17 July, the 362d fought desperately to hold its positions. Fighting in the sector of the 2d Battalion was especially severe. Three counterattacks of approximately company strength were repelled with serious casualties on both sides. Only through the dauntless courage of every member of the battalion and the extraordinary heroism of individuals, were the lines held intact.

When the first enemy counterattack struck hard at the leading elements of the 2d Battalion, Sergeant Carroll D. Eggers, machine gunner of Company H, was in position with his machine gun squad on the extreme left flank of the battalion. With complete disregard for his own personal safety, Sergeant Eggers carried his machine gun forward to a better firing position. Because his squad was composed of replacements and partially depleted by casualties, Sergeant Eggers manned the machine gun himself. Firing with great skill and accuracy, he helped repel two severe German counterattacks. However, a third enemy thrust succeeded in pushing back the left rifle platoon.

Sergeant Eggers remained where he was and continued to fire into the advancing Germans. Doggedly holding his position, he was wounded in the face and knocked away from his gun. He crawled back with the assistance of an ammunition bearer and resumed firing. By increasing his rate of fire, he was able to beat back the enemy attack. As the Germans withdrew out of range of his deadly fire, Sergeant Eggers was mortally wounded by a sniper's bullet, but his courageous action prevented the enemy from penetrating the left flank of the 2d Battalion.

While Sergeant Eggers was heroically repelling enemy thrusts against the left flank of the battalion, another example of extraordinary heroism in action was taking place in Company E. When the company radio operator was wounded, Private First Class Joseph O. Van Osdol, a member of company headquarters, volunteered to carry

"By increasing his rate of fire, Eggers was able to beat back the enemy attack."

and operate the radio set. A short time after he took over his new duties, he spotted a casualty struggling helplessly on a nearby exposed hillside. Realizing the extreme peril his wounded comrade was in, Private First Class Van Osdol, with his company commander's permission, turned the radio over to another member of his company, dashed to the aid of his fallen buddy, and assisted him through heavy artillery and mortar fire to the battalion aid station. Returning immediately through the deadly hostile fire, he resumed his duties as company radio operator. He had been back in the company command post hardly five minutes when he discovered that his set was not working properly. He moved his radio out of the draw in which the CP was located to an exposed position on top of the embankment, where he could get satisfactory reception. Enemy shells burst dangerously close to him, but he remained at his post.

Later in the day, during a severe enemy counterattack against Company E, Private First Class Van Osdol accompanied his company commander and another member of the company command group across two hundred yards of machine-gun swept terrain to a stone farmhouse, from which they were able to obtain unrestricted observation to the front. The Germans also had unrestricted observation of the stone building, and within fifteen minutes, the farmhouse was completely

"Van Osdol set up his radio in a small ditch, fully exposed to hostile mortar and artillery fire."

destroyed. By one of the freak chances of war, Private First Class Van Osdol and his company commander escaped the exploding enemy shells and flying debris and reached the safety of a nearby shell hole. The third member of their group had not been so fortunate. His muffled cries for help reached Private First Class Van Osdol, and for the second time that day, the gallant soldier left the safety of a sheltered position and ran to the assistance of a less fortunate comrade. Within a few minutes, he had removed the rocks and fallen timber which had pinned his battle mate inside the destroyed building and had dragged him to a covered position, where he administered first aid.

In the meantime, enemy opposition to the leading platoons of Company E had become increasingly stiff. Private First Class Van Osdol volunteered to go forward to the leader of the platoon hardest pressed by the enemy counterattack to provide direct communication with the mortar platoon. As he advanced, a burst of machine-gun ripped his canteen from his belt and cut his radio harness from his back, wounding him slightly. He continued forward to the platoon observation post, where he set up his radio in a small ditch, fully exposed to hostile mortar and artillery fire. He remained in this position in the thick of the fight, relaying the fire orders of the platoon leader to the mortar platoon. As a result the mortar platoon was able to place accurate fire on the attacking Germans and ward off a serious enemy threat to the whole company.

Throughout the day, Private First Class Van Osdol had exposed himself time and again to heavy enemy fire to provide his company with means of coordination and control. His valor, perseverance, and devotion to duty inspired his comrades to hold firmly to their positions in the face of severe and repeated enemy counterattacks and will never be forgotten by his buddies.

When darkness fell on 17 July, enemy resistance slackened until only occasional shell fire broke the stillness of the night. But there was little rest for the men of the 1st and 2d Battalions. In the daylong artillery pounding and the repeated enemy counterattacks, the battalions had suffered heavy casualties, and under cover of darkness the big task of evacuating these was begun. It did not stop until the last wounded soldier had been delivered to an aid station.

The 362d's attack was resumed at 0330 on 18 July, with the 3d Battalion replacing the 2d Battalion on the right. Enemy resistance was light, but the advance of the regiment was slowed by the rugged terrain. Shortly after daylight, the troops were checked by artillery fire from the Treggiaia area and by *Schü* mines,[1] the first the regiment had encountered. About noon the Germans were observed pulling their artillery across the Arno River.

The next day at 0800 the advance was again resumed, this time without enemy resistance. Terrain, demolitions, and minefields impeded the advance, but by 1500, the 362d had closed on its objective. One company from each battalion outposted the line, and patrols were sent to the Arno River.

Thus, after seven and a half days of bitter fighting, the Powder River men had accomplished their mission. Over 2,678 tons of ammunition had been fired by Division Artillery, with a peak of 323 tons in one 24-hour period of 16-17 July. The Division had suffered 904 casualties: 12 officers and 131 enlisted men killed and 37 officers and 724 enlisted men wounded or injured. The outstanding work done by the Medical Corps men and by the 316th Medical Battalion in caring for these battle casualties won the undying respect and appreciation of every man in the Division.

The Division was the first unit of Fifth Army to reach the Arno River and gain control of the high ground on its south bank. Major

---

[1]This antipersonnel mine was used extensively by the Germans. About half the size of an ordinary cigar box and filled with a half-pound of TNT, the *Schü* mine was armed with a non-metallic fuze which makes detection almost impossible. When stepped on, the hinged cover releases the detonator, which in turn sets off the main charge.

Brigadier General Williamson points out progress of the Task Force to the Fifth Army commander

General Willis D. Crittenberger, commander of IV Corps, wired General Livesay on 18 July: "Well done, 91st Division." That same day, in a general order, General Livesay commended the Division on its outstanding achievements. "I am highly gratified with the accomplishment of the Division," he said. "I have noted a spirit of determination and pride of service in all ranks that assures the further success of the Division."

The Arno River was reached on 16 July 1944, and during the closing weeks of the month, the 91st Division patrolled its banks and made a reconnaissance of enemy positions north of the river. In the last week of July, it relieved the 88th Infantry Division on its right flank, was assigned to II Corps, and prepared for the hard campaign to crack the Gothic Line.

The 91st had entered combat for the first time as a complete fighting unit in the face of stubborn German delaying action, and it had led the American forces in Italy in the drive to the Arno. To repeat General Crittenberger's words, it was "a job well done."

CHAPTER 7

TASK FORCE WILLIAMSON

THE final chapter in the story of the Division's advance to the Arno was written principally by the 361st and 362d Infantry Regiments, but the 363d Infantry, commanded by Colonel W. Fulton Magill, was the nucleus of Task Force Williamson and played the leading role in the capture of Leghorn and Pisa, 18-24 July.

In what was described as a spectacular "end-around play," the 363d Regimental Combat Team, reinforced, moved out of its assembly area at 1817 on 17 July under the command of Brigadier General Raymond E. S. Williamson, Assistant Division Commander. It organized in the 34th Division sector and at 0340, 18 July, knifed northwest through the gap between the 135th and 442d Infantry Regiments toward the great port of Leghorn.

At 2100 that night, the city, at whose gates the Fifth Army had been hammering for twenty-five days, fell to Task Force Williamson. The Germans were caught completely by surprise. They were hit when they were off balance, with their main forces deployed against the 34th Division. In a few hours the strongest enemy bastion south of the Gothic Line had fallen.

The 363d's 1st Battalion and the 2d Platoon of the 91st Reconnaissance Troop, striking from the high ground east of the port, were the first to enter Leghorn. The 2d and 3d Battalions moved in the following morning and reorganized for the attack on Pisa. Enemy resistance by this time was completely shattered, and the main German forces were withdrawing toward Pisa.

Principal obstacle to the advance on Pisa was a canal north of Leghorn cutting Highway No. 1. However, the 1st Battalion crossed the barrier at 1800, 20 July, and the battle for Pisa was underway. Enemy artillery was trained on the canal, and it was impossible to erect a bridge, so patriots were used as carrying parties to keep the forward troops supplied. The 2d Battalion joined the 1st on 21 July. The following day the 1st struck out for the south bank of the Arno River, where it established its positions that night. By 0530, 23 July, patrols had entered the city, and by 1245 they had grown to company strength. The 3d Battalion, infiltrating in small groups across fields, so disguised its movement that the Germans did not realize that a battalion was entering the city until it was too late. The 2d Battalion joined the 3d, and by late in the afternoon the two units held positions in the city. That night, however, mortar and artillery fire, directed from a German OP located in the famous Leaning Tower of Pisa north of the Arno,

Three members of the 363d take time for a little horseplay

Damage to the port at Leghorn was severe

was so heavy that General Williamson ordered one battalion to withdraw south of the city.

In the meantime, the 2d Platoon, 91st Reconnaissance Troop, with the mission of screening the left flank of Task Force Williamson, improvised a bridge across a canal south of Marina di Pisa, and during the night of 21 July succeeded in crossing two jeeps and pushed into the outskirts of the city. On the morning of 22 July, three more jeeps were sent across the canal, and with the assistance of two platoons from Company A, 363d Infantry, cleared the remainder of the city of all hostile resistance.

From 23 to 28 July, when it was relieved, Task Force Williamson was under constant artillery, mortar, and small-arms fire from the German lines across the river. At first enemy patrols came across in small boats to reconnoiter the American positions, but General Williamson put an end to this by establishing strongpoints at strategic positions. Throughout the entire period, Company C, 316th Engineer Battalion, worked hand in hand with the infantry. The Germans had outdone themselves in laying mines and setting booby traps and demolitions in the Leghorn area, and many a life was saved by the untiring efforts of the engineers in locating and neutralizing these death-dealing devices.

On the night of 28 July Task Force Williamson was relieved by the 45th AAA Brigade and withdrew to the south of Leghorn in preparation for movement to the east, where it was assigned the mission of

On the alert for snipers, men of the 363d patrol Pisa

Engineers detonate mines in Leghorn

screening the Fifth Army's right flank and maintaining contact with the 88th Division.

In its five-day attack, Task Force Williamson had made history. General Clark made a personal inspection of Leghorn after troops of the 363d Combat Team had occupied it. Dropping out of the sky in a Piper Cub, the general landed in the city at 0715, 19 July, hailed a jeep from Captain H. K. Chase-Dunn and Sergeant Everett Phariss, both of the 347th Field Artillery, and made an extensive tour of the bomb-gutted port city.

Later, in a commendation to the troops of the IV Corps for the campaign to the Arno and the capture of the city, General Clark singled out the Powder River Division when he wrote: ". . . I have been especially delighted over the performance of the 91st Division in its first major test." General Crittenberger of IV Corps added: "I consider it an honor and a privilege to have commanded such fine American troops of the caliber of the men of the 91st Infantry Division. The valiant deeds of these men, and their outstanding contribution in this Italian campaign, will go down in history as another great military achievement of American arms."

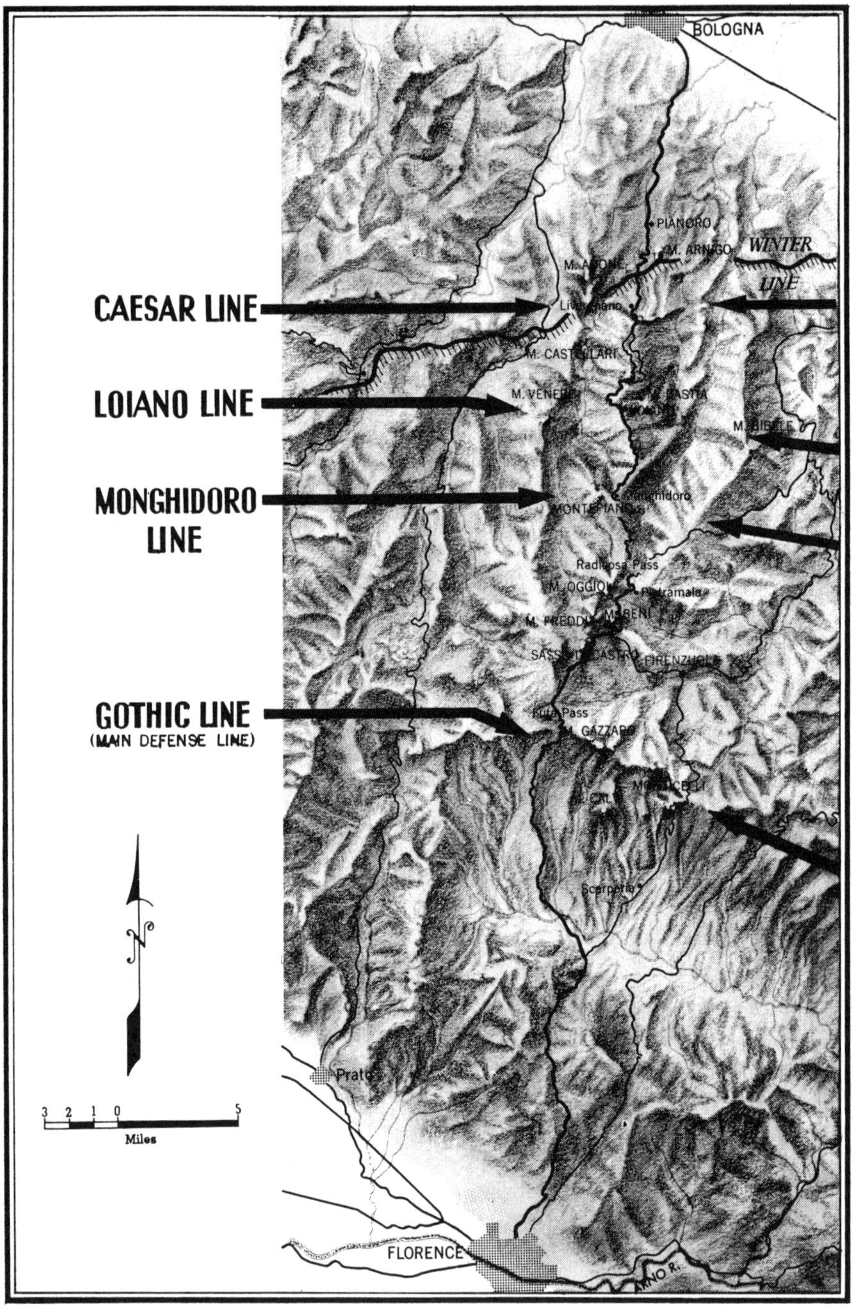

Map 11: From the Arno to Bologna

CHAPTER 8

THE GOTHIC LINE CAMPAIGN

ACTIVITY along the entire Fifth Army front, following the successful conclusion of the Arno River Campaign, was characterized by intensive preparation for the impending assault on the Gothic Line.

The extensive regrouping of forces which started in the latter part of July was completed by August, and the 91st Division assumed responsibility for the eastern flank of the Fifth Army, with Task Force Ramey on its left and the 8th Indian Division of the British Eighth Army on its right. The 362d Infantry, covering an eight-mile front running east along the south bank of the Arno from the small town of Buche, had organized defensive positions across the Division sector and was maintaining strong combat and reconnaissance patrols to the river. In addition, on the night of 30-31 July, the 2d Battalion, 363d Infantry, reinforced by the regimental Antitank Company and one platoon of the 91st Reconnaissance Troop and supported by the 347th Field Artillery Battalion and the regimental Cannon Company, relieved elements of the 350th Infantry, 88th Infantry Division, which were occupying positions facing north and east on the Fifth Army-Eighth Army border. The battalion's mission was to protect the right flank of the Fifth Army and to maintain contact between the Eighth Army and 362d Infantry.

In reserve were the 361st Infantry, relieved from attachment to Task Force Ramey at 0300, 1 August, and the remainder of the 363d Infantry. Division Artillery, less the 347th Field Artillery Battalion, was attached to Task Force Ramey, while the 178th Field Artillery Group —consisting of the 339th, 337th, 403d, and 936th Battalions—was in direct support of the Division.

The mission of the Division at this time was to establish a defensive line along the Arno River to protect the right flank of the Fifth Army; to screen the regrouping of the Fifth Army; and to maintain liaison with the 8th Indian Division.

Up to the time the 91st Division was relieved from the line on 17 August, the period between attacks was comparatively quiet. It was characterized by extensive reconnaissance and patrol activity, harassing artillery fire, and occasional patrol skirmishes. The enemy was sensitive to every move, endeavoring desperately to learn where the next blow would fall. During the day there was very little activity other than artillery duels. At night, however, enemy patrols, often forty to fifty men strong, crossed the river to probe the Allied lines. Sometimes

Riflemen of the 363d approach the outskirts of Pisa

these German patrols hid in houses south of the Arno by day and made reconnaissance forays by night. The enemy also made extensive use of observation planes and flares in an attempt to determine the disposition and intention of our forces. All ordinary patrol methods used by the Germans proved inadequate, however, and in a final effort to obtain information, they ordered their patrols to return with prisoners at all costs. In some sectors, it was reported, the Germans offered a reward of a two-day furlough, twenty Reichsmarks (approximately four dollars), and an Iron Cross Second Class for one Allied soldier brought in unharmed or slightly wounded.

The 362d Infantry, occupying positions along the river, had two primary missions: to learn as much as possible about the enemy's strength, positions, and movement, and to scout the river and its banks for information to be used in possible river-crossing operations later in the month. Its secondary mission was to screen the front of the Division and the Fifth Army and deny the enemy knowledge of the disposition and movement of our troops. To complete these missions, an average of five combat patrols, consisting of from eight to twenty men each, and fourteen reconnaissance patrols of four to eight men, with an officer leading each patrol, covered prearranged routes every night. Their success varied greatly.

In addition to the combat and reconnaissance patrols sent out by the infantry, the 316th Engineer Battalion sent out reconnaissance parties to gather information essential to a crossing of the Arno River. One such party reconnoitered the terrain for three nights and two days,

Machine gunners of TFW relax after Leghorn is cleared

18-21 August. The engineers waded and swam the river at many places to determine depths and widths of the stream and gathered other information concerning the banks and approaches.

From prisoners captured by combat patrols and from the reports of the reconnaissance parties of both the 362d Infantry and the 316th Engineer Battalion, it was possible to gradually build up an accurate and complete picture of the disposition of the enemy forces, as well as a detailed analysis of the Arno River and its banks.

The 362d's secondary mission—holding the sector against enemy attack and screening the movement and disposition of the Fifth Army troops—was complicated by the development of two weak points in the Division's defensive position. Constant harassment came from the enemy in the no-man's-land lying between the south bank of the Arno and the railroad running about a kilometer south of the river in this sector. During the day enemy patrols crossed the river and reorganized in comparative safety before proceeding on their missions. Another weak point was discovered in a narrow strip of land to the right of the Division's sector at the Elsa River, where some question of responsibility existed following the relief of the 8th Indian Division by the 2d New Zealand Division of the British Eighth Army on the night of 7-8 August. During the night of 11-12 August, however, Company B of the 363d Infantry took up positions running west from the Elsa River to the right flank of the 362d Infantry. This closed the gap between the 91st and the 2d New Zealand Division and effectively sealed out enemy patrols.

On the night of 12-13 August the 1st and 3d Battalions of the 363d Infantry relieved two battalions of the 362d Infantry on the line. The relief was completed at 0400, and at 0500 Colonel Cotton relinquished command of the sector to the 363d Infantry. During darkness on 13 August, the 2d Battalion of the 363d Infantry relieved the 1st Battalion, 362d Infantry, on the extreme left flank of the 363d's sector. The 363d now had all three of its battalions on the line.

Daylight reconnaissance was immediately undertaken to locate suitable positions for strongpoints in the area between the river and the railroad. By 2400, 14 August, five strongpoints were established and protected by mines and trip-wire flares. These measures made enemy reconnaissance much more difficult and materially strengthened the Division's defensive positions along the south bank of the Arno.

While the 362d was patrolling the Arno (1-13 August) the other two regiments and Division Artillery intensified training and the care and cleaning of equipment. On 5 August, training stressing marksmanship and physical conditioning was started in the 361st Infantry in an effort to bring the thousand-odd replacements who had come to the regiment since 2 June 1944 up to regimental standards. Instruction in scouting, patrolling, mines, and mine warfare and technical training for special units was also carried out. In the 363d essentially the same program of training was undertaken for those not actively engaged in the regimental mission. An opportunity was given every replacement to gain actual patrol experience under the guidance of experienced leaders. Division Artillery, in addition to activities similar to those carried on by the infantry regiments, concentrated on the care and cleaning of their equipment. The armament section of the 791st Ordnance Company, with the help of twelve men from the automotive section, carried out the semi-annual survey of the Division's artillery pieces.

On 13 August arrangements were begun by II Corps to move the Powder River Division to a rear area for specialized training. Movement of various units took place at night from 14 to 17 August. The 363d Infantry, which had relieved the 362d on the line at 0400, 13 August, was in turn relieved by elements of the 85th Infantry Division during the night of 17 August, and command of the sector was offi-

An American soldier pauses to watch an Italian laundry in operation

cially relinquished at 0445, 17 August. The Division assembled in the vicinity of San Gimignano and Gambassi to concentrate on river-crossing techniques, operations in mountains with mule supply, and the reduction of fortified areas.

During the remainder of the month of August the 91st carried out the training program outlined by II Corps. Originally scheduled for ten days, the training period was extended into the month of September. The 316th Engineers gave lectures and demonstrations on river-crossing techniques, and full use was made of the 11th Italian Mule Group for training in loading and using mules in mountainous terrain.

Such scenes were typical of rural Italian life

Firing ranges were set up by separate units, and further practice in marksmanship was held. Extensive drilling in night problems was also carried on.

Battle casualties of the Division were light during this period, and the 316th Medical Battalion, in addition to its normal duties, provided medical care for large numbers of Italian civilians. The Quartermaster Company engaged busily in re-equipping the Division. One of the most important jobs performed by the Quartermaster Company was the procurement and distribution of the Division's first beer ration. The 91st Signal Company maintained communications and conducted specialized technical training for communications personnel throughout the Division. The 91st Military Police Platoon had its hands full controlling traffic and keeping unnecessary motor movement in the Division area down to a minimum. The Division Staff made numerous reconnaissances and extensive plans for future operations.

This period was the first in which the majority of the organic units of the Powder River Division were in Fifth Army reserve. Despite the extensive training carried on, movies were shown, trips to nearby centers of historic interest were arranged, and every effort was made to furnish the troops with Special Service entertainment and recreation.

Prior to this time the personnel of the Division had had little opportunity to associate with the native population. The few contacts made with the Italians during the heat of battle were usually in quest of first-hand information concerning the enemy. The other main topic of conversation usually brought forth the stock answer—"*Niente vino; gli Tedeschi tutto portato via*" ("There is no wine; the Germans have taken it all away"). However, with the increase in pass quotas to Rome, San Gimignano, Volterra, and Siena and the frequent conversations with Italian families in trying to solve the laundry problem, something

James V. Forrestal, Secretary of the Navy, visits the Division area

new was being added to the GI vernacular. *"Bon giorno!"* was the usual morning greeting. Instead of "please" and "thank you," it was *"prago"* and *"grazie."* The Italian word for Germans, *"Tedeschi,"* replaced in common usage the better known terms, "Krauts" or "Jerries."

The quaint home life of the Italians was a welcome relief from the severe living conditions and uncertainties of combat. The rustic, well built Italian houses—those that were fortunate enough to survive a destructive war—were symbolic of man's everlasting struggle to preserve the peace and security of his home. The sad plight of Italian life in general was viewed with mixed emotions. Some said, "They brought it on themselves." Others, taking a more sympathetic stand, were unwilling to accept that the innocent should suffer with the guilty. The Italians themselves seemed to accept their fate with almost stoic calm. They went about the business of tilling their fields with slow-moving oxen. They gathered in hay and picked huge quantities of grapes for wine. The hard physical labor performed by the Italian peasant women and their uncanny ability to carry great loads on their heads were a source of wonder to the men of the Division. In appraising the qualities of a well built *signorina,* GIs would often remark, "Wow! What a swell ammo-bearer she'd make."

At the end of the month the men of the Division were rested—their

Robert P. Patterson, Secretary of War, congratulates an officer of the Division on winning the Silver Star for his part in the drive to the Arno

equipment cleaned, repaired or renewed. They had had a chance to correct the weaknesses which had shown up in their first extensive combat test and to learn new and valuable military techniques. Field Order No. 12, which outlined the Division's next attack, was issued on 25 August 1944. It would not be long until battle with the enemy would be renewed.

During August the 91st was visited by a number of high Army and Navy officials, and the second anniversary of its activation was celebrated. Within a week the Division met and entertained the Secretary of the Navy, the Honorable James Forrestal, and the then Under Secretary of War, the Honorable Robert P. Patterson.

Secretary Forrestal—accompanied by General Clark, General Crittenberger, and Major General Geoffrey Keyes, Commanding General, II Corps—inspected the Division command post on 9 August 1944 during a tour of the Fifth Army front. The party dined with General Livesay and the members of the General Staff.

On 14 August, Under Secretary of War Patterson arrived. He was accompanied by Lieutenant General Brehon B. Somervell, Commanding General, Army Service Forces; General Clark; General Keyes; General Crittenberger; Major General Vernon E. Prichard, Commanding General, 1st Armored Division; Major General Thomas B. Larkin,

The Secretary of War inspects a kitchen in the 361st area

Commanding General, Services of Supply, NATOUSA; and Colonel Forrest Conner, of the War Department General Staff. The party visited the regimental command post of the 361st Infantry, where Under Secretary Patterson presented decorations to six officers and enlisted men and greeted a guard of honor of fifteen men who had previously been decorated for heroism by the Division Commander. After reviewing the 361st's 2d Battalion and addressing the troops briefly, he was taken to the Division command post where he and his party were the guests of General Livesay at luncheon.

On 15 August, the Division celebrated the second anniversary of its reactivation. No formal ceremonies were held, but General Livesay, in a letter of greeting, expressed the quiet pride and satisfaction every member of the Division felt. He wrote, in part:

> The Division is now of age—it is no longer a division in training. It is a division that has met the enemy under the most trying circumstances of terrain and has driven him back with heavy casualties. I feel certain that the German High Command has this division registered as one of the first-line fighting divisions. The campaign to the Arno, the taking of Leghorn, and the investment of Pisa leave no doubt in my mind that I have the honor to command an organization of top-class fighting men.
>
> With all of my pride in you, I am still inclined to sound a note of warning. Let us steel ourselves to further, more definite, efforts. Let us improve ourselves in all of the things we have learned, so that nothing can stand successfully in the path of the Division.

CHAPTER 9

THE ATTACK: MONTH OF SEPTEMBER

THE history of the Powder River Division during the month of September 1944 is the story of the breaching of the Gothic Line. In this campaign, every resource of the Division, II Corps, and Fifth Army was thrown against the enemy's most formidable defensive positions in Italy. In twelve days our troops destroyed more than a year's work of thousands of men. Our attack advanced through elaborately constructed fortifications and over mountainous terrain, made doubly hazardous by rain and fog, with unflinching courage and unwavering determination. To the combat infantryman the climactic days, 12-22 September, were a "lifetime of mud, rain, sweat, strain, fear, courage, and prayers." And those days demanded no less of the artillery, the engineers, the medical troops, and every other unit of the Division. All of them supported the infantrymen without rest or reservation. With brilliant leadership and magnificent cooperation, the 91st completed its mission, and at the end of the month had come within sight of Bologna and the Po Valley beyond.

The September campaign falls into three phases: the advance from the Arno River to the Sieve River, 4-11 September; the breaching of the Gothic Line, 12-22 September; and the advance northward from the Santerno River, 23-30 September, which at the end of the month had reached a line just south of Monghidoro.

PHASE I

The first period was primarily one of maneuver for position. Contrary to expectation, the German high command did not elect to make a stand at the Arno, but withdrew to prepared positions north of the Sieve River. Thus, constant revision of plans became necessary, and an extraordinary number of field orders was issued to keep the Division abreast of the rapidly developing situation. On 25 August, while the 91st was in training in the Poggibonsi–Gambassi area, the first of these field orders, No. 12, appeared. One of the most detailed field orders ever issued by the Division, it prescribed every phase of the 91st's part in the Fifth Army's projected attack across the Arno to the Po Valley. On D-day the Division was to cross the Arno at six points east of Florence (Firenze) with two regiments abreast, the 362d Infantry on the left and the 363d Infantry on the right. After securing the initial bridgeheads, they were to push north to a line running northeast from Pantanice to V. Medici, approximately two kilometers north of the river. Successive objectives were: for the 362d Infantry, the area about

Bivigliano and Mt. Senario and the area from the Sieve River south and west to Cartone and Trebbio; for the 363d Infantry, Mt. Calvana and Highway 65 just south of San Piero. Although initially the Division's front was only a little over four kilometers wide, at about the 80th Northing it widened considerably until at the Sieve it was fifteen kilometers.

On the basis of intelligence reports, it was expected that enemy opposition would be four divisions in the line, estimated to number about 12,600 men, with at least one division, of 2100 men, in reserve in the vicinity of Prato. The first delaying position was expected to be a line—consisting principally of rifle pits, temporary gun positions, and minefields—which was anchored on the west flank by Mt. Morello and on the east flank by Mt. Senario. It cut across Highway 65 at Pratolino. North of this line lay the Green Line, where an extended stand was anticipated. The Green Line ran eastward from Fontebuona, through Ferraglia, Bivigliano, and Mt. Senario, to Il Poggiolo.

In assuming command of the sector assigned to it by II Corps, the 91st Division took over an area previously held by the 8th Indian Division. Utmost care was exercised in moving the Division east, with the British screening the movement. By 0455, 5 September, the 363d Infantry had closed in an area 3,500 yards south of Compiobbi. The 362d closed in an area southeast of Badia-a-Candeli at 0500 the same day.

While the British were screening the Division's movements, they found, contrary to expectation, that the enemy had decided not to make a stand at the Arno, but had begun to withdraw. Accordingly, Division Field Order No. 13 was issued at 2000, 5 September. So materially had the situation changed that Phase Line D of Field Order No. 12, running between the 77th and the 78th Northings, eight kilometers north of the river, became the line of departure of Field Order No. 13. The main effort was to be made on a maximum frontage with rapid and continuous progress toward the Orange Line, running in an arc approximately 2,000 yards beyond the Sieve. No change was made in either the Division's boundaries or objectives. The sectors of the 363d Infantry and the 362d Infantry were changed somewhat, however, by shifting the boundary between the two eastward, north of the 81st Northing.

The 8th Indian Division, under the operational control of the 91st Division, patrolled constantly in an effort to maintain contact with the withdrawing enemy. By 6 September, it was found that the Germans were located generally across the 79th Northing. Thus the Division

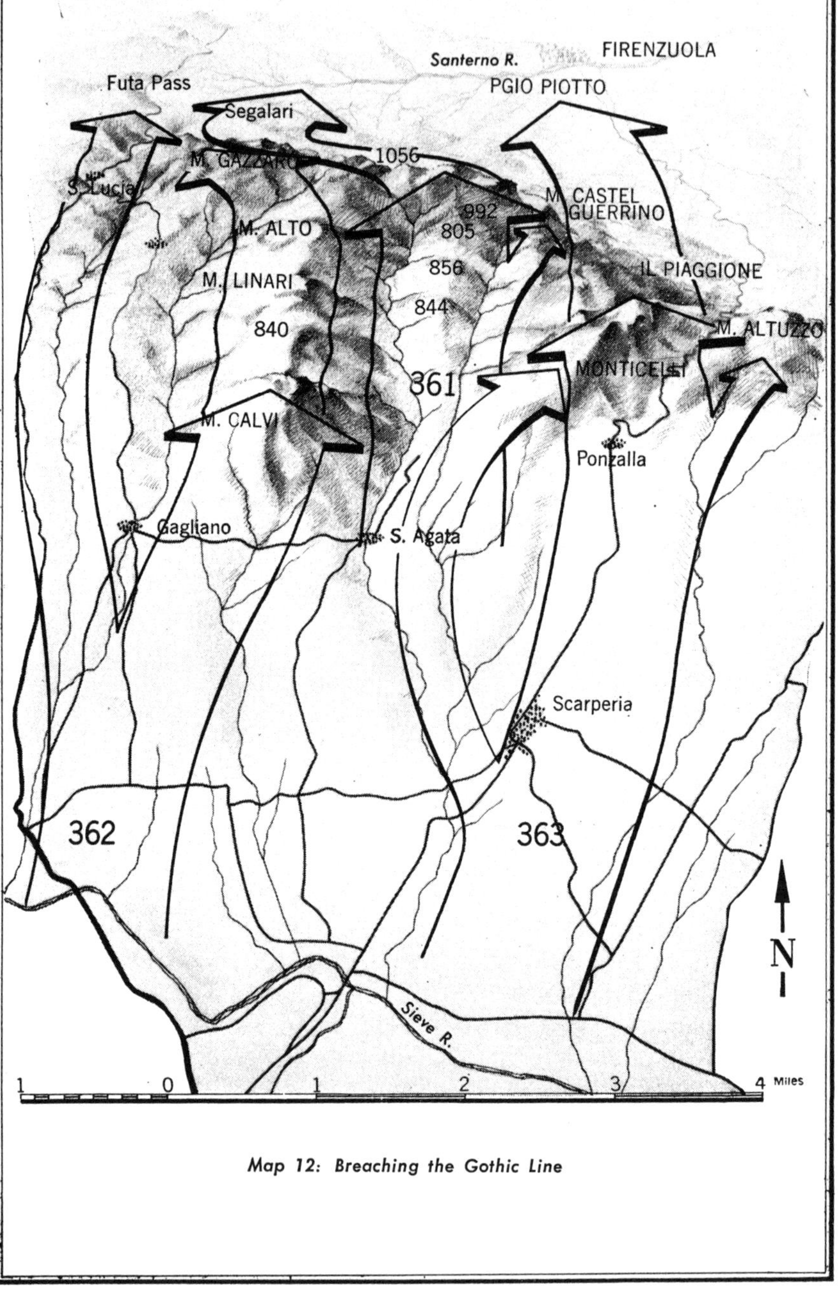

Map 12: Breaching the Gothic Line

was able to move unimpeded across the Arno and to establish, on 6 September, a line just south of the 74th Northing.

On 8 September, British patrols reached Farraglia, Bivigliano, Mt. Senario, and Mt. Calvana along the 80th Northing, which had been designated as the Green Line in Field Order No. 12. No enemy was discovered, and the British 1st, on the left, and the 8th Indian Divisions moved forward to occupy the line. Although at 1420, 8 September, both the 362d and the 363d Infantry Regiments were alerted for a possible move into position for the attack, II Corps, at 1655, 8 September, postponed the movement for at least another twenty-four hours.

New plans, drawn up in light of the continued withdrawal of the enemy, were embodied in Field Order No. 14, issued at 1200, 9 September. Essentially a restatement of the later phases of Field Order No. 12, No. 14 retained the Orange Line as the main objective of the Division. The line of departure for the attack was the lateral road running from a point south of Vaglia to Bivigliano. The major change occurred in the Division's boundaries. These were shifted some six kilometers to the west so that the Division would attack directly north of Florence along Highway 65, the main road from Florence to Bologna. In case a breakthrough was achieved, the 91st Reconnaissance Troop was to be prepared to push rapidly forward on the entire Division front and to maintain contact with the enemy. Likewise, the 361st Infantry was to be ready to exploit any weakness or gap in the enemy defenses. The time of the attack was set for 0530, 10 September.

In compliance with the field order, units of the 91st moved into position during the night of 9 September, hampered by mined roads and fields. Company G, 362d Infantry, lost 9 killed and 29 wounded during the night. The 362d relieved the 2d Brigade of the British 1st Division in the vicinity of Vaglia, and the 363d, moving through the 3d Brigade, closed at 0100, 10 September, in its forward assembly area just south of Bivigliano. The Division Artillery displaced northwest of Grigliano to the vicinity of Pratolino. By 1945, all artillery had registered, and preparations were completed for the attack just before daybreak.

The attack jumped off according to plan. No resistance was met during the morning until, at about 1000, the 2d Battalion, 362d Infantry, received enemy artillery fire near Labbia. From then on both regiments were subjected to harassing artillery and small-arms fire from enemy positions north of the Sieve River. Although the extensive use of mines along the banks and streambed of the river and our

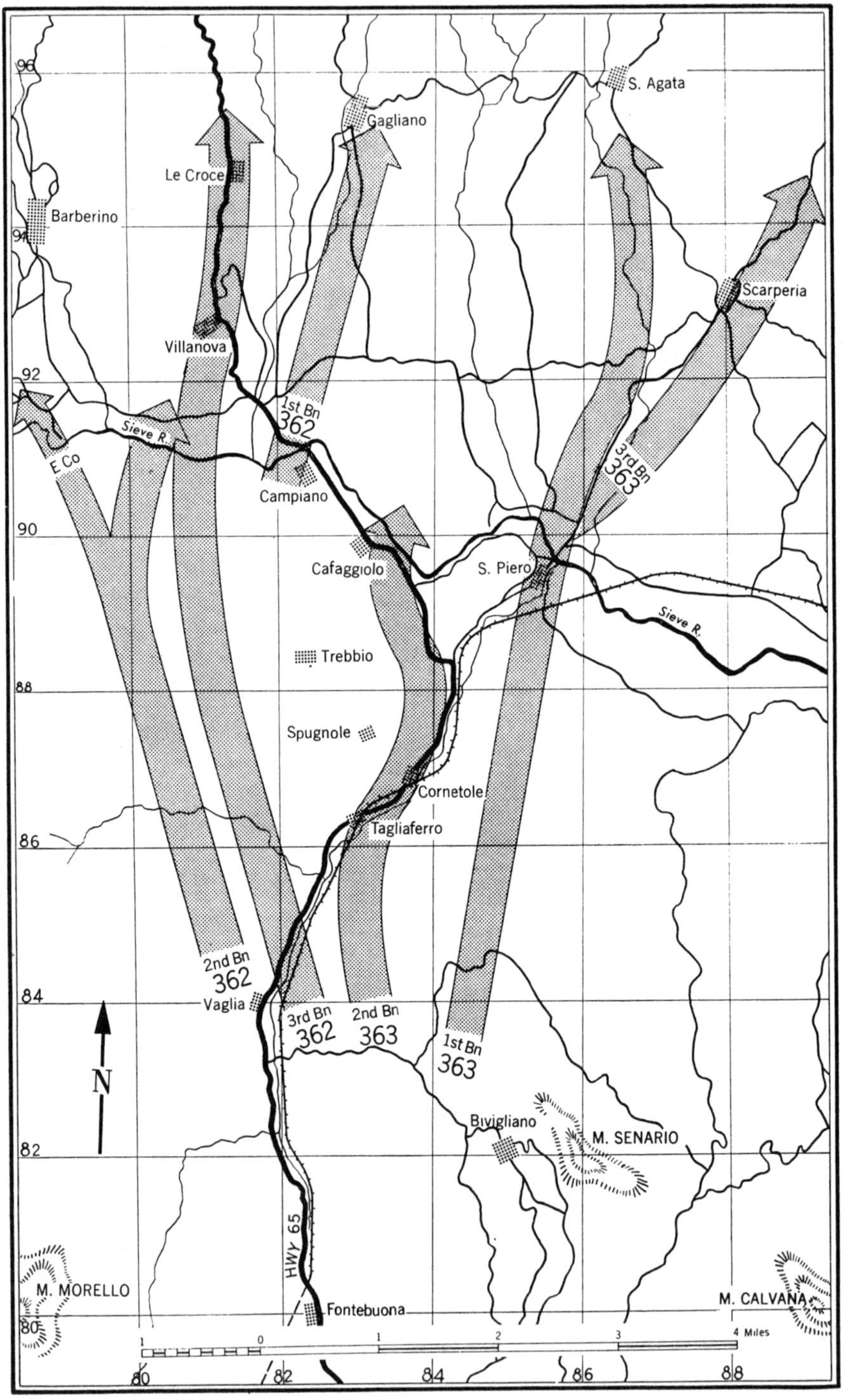
S. Agata
Gagliano
Le Croce
Barberino
Scarperia
Villanova
1st Bn 362
Sieve R.
E Co
Campiano
3rd Bn 363
Cafaggiolo
S. Piero
Trebbio
Sieve R.
Spugnole
Cornetole
Tagliaferro
2nd Bn 362
Vaglia
3rd Bn 362
2nd Bn 363
1st Bn 363
N
Bivigliano
M. SENARIO
HWY 65
M. MORELLO
Fontebuona
M. CALVANA
4 Miles

Map 13

troops' lack of knowledge of enemy dispositions on the north bank made this crossing a major operation, it did not halt the advance. By 0250, 11 September, the 363d's 1st and 2d Battalions had crossed the river. Company E, 362d Infantry, completed its crossing at 0315. Throughout the night, troops of the 1st and 3d Battalions, 362d Infantry, and the remainder of the 363d waded the river and occupied the Orange Line, the Division's first objective.

During the day of 10 September the advance was so rapid that Field Order No. 15 was issued giving new objectives for the Division. Once again its sector was narrowed. Although the eastern boundary remained the same, the western boundary was moved, at some points as much as three kilometers. The 2d Battalion, 362d Infantry, went into regimental reserve. The Orange Line positions were designated as the line of departure. Further objectives of the 362d were to be Mt. Galvi, Mt. Alto, Mt. Gazzaro, and Futa Pass. The objectives of the 363d were Mt. Altuzzo and Mt. Castel Guerrino. A Red Line, running from S. Lucia through Mt. Gazzaro to a point a kilometer south of Casanova, was also set up as the general divisional objective. All units of the Division were instructed to maintain contact with the enemy at all times. Reserves were to be kept well forward to permit their immediate employment to exploit any enemy weakness or withdrawal. Towns and strongpoints were to be bypassed and subsequently reduced by reserve elements.

On the morning of 11 September the two regiments pressed the attack under cover of an aerial assault on the Gothic Line by four hundred medium bombers. The 362d Infantry, moving forward with the 1st Battalion on the right and the 3d Battalion on the left, made considerable gains. Enemy opposition was light, since the Germans had withdrawn from their outpost line on contact, but the mountainous terrain and enemy minefields slowed the advance. By the end of the day the 1st Battalion had reached positions just north of Gagliano, and the 3d Battalion was at the 96th Northing—an advance of approximately four kilometers. The 363d, attacking on the right, also had made excellent gains, reaching Sant' Agata.

With the jump-off at 0600 on 12 September 1944, the Division began its attack on the immediate approaches to the Gothic Line. The 363d Infantry, moving forward from Sant' Agata, advanced on Monticelli, its first main objective. Although there was some small-arms, mortar, and artillery fire, terrain was still the greatest obstacle. Progress was good, however, and after an attack by the 1st and 3d Battalions at 2115, elements of the 3d, on the right, had reached the 98th Northing.

Members of the Division Ordnance Company inspect a captured two-man enemy tank

On the left, the 362d Infantry Regiment advanced toward Mt. Calvi.

At 2000, General Livesay issued Field Order No. 16, which made significant changes in the Division's plans. The decision of II Corps to commit the 85th Division on its right flank brought about a narrowing of the 91st's sector. Mt. Altuzzo, Barco, Mt. Frena, and Firenzuola were no longer Division objectives. Instead, the 363d Infantry was ordered to facilitate the movement of the 85th Division into position by securing Mt. Altuzzo. Its mission then was to seize Monticelli and move northward to Mt. Castel Guerrino. Both the 362d and the 363d were to continue their drive northward until they reached Mt. Oggioli, the Divisional objective on the White Line, and to exert special pressure at 0600, 13 September, to facilitate the movement of the 85th Division. According to plan, the 363d Infantry consolidated its positions before Monticelli. While the bulk of the 3d Battalion remained in position, elements of Company L seized Mt. Altuzzo during the darkness of early morning, but were unable to hold the position. After this attempt, since the 85th Division had assumed control of the sector, the 363d Infantry stopped its attack on Mt. Altuzzo and directed all its efforts toward Monticelli.

At 1500, 13 September, General Livesay committed the 361st Infantry, which up to this time had been in Division reserve. The regiment was to attack at 0600, 14 September, with two battalions abreast, between the 363d and 362d Regiments. On the right, the 363d, after securing Monticelli, was to revert to Division reserve, with the 361st taking over of the right half of the Division's sector. The 362d, on the left, was to secure Mt. Calvi and then proceed to its subsequent objectives, Mt. Faggio, All Ombrellino and Mt. Gazzaro. Until the 363d reverted to reserve, the 91st Division was to have nine battalions on the line: three on the left, one moving north along Highway 65 and two attacking Mt. Calvi; three in the center attacking hills 844 and 856; and three on the right attacking Monticelli. The great drive on the main defenses of the Gothic Line was underway, with the 91st Division making the main II Corps effort on a broad, twelve-kilometer front (the entire II Corps front was only twenty-four kilometers wide along the main defenses of the Gothic Line).

Phase II

Monticelli, the 363d's objective, was one of the most important positions in the Gothic Line. It overlooks Il Giogo Pass, through which Highway 67 runs. In terrain where the problems of supply were extremely knotty, the freeing of this pass was of great importance to the 85th Division, fighting on the right flank of the 91st. The position constituted the anchor of the Gothic Line in the 91st's sector. Since, as it later appeared, the enemy had no positions of comparable strength between the mountains and the Santerno, the taking of Monticelli unhinged the Gothic Line in the east. It made possible the taking of Mt. Altuzzo by the 85th Division and assisted the drive of the 361st Infantry on the west.

Monticelli is a rocky, cone-shaped peak, 3,000 feet high, wooded three-fourths of the way up, but devoid of any cover or concealment in the last six hundred feet before the summit. On its sides pillboxes and dugouts were built in such a way as to afford protection for each other. They were camouflaged so carefully that they were invisible to the naked eye. One typical pillbox, large enough to accommodate five men, was constructed of concrete with a roof covered with three feet of logs and dirt. In front was a firing slit six inches high and three feet long. Row after row of barbed wire, one foot high and twenty-five feet wide were placed at 100-yard intervals to the top of the mountain. In the two ravines which led to the top of the mountain the enemy had laid small minefields. On the reverse slope of the peak elaborate dugouts were constructed. They were dug straight back into the mountain to a distance of seventy-five feet and were large enough to hold up to twenty men. On a hill 300 yards north of Monticelli was found a huge dugout blasted out of solid rock. Shaped like a U and equipped with

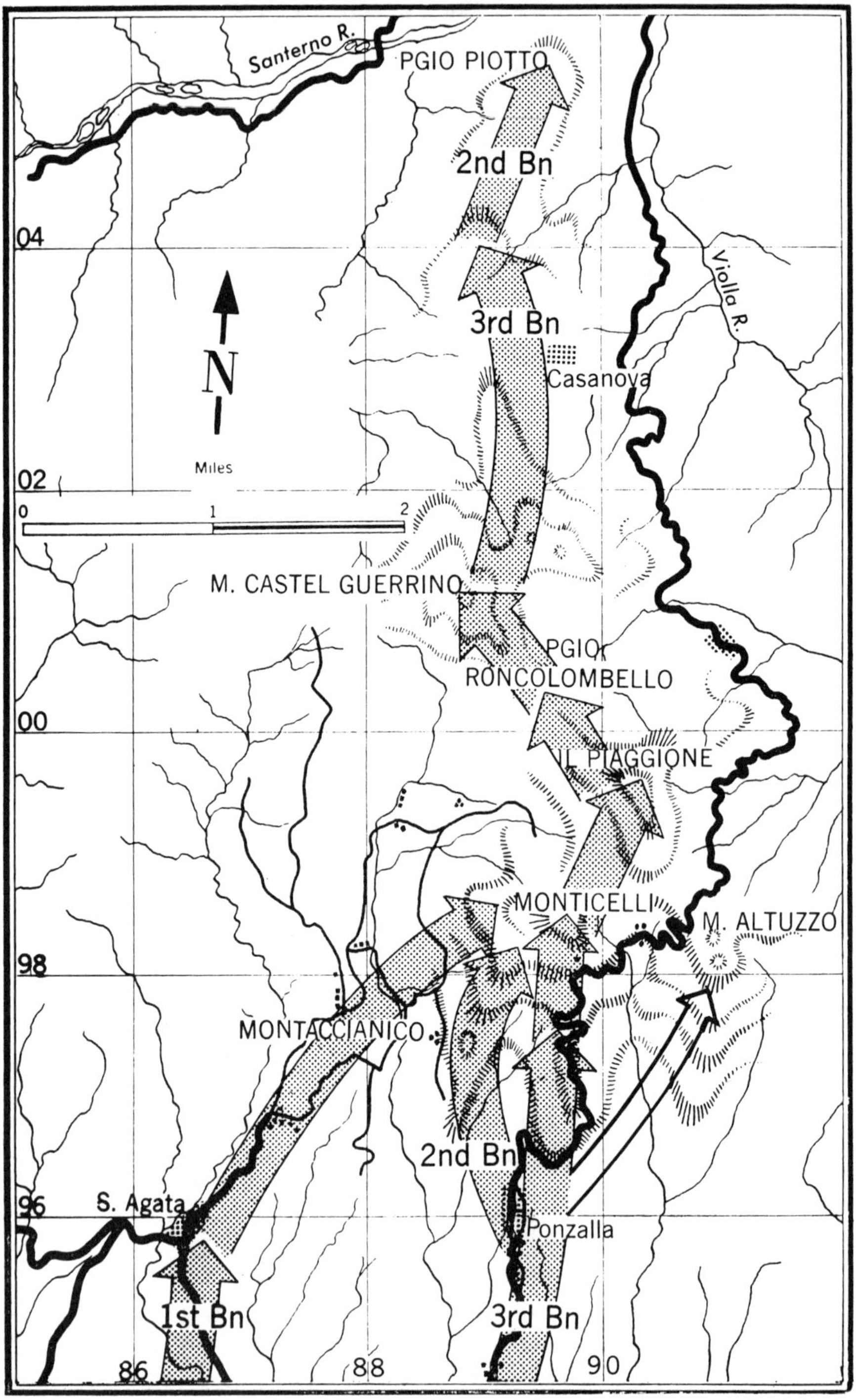

Map 14: The 363d Infantry on the right

cooking and sleeping quarters, it was large enough to accommodate fifty men.

On 13 September the 1st Battalion, 363d Infantry, prepared to assist the 361st Infantry's move into position on its left. At 1625 it was reported that the 363d's 2d Battalion, committed for the first time, had passed between the 1st and 3d Battalions and, with Company F in the lead, pushed to within 600 yards of the crest of Monticelli under cover of a smoke screen. It was subjected to a very heavy counterattack at daybreak, however, and withdrew from its exposed positions. At 1100, 14 September, the 1st and 2d Battalions advanced against heavy small-arms and mortar fire and made slight gains.

On the morning of 15 September the attack continued. Working forward by fire and movement, elements of Company B succeeded in destroying an enemy machine-gun position and partially flanking Monticelli by advancing to the crest of a ridge extending west from the mountain. The 1st Battalion was ordered to reinforce Company B and succeeded in part. Company G, attached to the 1st Battalion, took up positions on the right rear. At the same time, on the right of the regimental sector, the 2d Battalion, with Company F in the lead, continued to reduce pillboxes one by one on the forward slope of the mountain. During the night of 15-16 September, Company B was subjected to four separate counterattacks, all unsuccessful due to the courage of the officers and men and their determination to hold all ground gained.

It was during this action on Monticelli Ridge that Sergeant Oscar G. Johnson, Jr., of Company B, killed 40 Germans, captured 24 more, and destroyed three enemy machine guns.

A private first class at the time, Sergeant Johnson was acting leader of a mortar squad. The mortar platoon had fired its entire supply of ammunition earlier in the day, so, when the Germans counterattacked for the first time at dusk, Sergeant Johnson fired his Garand rifle, but because of the brush, was unable to see the effect of his fire. The counterattack, in which the Germans used hand and rifle grenades, failed, but Sergeant Johnson's platoon was obliged to take cover under the ridge for the night. Taking advantage of the darkness, Sergeant Johnson and a comrade made their way to the base of the hill, while a German artillery concentration pounded about them. They returned early the next morning with rifle ammunition, a litter, and drinking water.

The Germans fired mortars all through the second day, but did not

… from the top of Mount Altuzzo

Infantrymen rest outside a German dugout, typical of the many taken on the Gothic Line

counterattack. During the day Sergeant Johnson forced the surrender of twenty-four Germans.

That night, going down again for water and ammunition, this time with two comrades, Sergeant Johnson encountered three Germans by a stone wall and fired on them, but could not see whether he had hit them or not. He returned to the ridge to sweat out a night-long concentration of German mortar fire.

The area was drenched with enemy mortar fire early the next morning in preparation for continuous counterattacks which were to last until late in the afternoon. Of three men in a foxhole protecting the company's left flank, two were wounded seriously and the third moved in toward the center of the company, leaving the left flank unguarded.

Sergeant Johnson, carrying a rifle and a pistol, moved up to the left to protect the flank. In a foxhole overlooking the ridge and commanding a long field of fire to a crest 200 yards beyond, he found a tommy gun, an automatic rifle, two more Garand rifles, and two magazines of automatic-rifle ammunition.

At least three German machine guns were firing from the crest beyond Sergeant Johnson's hole, and four others, on the high ground to

"Johnson perched himself on the edge of his foxhole, exposed to enemy fire from the front and left."

his left, enjoyed flanking fire along the ridge. Two officers to his right were killed when they attempted to stand up and fire.

Disregarding his personal safety, Sergeant Johnson perched himself on the edge of his foxhole, exposed to enemy fire from the front and the left, and remained there until the counterthrusts abated toward evening of the third day. Firing every weapon he had, and miraculously surviving, he silenced all the machine guns in front of him and cut down twenty enemy infantrymen as they charged a friendly machine gun behind the ridge. He also gave water and first-aid treatment to two wounded men.

The counterattacks subsided late in the afternoon, but heavy mortar fire fell for some time afterward. Then the Germans pulled out, and early the next morning, the 361st Infantry moved in to take the high ground to Company B's left. Of the scattered German dead in the area, Sergeant Johnson was officially credited with forty. Later he was awarded the Medal of Honor by General Clark, then 15th Army Group commander, at a Division ceremony in Gorizia, Italy.

On the morning of 16 September, the 1st Battalion, 363d Infantry, remained in position, securing its gains and resupplying its companies. The 2d Battalion maneuvered in an attempt to reduce the pillboxes which had halted Companies E and F, but progress was slow. At 1125 the 3d Battalion, which had been in reserve since 14 September, attacked, with Company L and Company K in the lead. By 1425 it was on its way up Monticelli, receiving heavy mortar and artillery fire and encountering many *Schü* mines.

Company K was finally forced to withdraw from its forward positions by intense mortar and machine-gun fire, but the attack was continued through the night of 16-17 September by the 3d Battalion and elements of the 2d Battalion. Some progress was made by both battalions, and the 2d captured 26 prisoners in pillboxes. The 3d, however, encountered heavy machine-gun fire, which forced it to withdraw for reorganization.

On the morning of 17 September, General Livesay laid the plans, and supervised the preparations, for the final assault. Every resource was marshalled for the effort. Company F, the right-flank unit of the 361st Infantry, was attached to the 1st Battalion in exerting maximum pressure on the left flank of Monticelli. The 3d Battalion maintained maximum pressure on the right of the regimental sector. The 2d Battalion, plus Company A, launched its attack at 0545, and at 0925 Companies E, F, G, and K were moving up the mountain, reducing pillboxes as they went.

"Captain Conley led his men in the assault of several fortified positions."

It was a climactic day for the troops of the 363d Infantry. The slopes and crest of Monticelli were a blazing inferno of every type of mortar, artillery, and small-arms fire the dogged enemy forces could amass in a final effort to hold this key terrain feature and shatter the forward surge of the attackers. The Germans succeeded in slowing down the forward movement of units, but they could not stop the attack because they had not reckoned on the supreme courage and indomitable will of individuals.

Captain Edward J. Conley, Company G commander, led his men in the assault of several fortified positions, until they were stopped after sustaining heavy casualties. After reorganizing his depleted company, Captain Conley led it in repulsing an enemy counterattack, during which he was painfully wounded in the arm. Making his way to report to his battalion commander, he was knocked unconscious by a shell burst. He soon regained consciousness and learned that the adjoining units, companies E and B, had lost all their officers and had suffered heavy casualties. He made his way to both companies, reorganized them, and remained in command of all three units until the objective was secured.

Meanwhile, when the attack of Company F, on the right of the battalion, began to bog down, Lieutenant (then 1st Sergeant) Russell M. McKelvey volunteered to lead an assault platoon against a strong-

"Lieutenant McKelvey threw a Molotov cocktail into the embrasure of a pillbox."

point holding up the company's advance. Moving up the ridge on which five enemy emplacements were known to be located behind barbed-wire entanglements and minefields, Lieutenant McKelvey repeatedly exposed himself to enemy fire in his efforts to keep his platoon organized and in moving from man to man with words of encouragement.

Realizing the seriousness of the situation, he secured several Molotov cocktails[2] and proceeded alone through a hail of enemy machine-gun fire toward an enemy strongpoint. A sniper's bullet pierced his helmet while he was disarming a German S mine, but he went on to a point within ten yards of his objective and threw a Molotov cocktail into the embrasure of a pillbox. Dashing into the emplacement, he flushed five Germans out of a sub-cellar and captured them and their weapons. Then, leading his platoon in a flanking attack, he knocked out the remaining enemy emplacements, and the entire battalion was able to advance.

In the meantime, Company K, commanded by Captain William B. Fulton, was spearheading the 3d Battalion's advance on the right and had moved to a point within three hundred yards of the crest of the mountain. There it was pinned down by heavy mortar and artillery

[2]Molotov cocktails were made by filling any type of bottle with an inflammable mixture and closing the mouth with an oil-soaked rag to be used as a wick. When tossed against a tank or concrete pillbox, the bottle broke, causing the contents to burn fiercely.

"Advancing over barbed-wire entanglements and through minefields, Captain Fulton led them to the objective."

concentrations. In order to continue the advance, it was necessary for the company to cross a draw swept by bands of machine-gun and rifle fire. Telling his men that they must take the crest of the hill, Captain Fulton crossed the draw and, completely exposed to concentrated enemy fire, began the perilous climb up the precipitous mountain slope. Eight of his men followed him. Advancing over barbed-wire entanglements and through minefields, he led them to the objective. He quickly secured two concrete pillboxes, deployed his eight followers as they reached the top of the hill, and organized the position for defense.

The enemy immediately laid an intense artillery and mortar concentration on the crest and directed machine-gun and sniper fire on the position held by Captain Fulton and his men. By radio Captain Fulton adjusted artillery fire on the Germans and led his men in repulsing three fierce counterattacks made by the enemy in trying to regain the position. When reinforcements arrived that night, Captain Fulton had so planned and organized the defense of the hill that they moved directly to their positions without confusion or loss of time. Finally, by 2240, Colonel Magill was able to report that his men were well dug in and prepared to hold the mountain against any counterattack. At 0530, 17 September, Company F, 361st Infantry, assaulted pillboxes which had been taken previously by Company A and Company

B and found untenable. Company F took two prisoners, the other pillbox occupants having withdrawn before the attack.

During the night two batteries of the 347th Field Artillery Battalion laid a ring of steel around Monticelli, firing a volley every three minutes. No counterattack developed, and by 0820, 18 September, 300 men of the 2d and 3d Battalions were on Monticelli. The position was organized, and supplies were brought up.

Monticelli had been won by the courage and daring of the infantrymen of the 363d Infantry Regiment and the superb support of the 347th Field Artillery Battalion and its associated units. On two separate nights, when the infantrymen had difficulty orienting themselves because of the nature of the terrain, the artillery laid smoke shells on the top of the mountain. On 17 September, the artillery fired 4,000 rounds to prevent serious counterattacks on the men of Company K before the crest could be occupied in strength. The effect of this constant hammering of the enemy by the artillery was disclosed in reports of the 1st Battalion, which, while occupying the ridge running west from Monticelli, had directed fire on surrounding areas for four days. After the ridge and Monticelli were made secure, the 1st Battalion commander counted 150 dead in the area the artillery had battered. One of the targets of the all-night firing, 17-18 September, was a battalion command post thirty feet wide and dug one hundred yards into the side of the mountain. Thirty-three prisoners were taken from the dugout, and all were dazed and shaken by the pounding they had received. The artillery had run the enemy into his holes; the infantry had dug him out, and Monticelli was ours.

General Keyes, Commanding General of the II Corps, expressed his satisfaction on the capture of Monticelli in a wire to General Livesay:

> CONGRATULATIONS UPON THE CAPTURE OF MONTICELLI STOP THE SUCCESSFUL ACCOMPLISHMENT OF THIS TOUGH ASSIGNMENT IS FITTING TRIBUTE TO THE DOGGED DETERMINATION AND COURAGE OF THE 91ST STOP

Division plans contemplated that the 363d Infantry would be pinched out by the 361st after it had taken Monticelli. However, at 1030, 18 September, II Corps, desiring to exploit the capture of Monticelli as rapidly as possible, directed that the 363d send forces to Il Piaggione and thence to Mt. Castel Guerrino. Accordingly, a patrol, to be followed by a company, was sent by the 3d Battalion. Little resistance was met, and by 1735, 18 September, Company K and the patrol were consolidating positions just south of Il Piaggione.

Doughboys of the 363d move forward after bitter fighting through the Gothic Line

After a quiet night, in which no enemy activity was reported in the 363d's sector, the 3d Battalion jumped off at 0530, 19 September. The advance was slowed by the rugged terrain and some enemy resistance, but it continued to the approaches of Casanova, where really stubborn opposition was met. All the next day stiff fighting went on for Casanova, and at 1935 Colonel Magill decided to relieve the 3d Battalion with the 2d. The next morning, 21 September, Company F relieved Company I, and Company E relieved Company K. During the night the enemy had withdrawn across the Santerno River in this sector, and the 2d Battalion advanced rapidly. By 1018, one platoon of Company F and one platoon of Company E had occupied Poggio Piotto. Patrols were immediately sent out to the Santerno, but no enemy was found. The 2d Battalion then organized Poggio Piotto, maintaining contact with the 85th Division on the right and the 361st Infantry on the left. Strong patrols were sent across the Santerno to maintain contact with the enemy. The 1st and 3d Battalions, which had assembled in the vicinity of Casanova, began at 1400, 22 September, to move by truck to Villanova for a brief rest and rehabilitation.

During the advance across the Sieve River, the 361st Infantry had been in Division reserve, but at 1531, 13 September, the regiment was directed to prepare for movement into the line in a sector between the 362d on the left and the 363d on the right. The 361st was to pass through the forward elements of the 363d, and when the latter had secured Monticelli, it was to pinch it out. The 361st moved immediately to an assembly area one mile northwest of Scarperia and prepared to attack at 0545, 14 September.

The situation facing the 361st Infantry was much different from that of the 363d. The 363d was concentrating all its forces on taking a single important terrain feature, Monticelli, capture of which unlocked the door for a steady advance northward. The sector assigned the 361st, running north and northeast from Montepoli, was a bowl surrounded on three sides by a mountain range shaped like a horseshoe. At the right point lies Monticelli; at the left, Mt. Calvi; to the front, Mt. Faggio, All Ombrellino and Mt. Castel Guerrino, the latter one of the highest points in the Apennine Range. The floor of the bowl was cut by ridges running north and south. The whole area was under excellent enemy observation from the mountains on all three sides. For an understanding of the achievements of the 361st Infantry in the Gothic Line Campaign, these facts are of paramount importance. Somebody put it well when he said that the enemy was looking down the throats of the 361st. From their prepared positions the Germans were able to place terrific machine-gun, mortar, and artillery fire on our troops as they advanced north.

Another problem, which all regiments of the 91st Division faced but which was an especially tough one for the 361st Infantry, was supply. On the left and right, roads were available at least part of the way north, but in the 361st's sector the only road of any size stops at Casal. By ceaseless effort, however, the engineers rapidly extended to Coppo a trail adequate for quarter-ton trucks. This was one of the most difficult jobs of the month for the engineers. The trail was so rocky that it was impossible to scrape a road out of the mountainside and so steep on the outside that it was equally impossible to bank it up to a passable width. Yet, by blasting and chipping the rock wall and base, Company A, 316th Engineer Battalion, using its platoons in succession and working night and day, succeeded in widening the trail into a road. It was a dangerous road, especially in the dark when drivers could not see the tracing tape and had to be led by a convoy officer, but it was usable. Above Coppo there were only mule trails. For days every drop of medicine, every round of ammunition, every ounce of food and water was carried up from Coppo by mule. The trail was so narrow and dangerous that traffic-control points had to be set up along the way to enable litter bearers bringing out wounded to pass mule trains bringing up supplies. Initially the problem of supply was not as serious as it was after the regiment had advanced into the high ridge line of the Apennines. Once they had made the advance, however, it was acute until they had reached the Santerno and Futa Pass had been cleared.

There were, on the other hand, excellent reasons for attacking at this point. Within the Division sector it was possible to attack here or at Futa Pass. Futa was the most heavily defended position in the Gothic Line and had the further advantage of being easily supplied down Highway 65. The section of the Gothic Line in the 361st Infantry's zone, although heavily fortified, was not prepared in depth and was difficult for the enemy to supply. When the 361st broke through the main line of resistance in its sector, it found that the enemy had failed to solve the supply problem. Most of the prisoners captured had had no food for three or four days, and their ammunition was very low. Thus, although the sector presented great difficulties to an attack, it presented equal difficulties to the defense. The wisdom of committing the regiment in this sector was borne out by the subsequent success of its drive.

Thus, as in the lines of Tennyson, with "cannon to right of them, cannon to left of them, cannon in front of them," the 361st jumped off at 0545, 14 September, from a line north of Montepoli, with the 3d Battalion on the left and the 1st on the right. Almost immediately they met fire from all three sides. Throughout the day fire from Poggio Roncolombello was especially troublesome, and many artillery concentrations were directed there in an effort to neutralize the enemy fire. Fire was also heavy from Apparita, near the head of the bowl, and from Mt. Calvi. The capture of the latter by the 362d Infantry at 0800 somewhat relieved the pressure from the left flank however.

In the 3d Battalion sector, Company I, working directly north, was held up by small-arms fire, and companies K and L were brought around the right flank to pinch it out. The 1st Battalion advanced rapidly during the morning, but in the afternoon was slowed down by mortar and small-arms fire. By 1700 it had stabilized its position in the vicinity northeast of Nicchi along the 98.5th Northing, and the 3d Battalion, echeloned to the left rear, was stabilized between the 98th and 97th Northings. The 2d Battalion was brought up in close support to repel a possible counterattack.

It was clear from the first day's fighting that extensive use of mortars and machine guns would be necessary if any marked advances were to be made, and when General Livesay visited the regimental command post late in the day, he said, "Fire all the ammunition you can haul."

No enemy activity was reported by the battalions during the night except intermittent artillery fire, and at 0600, 15 September, the second

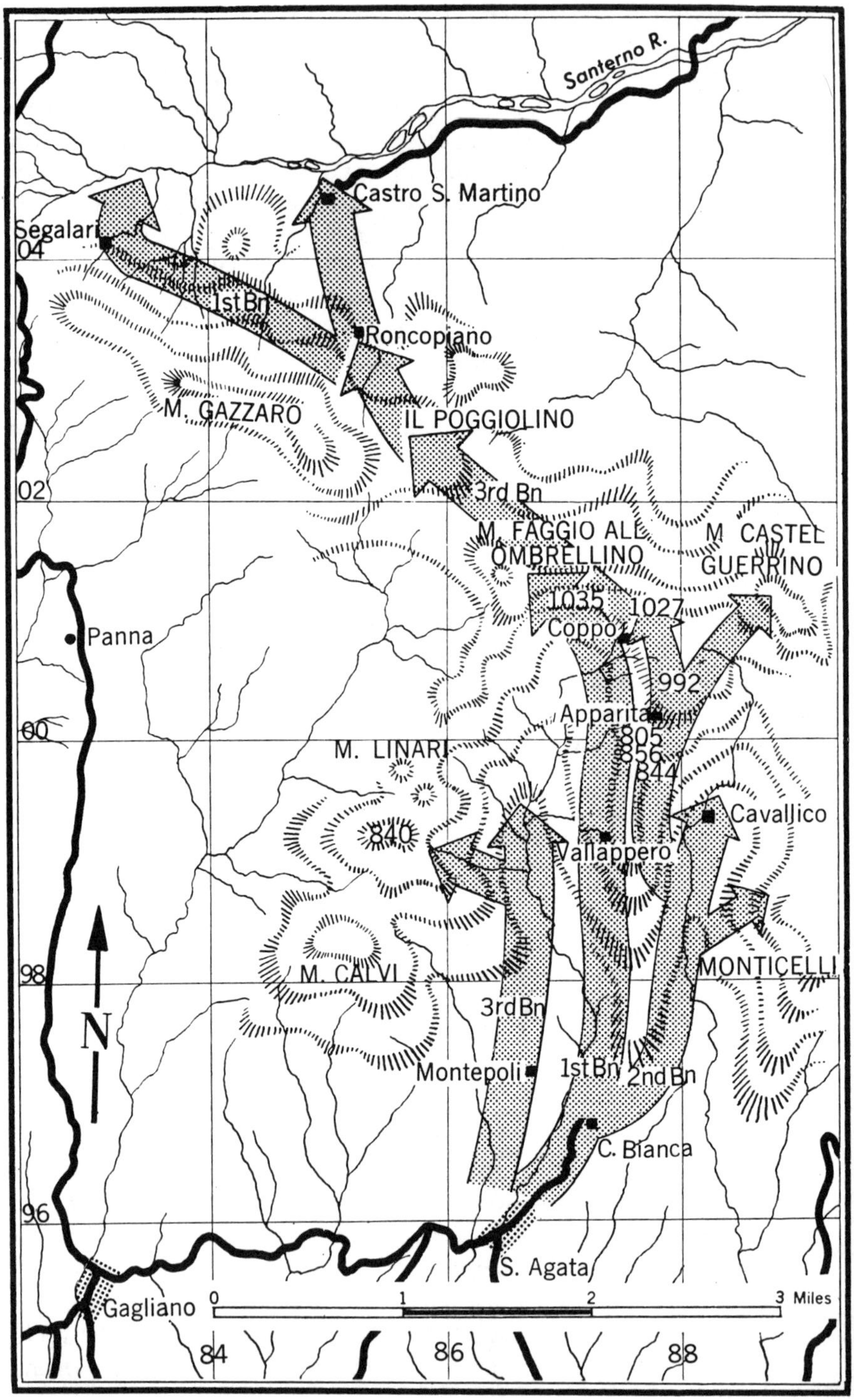

Map 15: The 361st Infantry in the center

day's attack jumped off. Both battalions advanced slowly, hindered by barbed-wire entanglements, pillboxes, dug-in positions, and heavy fire of all kinds. At one point the 3d Battalion reported that in front of it were "two banks of wire, each fifteen to twenty feet wide, with a space of twenty feet between which was undoubtedly heavily mined." Even 105mm artillery shells could not breach the obstacles, and at the end of a day of bitter struggling against such difficulties, covered always by terrific fire, very little progress had been made.

At 0600 on 16 September, the attack was again resumed, with the 2d Battalion committed on the right and ordered to attack to the northeast, capture Mt. Roncolombello, and so reduce the fire from that high point. The 1st and 3d Battalions were to continue their advance northward. As they jumped off, all three battalions were subjected to the fiercest resistance. Three times during the day severe counterattacks were received and repelled with heavy losses. One from the east, directed at the 1st Battalion, was repelled by mortars. Another, at 1345, in which the Germans drove in force from the northeast toward Vallappero in an attempt to drive a wedge between the 1st and 3d battalions, was turned back with the help of Company B, heretofore held in regimental reserve, and the 81mm mortars of Company M.

The 3d Battalion, in an attempt to neutralize the fire which had been harassing it from the west, sent Company I, followed by Company K, up a draw toward Hill 840 between Mt. Alto and Mt. Calvi.

Although losses were heavy in the day's fighting, some progress was made, especially in the 3d Battalion sector, and the line was stabilized on the 99.2th Northing.

The advance was continued next day, again in the face of very heavy mortar, machine-gun, artillery, and small-arms fire. In the 3d Battalion zone, Company M remained near Mt. Alto, which had not yet been captured by the 362d Infantry, prepared for a counterattack. Companies I and K pushed forward through barbed wire and small-arms fire to Hill 840, and here Company I repelled a determined enemy counterthrust, sustaining heavy casualties in the process. In the center, Companies A and G advanced toward Hill 856.

During the night of 17 September the regiment maneuvered the companies into better positions. Companies I and K were pulled back from Hill 840 to positions in the rear of the 2d Battalion, and Company F was released to control of the 363d Infantry to assist in the capture of Monticelli.

Three days of bitter fighting—in which pillbox after pillbox had been methodically reduced, minefields breached, and barbed-wire entanglements blown up by artillery shelling and bangalore torpedoes—resulted in remarkable successes on 18-19 September. The fighting that day was just as bitter as it had been on the previous days, but the enemy's losses in men and positions began to tell, and his attempts to reinforce his lines were only moderately successful.

This was noted along the entire Division front. Our terrific artillery and mortar concentrations and the constant drive of our infantry had taken their toll. Replacements were brought up by the enemy as early as 13 September, but they were adequate neither in numbers nor in combat training. New organizations—the Lehr Brigade, the Aufstellungstabe Brehde, and the 715th Division, as well as men from the rear-echelon units of the 4th Parachute Division—began to appear among the prisoners of war. Getting replacements was no small task for the enemy. A report on one prisoner's statement gives an idea of this:

> After being stationed in the Santa Lucia area for some time the 4th Company left for their present position 14 September. They were repeatedly attacked by our [American] bombers on the way and suffered heavy casualties. Company was committed night of 16 September. PWs state that the personnel of the company consists mostly of very young recruits and that morale is very low. Many men lost their weapons on the march to the main line of resistance because they were too exhausted to carry them.

By 19 September enemy disorganization was mounting rapidly, and captives flowed through the prisoner-of-war cage. Of the 896 prisoners captured between 9 and 30 September, 502 were taken in the four-day period 18-21 September. The preamble to the "Interrogation of Prisoners of War Report" of 19 September describes the situation thus:

> PWs present a good picture of the disintegration of the German lines. Most of them are stragglers who were not able to withdraw with their units. Others belong to reserve units which were committed only one or two days ago to stem our advance; none of the fresh units were in an orderly fashion caused by our tremendous and accurate artillery fire. Casualties were very high but exact figures could not be obtained due to the general confusion of the German troops. During the night from 17-18 September an order for a general withdrawal was apparently given. Many units, however, could not be reached and simply were left behind.

Thus, although much hard fighting lay ahead, the enemy had begun to crack. In every regimental sector the deterioration was noticed, and the tempo of the advance picked up.

This was especially true in the 361st Infantry's zone. By 0650, Companies A and G were reported on Hill 856, and at 0811 Company E reported Hill 844 completely occupied. The capture of Hill 844 was particularly important, for it had been the most strongly fortified hill facing the regiment, and its loss seriously crippled the enemy's defensive position. Companies A and G, working up the left side of the ridge toward Hill 856, had bypassed the heavily fortified pillboxes facing south and southeast on the broad shelf halfway up Hill 844. An elaborate system of communication trenches connected these positions to the rear. With the reduction of the defensive positions on Hill 844 and the capture of Hill 856, the position became untenable, and the Germans were forced to retreat. So determined were they to deny the use of this ridge to our troops, however, that almost immediately following its capture the entire area was literally sprayed with 40mm AA and 88mm artillery fire. Our lines did not waver, however, and after a brief period, the advance continued.

At 1000, 18 September, General Livesay designated new objectives for the 361st. The 1st Battalion was to push north to Hill 805 and Hill 992, and the 2d Battalion was to capture hills 997 and 947. Once the hills were secured, both battalions were to drive to the ridgeline at hills 1022 and 1027, and Mt. Castel Guerrino (Hill 1117). As the afternoon wore on, the disorganization of the enemy became more and more apparent. They took hasty positions for brief stands and then ran back to others. By 1530, Hill 805 had been taken and organized by Companies B and G. Bombing missions were flown against hills 1022 and 1027 four times in the late afternoon in a further effort to disorganize the German withdrawal and prevent a successful stand on the ridgeline.

In accordance with Field Order No. 18, issued at 1630, 18 September, the attack continued at 0500, 19 September, under a tremendous rolling barrage. Both the 1st and 2d Battalions advanced rapidly, and hills 992 and 947 were soon taken. Companies B and G pressed on to take Hill 1022 and Hill 1027, while Company E pushed toward Hill 1117. By 1525 the 361st Infantry had secured its portion of the Division objective, with the exception of Hill 1035. Company M continued to assist the 362d in its drive on Mt. Alto, and Company E had established contact with the 363d on Mt. Castel Guerrino.

Since the 363d had broken through Monticelli and was advancing north to secure the Division's right flank, the 361st was ordered to follow the ridgeline to the northwest. Company C was to secure hills 1035 and 1056, and the 3d Battalion, after passing through C Company, was to seize Il Poggiolino. Both units, advancing across extremely rugged terrain, made more difficult by rain, moved forward against light resistance, and the 3d Battalion occupied Il Poggiolino with two companies at 1845, 20 September. At the same time the 1st Battalion, 362d Infantry, advanced and seized Mt. Gazzaro. With its left flank protected, the 3d Battalion of the 361st moved northwest to seize Roncoppiano and then sent patrols to Castro San Martino. Without opposition the battalion occupied hills 763, 613, 614, and 617, the high ground overlooking the Santerno. The 1st Battalion, which had been closely following the 3d, occupied positions from Segalari east to Hill 705. The 2d Battalion, meanwhile, originally drawn back for possible use with the 362d Infantry in its drive on Futa Pass, remained at Panna occupied with rehabilitation of its troops.

In their positions along the Santerno, the 1st and 3d Battalions were very nearly cut off from their supplies. Mule trains from Coppo and Panna constituted the only means by which ammunition, medical supplies, and food could be brought up, until the lateral trail running east along the Santerno from Futa Pass could be used. The opening of this road depended entirely upon control of the pass, the north approaches of which had not been secured on 21 September. Even litter cases still had to be evacuated by pack mule down the long and treacherous trail to Panna, but by 1600, 22 September, the 1st Battalion, 362d Infantry, was well north of the pass.

During 23 September Company B took positions immediately east of the road junction in Futa Pass, covering it with machine guns. Other elements of the 361st's 1st Battalion sent patrols across the Santerno, but they were driven back by enemy fire. As the line south of the river stabilized, every effort was made to rehabilitate the troops; to supply hot food and clean clothes; and to allow as much rest to the weary men as the tactical situation permitted.

While the 363d Infantry was making an all-out assault on Monticelli on the left and the 361st was smashing at the center of the line toward the high mountains on the north, the 362d was fighting along Highway 65 toward Mt. Calvi and Futa Pass. This area was highly organized, and fields of fire had been cleared by the enemy, making it necessary to advance in many places without cover. On 12 September the 1st Battalion, on the right, jumped off for Mt. Calvi and made good progress. On the left, the 3d Battalion, along Highway 65, met considerably stronger opposition. At the end of the day the 3d stablized its positions along the 97.5th Northing on Highway 65, and the 1st was on the southern slope of Mt. Calvi. During the night the 2d relieved the 1st for the attack at 0600, 13 September.

The fighting during the next three days was very bitter. Progress was slow, and losses were heavy. The 362d Infantry, like the other two regiments, fought its way with unwavering determination and dauntless courage from pillbox to pillbox, through barbed wire and minefields, and always the enemy had excellent observation and prepared fields of fire. During daylight of 13 September both battalions received very heavy artillery, mortar, and small-arms fire. By nightfall the leading company of the 3d Battalion had dug in at the 97.6th Northing after an advance of only 800 yards, and the 2d had stabilized its positions along the 97.7th Northing, on the slope of Mt. Calvi.

The 2d Battalion attacked again at 0400 the next morning and at 0635 one platoon of Company G had fought its way through the defenses of Mt. Calvi's southern slope and was fifty yards from the crest. By 0800 Companies G and E had passed over the crest and were climbing down the steep northern slope.

In the saddle at the bottom of this slope was the beginning of the enemy's second or switch line, from which the ground rose rapidly to a point 800 yards north, which was 120 feet higher than Mt. Calvi itself. Fire from these defenses forced the 2d Battalion to hold up its attack on hills 821 and 840, immediately to the north, during daylight hours.

The 2d Battalion had taken Mt. Calvi the hard way, by direct frontal assault in the night. It was the only way, however. Ground for a flanking attack was not yet in our possession, and even if it had been, the enemy had prepared his defenses with a view to stopping an attack around either flank of the mountain. The defending paratroopers said they had expected to stay there until spring.

The capture of Mt. Calvi was the first rupture by any element of the Fifth Army of any portion of the Gothic Line. It deprived the enemy of very superior observation and took a tremendous amount of pressure off the 361st and 362d Infantry Regiments. The attack on Mt. Calvi was pushed forward to success by individual squads and platoons with bayonets, rifles and grenades. The surprise of the attack hour, 0400, without preliminary artillery fire, helped at the start.

On 14 September the 3d Battalion advanced 800 yards nearer Futa Pass, while the 1st cleaned out enemy pockets in the rear of the 2d and secured the western approaches to Mt. Calvi. A platoon of Company C was attached to the 2d Battalion to garrison the crest of

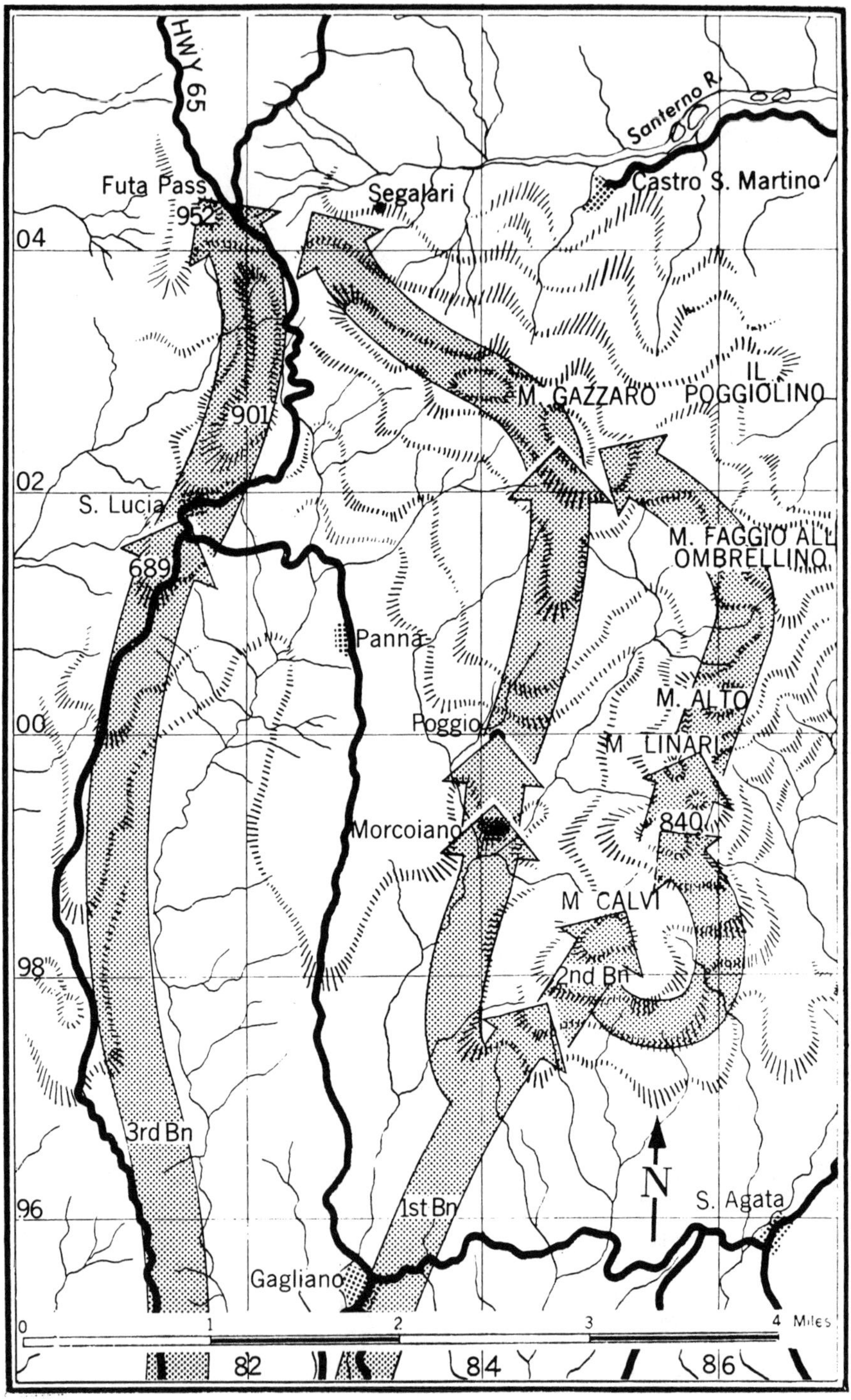

Map 16: The 362d Infantry on the left

German antitank ditches wind across the Italian countryside south of Futa Pass

the mountain. Later, all of Company C (less one platoon) was ordered to the top of the mountain and, due to the nearness of enemy small-arms fire and the bareness of the crest, had to organize a reverse slope defense, just in the rear of the mountain top.

The 15th of September was spent mainly in maneuvering for better positions. Only on the left, in the sector of the 3d Battalion, was the opposition light, and appreciable gains, to the 99.2th Northing, were made. The 2d Battalion consolidated its positions on Mt. Calvi and maneuvered into position under heavy machine-gun, mortar, and artillery fire from the ridge to the north. After nine tremendous artillery concentrations had been placed on the enemy within fifteen minutes, the 1st Battalion attacked at 1230, 15 September, according to a plan worked out with General Clark and General Livesay at the command post. Skirting the left flank of Mt. Calvi, the 1st Battalion struck north toward Morcoiano, and at 1732 had reached a point about a kilometer and a half below the town.

At 0430, 16 September, after a very heavy artillery preparation, the attack was continued, with all three battalions engaged. Again opposition on the left was light, and the 3d Battalion moved rapidly north until it reached a gigantic antitank ditch across Highway 65. The ditch was about 8,000 yards long, with an average depth of eight feet, and fifteen feet across at the top. On either side of it, barbed wire had been strung. Mines had been laid in and around it, and covering minefields lay both in front and in the rear. The Battalion stopped here and sent patrols as far as Santa Lucia to secure information, maintain contact, and probe the nature and extent of the enemy defenses.

In the 2d Battalion zone some progress was made toward Hill 840, but barbed wire, covered by machine guns, made the advance most difficult. Moving from Mt. Calvi to Hill 821 and Hill 840 involved some of the most difficult fighting the regiment had yet encountered.

Mt. Calvi has a gentle, bare forward slope, which rises to a rounded top 2,454 feet high. It then drops abruptly to the foot of Hill 840. At some points the drop is as much as 500 feet in 200 yards. The valley between Mt. Calvi and Hill 840 was so steep and narrow that even high-angle artillery fire was unable to reach the Germans dug in below. Because of this, the 346th Field Artillery, in direct support of the 362d Infantry, was only partially successful with its fire, and the infantrymen met much stiffer resistance than in areas where the artillery could reduce pillboxes and concrete emplacements.

Descent of the precipitous northern slope of Mt. Calvi, breaching the barbed-wire entanglements and numerous minefields ahead and getting through an area completely zeroed in by enemy mortar, artillery, machine-gun, and sniper fire called for the combined qualities of a mountain goat and Superman. However, according to plan, Company G, from its position to the left of Company E, moved forward to seize Hill 821, and Company F, from its reserve position in the rear, passed through the right elements of Company E toward Hill 840. Under cover of darkness excellent progress was made by both attacking companies.

Daylight found Company F halfway up Hill 840, with its leading elements within 200 yards of the crest. The heaviest enemy defenses had been bypassed in the darkness, and hostile fire was being received from all directions. During the ensuing fire fight, the weapons platoon became separated from the company, and its leader became a casualty.

"Westfall promptly killed all three with one burst from his submachine gun."

Staff Sergeant Welton C. Westfall immediately assumed command and, under blistering automatic fire, reorganized the platoon, led his men to a better position, and deployed them in vantage points from which they could return the enemy's fire. After regaining contact with his company, he directed his platoon against two German counter-attacks. He moved from one gun position to another, adjusting fire and encouraging his men, unmindful of the bullets which were barely missing him. Later in the day, when several men were wounded in a minefield, Sergeant Westfall entered the dangerous area, personally carried some of the wounded to safety, and guided litter bearers into the field to evacuate others. At another time during the company's attack, he was confronted suddenly by three heavily armed Germans determined to impede the advance. He promptly killed all three with one burst from his submachine gun. During the entire engagement, he performed heroic deeds, minimizing casualties in his platoon and facilitating the advance of Company F.

To the left of Company F, Company G had experienced the same kind of vicious fighting in its attack on Hill 821. Discouraging news was received at 1000: the company commander was seriously wounded, and all other officers were gone. Technical Sergeant John C. Brooks took command.

In the meantime, the main enemy positions bypassed by Companies F and G during darkness were engaged by Company E in a fierce fire

"Hamilton continued forward and killed six Germans in the ensuing fire fight."

fight. In an attempt to improve the company's position, several men tried to flank the enemy, but were checked by heavy automatic fire. Private First Class Harry E. Hamilton, an automatic-rifle man, boldly made his way to a forward position and opened fire on the Germans. Silencing an enemy machine gun with several bursts from his BAR, he continued forward and killed six Germans in the ensuing fire fight. Three Germans attempted to man an abandoned machine gun, but Private First Class Hamilton disposed of them quickly. Later, he repulsed a counterattack by thirty Germans with a heavy volume of fire from his automatic rifle. Following this unsuccessful attempt to dislodge Company E from its position, a force of 75 Germans launched a counter-thrust against the company's flank. Locating an enemy machine gun supporting the attack, Private First Class Hamilton arose and fired at it from his hip. This inspired his comrades to force the withdrawal of the enemy. Private First Class Hamilton was wounded severely during the fight, but he continued to fire at the enemy the remainder of the day. Toward evening the remnants of Company G were in position alongside of Company E, and Company F was ordered to hold its dearly won position on Hill 840.

In the light of later events, it is known that the 2d Battalion's attack on 16 September was a partial success, for one platoon of

Company G had actually reached the top of Hill 821, where it held out for two days against heavy enemy fire and counterattacks. The platoon's radio was put out of action during the passage of the first defensive belt, and runners sent back were not able to get through the enemy lines while their information was still of value, so the platoon's success was not known for many days.

Part of Company F (some forty men) arrived at a point about two hundred yards from the crest of Hill 840 and held until nearly noon, when, after reporting (by radio) that the enemy had almost surrounded them, they were ordered to withdraw down the long southeast nose of Hill 840 into the 361st sector. They arrived with ten prisoners about dark on 16 September, after suffering more casualties in the minefields, including several of the prisoners. Company E and the remainder of Companies G and F had meanwhile been fighting in the first defensive belt, but found they were not strong enough to put it out of action or to penetrate it. When the commander of Company G was wounded by an enemy grenade which burst in his face, they were ordered back to their original positions and started to infiltrate up the northern slope of Mt. Calvi back into our lines. The enemy positions were all in well timbered bunkers and emplacements, and it was impossible for our artillery to furnish adequate fire support because of the steep grade. By dawn, 17 September, between 40 and 50 men from each of these companies had arrived back under battalion control. The 2d Battalion commander and his staff were making plans and reconnaissance to take Hills 821 and 840 by moving through the 361st area and attacking again over the same ground.

On the left of the 2d Battalion, the 1st, which had been ordered to jump off at 0600, was unable to move on schedule because of the sharp counterattack directed at Company B, but after reorganizing, it began its attack and throughout the day was subjected to an unmerciful pounding by enemy mortars, artillery, and machine guns. However, the battalion reached a point just below Morcoiano and held there for the night.

As the situation developed in the 362d's sector, it became apparent that the 3d Battalion, considerably detached from the rest of the regiment and advancing north along Highway 65, was expected to maintain constant pressure, but, in General Livesay's words, its members were not to try "to win the war all by themselves." The main effort was to be made along the high ridgeline which constituted the boundary between the 362d and 361st Regiments, running north from Mt. Calvi, through hills 821 and 840, Mt. Linari, and Mt. Alto, to

Mt. Faggio All Ombrellino, where it joined the main axis of the Apennines and swung northwest along Mt. Gazzaro and Il Poggiolino to Futa Pass. This disposition of troops, however, left a gap in the center, on the low ground along the Gagliano–Panna road. To protect it, on the night of 16 September, elements of the 91st Reconnaissance Troop, reinforced with tanks, formed a defensive line approximately two kilometers north of Gagliano.

The 1st Battalion attacked at 0500, 17 September, with Morcoiano as its objective. By noon Company B was in position to assault the town, with Company A, supported by tanks, flanking to the left. During the afternoon, Company I was relieved from the 3d Battalion, since the 3d was expected merely to hold at the antitank ditch, and attached to the 1st Battalion. The 2d Battalion continued its dogged fight for Hill 821, with Company F leading.

At 0340, 18 September, Company B was fighting in the southern outskirts of Morcoiano, with Company A to the west. Company I, brought from the 3d Battalion area by truck, took up positions behind Company B. After Companies A and B had secured Morcoiano, Company I passed through them and pushed north toward the town of Poggio. Shortly after midnight it encountered strong enemy resistance, and at 0410, 19 September, when it attempted to take the town, the resistance became so strong that it withdrew and called for artillery fire. Meanwhile, on 18 September, the 2d Battalion continued to try to take Hill 821. Since the advance from the southwest had been disappointing, most of the day was spent in maneuvering to the right so that Hill 840 could be attacked from the southeast. At 0705, 19 September, the 2d was moving up the southeastern slope of Hill 840.

At 0700, 19 September, behind a rolling barrage, Company I assaulted Poggio. The Germans had fortified the town, making it a strong-point that had to be cracked before the Division objectives could be taken. The barrage laid down on it by the 346th Field Artillery Battalion was considered the most nearly perfect of any ever fired by the battalion. Fire was adjusted across the 1st Battalion front by Captain Robert E. Scott, Jr., liaison officer, who watched the infantry's advance from his observation post and moved the barrage forward appropriately. The artillery fire did not smash the fortifications, but it forced the defenders to seek cover and button up completely. When the fire moved past a given point, before the enemy could jump out of their holes to man their weapons, the infantry, a scant three hundred yards behind the barrage, was on them. Two hundred prisoners were taken, and Poggio fell. The attack literally walked through a strongpoint that would ordinarily have been the scene of bloody and prolonged fighting. The rest of the day was spent in cleaning out the town, reorganizing the companies, and preparing to attack Mt. Gazzaro, the next objective.

Meanwhile the 2d Battalion continued to move over Hill 840 to Hill 821, both of which were secured. Patrols were sent to Mt. Linari, and, advancing rapidly to keep contact with the enemy, now driven from his main line of resistance, the 2d occupied Mt. Alto during the night of 19-20 September. Here it reverted to regimental reserve and spent the next two days cleaning out the draws and valleys and mopping up isolated pockets of resistance which had been bypassed.

Behind successive concentrations of artillery fire, the 1st Battalion pushed off at 0600, 20 September, in its assault on Mt. Gazzaro. The approaches to this key terrain feature, as was characteristic throughout the entire main defenses of Futa Pass, were heavily guarded by a series of mutually supporting, well fortified emplacements. Company B, pushing forward on the right of the 1st Battalion's zone of attack, encountered several of these enemy positions shortly after the attack jumped off.

Private First Class John Czinki, Company B, noticing enemy movement near a draw on the right flank of his company, moved up to investigate. Armed with his rifle and two grenades, he approached a dugout from which a German soldier was emerging. He fired and wounded the German and then continued his bold advance toward the fortified emplacement. As he reached the entrance to the dugout, several Germans appeared and surrendered to him.

He approached an adjacent pillbox with his prisoners in front of him and ordered its occupants to surrender. When they refused, he threw a grenade, without pulling the pin, into the enemy position. Fifteen Germans emerged and surrendered. Keeping his prisoners under control, he methodically visited three other fortified emplacements in the vicinity, each time approaching the entrance and ordering surrender. He neutralized five enemy positions and captured forty-three prisoners, eliminating a decided menace to his battalion's right flank and greatly facilitating the early capture of Hill 896.

Later, when the advance of the 2d Battalion was seriously threatened by increased resistance, Private First Class Czinki volunteered to act as company observer in a building under heavy mortar, artillery, and direct self-propelled-gun fire. While direct hits threatened to level the building, he remained at his post for an extended period and furnished his company command post with valuable information on

"Czinki neutralized five enemy positions and captured forty-three prisoners."

enemy dispositions. During one of his trips to the command post to report, an enemy self-propelled gun scored a direct hit on a corner of the building. Rushing toward his observation post to locate the hostile weapon, Private First Class Czinki was mortally wounded by shell fragments. His courage and fighting spirit made him an outstanding example of the individual initiative and selfless devotion to duty that contributed so materially to the destruction of the crack German forces opposing the Division.

Progress was slow for the 1st Battalion following the capture of Hill 896, and by nightfall the general line was stablized on the 02.3d Northing, along the southern slope of Mt. Gazzaro. Two platoons however, were on top of the mountain, whose 3,714-foot elevation made it the highest point in the zone of the 91st Division.

After a quiet night the battalion moved off at 0430, 21 September. By 1015 three companies—B, E, and I—were on Mt. Gazzaro. Company I advanced without pause along the ridgeline to the northwest to Futa Pass. Advances were rapid, the extremely rugged terrain being the most important delaying factor. By 2130 Company A had reached positions on the high ground overlooking the pass from the southeast and had machine guns trained on the pass. Company I, ready to rejoin its battalion, was in position on the left of Company A, and Company C had returned to the 1st Battalion when the 2d arrived on Mt. Gazzaro.

One of the Tiger tank turrets guarding Futa Pass blasted from its concrete emplacement by Division Artillery

While the 1st and 2d Battalions were advancing on the right, the 3d Battalion on the left flank of the 362d had made substantial gains along Highway 65. Operating almost alone—with the closest unit more than a thousand yards away—the 3d had battled its way to within 1,500 yards of the formidable Santa Lucia position, which guarded the southern entrance to Futa Pass. On the morning of 16 September Company L seized strategic Hill 588 on the left of Highway 65, and Company K captured Casa Sodi on the right. Both were high-ground positions from which artillery could be directed against the German line. The defenses consisted of the anti-tank ditch the 1st Battalion had met on the right and interlocking fields of fire hinged on prepared strongpoints on the forward slopes of Hill 689. One Tiger tank's 88mm gun and turret was emplaced in a steel tunnel on the summit just off Highway 65, and other concrete casemates and pillboxes were situated on the right. There were also several self-propelled artillery pieces in the rear.

Against the classic maneuver of attack through draws, the kind of defense for which it was designed, this organization was probably invulnerable. But against a drive along the high ridgeline to the west, it proved defective.

The first problem which faced the battalion as it stood on the threshold of Futa Pass was to determine whether or not the position was manned in strength. Lieutenant Colonel George R. White, commander of the 3d Battalion, sent out two patrols early in the morning of 17 September. Company K's patrol, moving forward under the cover of darkness, succeeded in reaching the antitank ditch, where it heard Germans on the other side and encountered numerous minefields and barbed-wire entanglements. Company L's patrol, leaving at daybreak, drew so much fire from the left flank that it was unable to advance.

Realizing that the enemy was determined to hold the position, Colonel White established two observation posts, and for two consecutive days, 18 and 19 September, the 346th Field Artillery, reinforced by Corps artillery, pounded Santa Lucia to soften it for the attack. The artillery fired on every movement attempted by the enemy. The Tiger tank turret on Highway 65 was blasted from its mount. Two 105mm self-propelled guns were also destroyed.

When Company I was attached to the 2d Battalion on the night of 17 September, one platoon from Company K and one from Company L were placed in battalion reserve. The battalion consolidated its lines so that it could act as its own protection against harassing attacks directed at its dangerously exposed flanks.

On 20 September the 3d Battalion attacked north to seize Futa Pass. The plan was based on an idea that the Division's drive on the right flank, coupled with the enemy's confidence in his prepared defenses, would have thinned the German forces opposing the attack. Also it was felt that in attacking over the ridges instead of through the draws, an element of surprise would be achieved. Both possibilities materialized, and the surprise was so complete that one German officer was captured while taking his morning walk.

By 1230 Company L had reached the ditch. It pushed across during the afternoon against medium resistance, and by 1700 it had battled up the forward slopes of the important Hill 689, 350 yards southeast of Santa Lucia. The rolling barrage which accompanied the attack was so heavy that the infantrymen were able to move in and wipe out the enemy emplacements before the Germans could fully man their weapons.

That night the battalion held its position on Hill 689. It attacked again at 0600, 21 September, behind another tremendous rolling barrage. Company L moved forward on the left, while Company K crossed from the west to seize the town of Santa Lucia. German 120mm mortar crews kept up a running fight, delivering heavy concentrations all day. Strong small-arms resistance was also met, but, again under the cover of artillery fire which was seldom more than 300 yards ahead of the front-line troops, the two companies moved ahead, and by 1415 they had seized Hill 901, key position along the western ridge of the pass. That night they outposted inside Futa Pass in preparation for a final all-out assault against Hill 952 the following day.

Company I, which had been fighting with the 2d Battalion since 17 September, moved over the eastern ridge to rejoin the 3d for the attack against Hill 952. Attacking in the early morning of 22 September, with three companies abreast, the 3d fought all day against heavy artillery, mortar, machine-gun, and small-arms fire. Under the protective cover of close-supporting Division and Corps artillery, it slowly inched its way up Hill 952 and by nightfall had outposted positions on the summit.

The capture of Hill 952 was of tremendous strategic importance to the Division because it commanded the vaunted Futa Pass position and made untenable the great fortifications there. It also was the culmination of the Division's twelve-day battle to crack the Gothic Line. With the fall of Futa Pass, the Gothic Line, already outflanked and turned by the drives of the 361st and 363d Infantry Regiments, was broken. The taking of Hill 952 was the end of the second phase of the month's campaign, and the Division was ready to begin the third phase—the drive for the great city of Bologna.

After the 362d's capture of Futa Pass, General Livesay received this wire from General Keyes:

CONGRATULATIONS UPON THE CAPTURE OF FUTA PASS AND THE STEADY PROGRESS OF THE NINETY FIRST UPON THE CASTRO–FREDDI HILL MASS. THE NINETY FIRST HAS INDEED WON ITS SPURS.

On 22 September 1944, then, the Powder River Division had occupied an outpost line approximately one kilometer north of the Santerno River and extending from Futa Pass to the Violla River. Five battalions were deployed on the line.

It was the end of a brilliant campaign, which had completely destroyed and overrun the main defenses of the Gothic Line in twelve days—and the German *Organisation Todt* had worked for over a year to prepare them! Pillboxes; concrete emplacements, some so thick 105mm shells bounced off them like peas off armor plate; barbed wire; tank guns mounted in concrete turrets; minefields; elaborate antitank ditches, and precipitous mountain terrain—this was the Gothic Line. The position of each fortification had been most carefully planned so that bands of fire interlocked, every probable approach was cov-

Two infantrymen of the 362d look back over the area they fought to take Futa Pass

ered, and every emplacement was mutually supported by another. Near each firing position, caves had been blasted from rock or concrete rooms had been built as dormitories, storage rooms, or ammunition dumps. Acres of timberland had been cut and buildings razed to make unbroken fields of fire. Distance and azimuths had been recorded to every probable target within range of the weapons.

All the fortifications had been constructed in mountainous terrain which in itself would have constituted a serious hazard to an advancing enemy. Rocky and precipitous, unsuited to maneuver, and without natural avenues of supply, the Apennines were a formidable barrier even without man-made embellishment.

Since the stakes were high, the Germans did not depend solely on material obstacles, however. To man the fortifications they sent the 4th Parachute Division, one of Hitler's favorite divisions in Italy. Numbering approximately three thousand men, of which very few were *Volksdeutsche,* the division was strengthened between 25 August and 4 September with approximately 680 replacements. Later, after it had been seriously cut up by casualties and captures, it received numerous replacements from the Lehr Brigade, the 715th Division, the 305th Reconnaissance Battalion, and the 1060th Regiment. So the 91st Division had smashed Hitler's best troops, manning his most carefully constructed line of defense in Italy.

An All-out Fighting Team

This period in the history of the 91st Division cannot be closed without more detailed mention of the artillery, the engineers, the medical personnel, and the other supporting units who so magnificently fulfilled their roles in the campaign. Truly it could be said that the *Division* cracked the Gothic Line, because each unit of the Division contributed nobly to the accomplishment of the task. This period exemplified the finest possible application of the power of teamplay.

The 316th Medical Battalion, its equipment and staff strained by 3,500 casualties, did magnificent work. Litter bearers carried patients over narrow, slippery mountain paths, through minefields and barbed-wire entanglements, and over streambeds. The good condition of the patients when they arrived at aid stations was a tribute to the excellent care rendered by the company aid men and the painstaking efforts of the litter bearers. To evacuate wounded to the rear, ambulances operating under direct enemy observation made seven hundred trips—10,800 miles of driving. One hundred and forty men, wounded in small mountain draws and other inaccessible places, were carried by litters two miles and more over dangerous rocky paths. Yet without thought for themselves, the medical men worked to treat the wounded and evacuate them from the battlefields. Despite the enemy, the terrain, and the weather, the battalion evacuated all casualties through the Division clearing station in the rapid average time of six hours.

For the 316th Engineer Battalion the drive from the Sieve to the Santerno was a continuous nightmare. The roadnet in the Division sector was poor, and, damaged by shelling, demolition, and rain, what roads there were became almost impassable. By working night and day, however, the engineer battalion made possible the flow of supplies to the forward elements. They swept and maintained 99½ miles of road and built 23 bypasses and 19 culverts. They built bridges, sometimes abreast, or ahead, of the front lines, and roads where there were no roads before. Their greatest task and their greatest success was in keeping Highway 65, the Corps' main supply route, open. Giant craters, one of which required 75 truckloads of rubble and rock to fill, were bypassed or filled. Bridges were built, and when they were washed away by rain-swollen streams, they were rebuilt. Because of the engineers' untiring efforts, ammunition, medical supplies, and food reached the front-line troops.

Much of the credit for breaching the Gothic Line and cracking the spirit of its defenders must be accorded Division Artillery, augmented by II Corps artillery. During the campaign the Division controlled the fire of 168 guns. From 11 to 22 September, inclusive, 94,379 rounds were fired, and during a single 24-hour period on 15 September, 14,321 rounds were fired. On 14 September, two days after the campaign began, a prisoner from the enemy's 11th Parachute Regiment testified to the effect of this crushing artillery upon enemy troops. The intelligence report on him says:

> The morale effect of our artillery fire is high and prevents them from coming out of their foxholes. Casualties, however, are very few, as their positions are well prepared and afford sufficient cover. Phosphorus shells excite their anger but are generally preferred to fragmentation shells, as "they would rather be burned than killed."

As has been noted before, prisoners again and again were dazed and stunned by the artillery fire to which they had been subjected. The heavy artillery fire held the enemy helpless in their emplacements, unable to ward off death or capture by the infantrymen with grenades and automatic weapons who swiftly followed up the concentrations. The extensive use of rolling barrages, especially by the 362d Infantry, is a noteworthy application of this technique of advance.

On numerous occasions accurate artillery fire reduced positions which would have been extremely difficult for the infantry to take alone. Much of the trouble experienced by the 1st and 2d Battalions of the 362d Infantry may be explained by Mt. Calvi's masking of artillery fire. Here the infantry was forced to slowly reduce positions which normally artillery might have knocked out with a few well placed rounds. In other sectors, emplacements too large for Division Artillery were rendered helpless by the power of the Long Toms, 8-inch guns, and 240mm howitzers of II Corps. Finally, after the 4th Parachute Regiment had been seriously depleted and the enemy was trying to bring up replacements, the artillery directed telling fire on the men as they were being moved into line. In one night two hundred enemy replacements were killed by artillery fire as they neared the lines. Thirty-six hours after the Lehr Brigade entered the line, three or four company commanders had been killed, and the unit was almost completely disorganized. Throughout the campaign the artillery supported the infantry superbly with accurate and effective fire.

All along the line, this same spirit of unified effort prevailed. The 791st Ordnance Company, the 91st Signal Company, the 91st Quartermaster Company, and the MP Platoon carried on their tasks of backing up the forward echelons with untiring devotion.

The 91st Division was a coordinated fighting unit, strong in the combined strength of its component parts. It was the Division as a whole which captured Monticelli and Mt. Calvi and fought bitterly for hills 840 and 844. It was the Division as a whole which advanced through rain and fog over steep and rocky terrain along the ridgeline of the Apennines to the Santerno River. It was the whole Division which refused to be a holding force and swept northward along Highway 65 to capture Futa Pass, the most strongly defended position in the Gothic Line. Each man in the Division had acted as if he wanted "to win the war all by himself," and the tales of heroism and gallantry are legion. In twelve days' time, the Division had reduced to nothing a year's work of thousands of impressed laborers and had decimated the best troops Hitler could put in the line against it.

Phase III

The third phase of the 91st Division's operations during the month of September began at 0530 on the 24th. By 22 September the outpost line, approximately one kilometer north of the Santerno River, had been manned by five battalions from Futa Pass to the Violla River. During 22 and 23 September all regiments made efforts to rest and reorganize before starting the northward drive to Bologna. Vigorous patrolling maintained contact with the enemy, and contact was established with the 34th Division on the left and the 85th Division on the right. The engineers continued to work on the roadnet and improvement of Highway 65, while the Reconnaissance Troop reconnoitered the roads to the north.

The next defensive position in the II Corps zone was expected to be anchored on Mt. Bastione in the 34th Division sector, Mt. Oggioli in the sector of the 91st Division, and Mt. Canda in the sector of the 85th Division. South of this line lay Mt. Freddi and Mt. Beni, both just west of Highway 65. The character of the area was quite different from that of the Gothic Line. Mt. Canda, Mt. Oggioli, and Mt. Bastione dominated the surrounding hilly country and furnished excellent observation over the entire front. The approaches to these heights and to Mt. Freddi and Mt. Beni were open, and any advance could easily be seen by the enemy. Throughout this period the Division's flanks were exposed to artillery and mortar fire from both east and west, directed from Mt. Canda and Mt. Bastione. The mountains themselves were sheer, rocky peaks, with little cover and no concealment, from which the enemy could be routed only by the hand-to-hand combat of

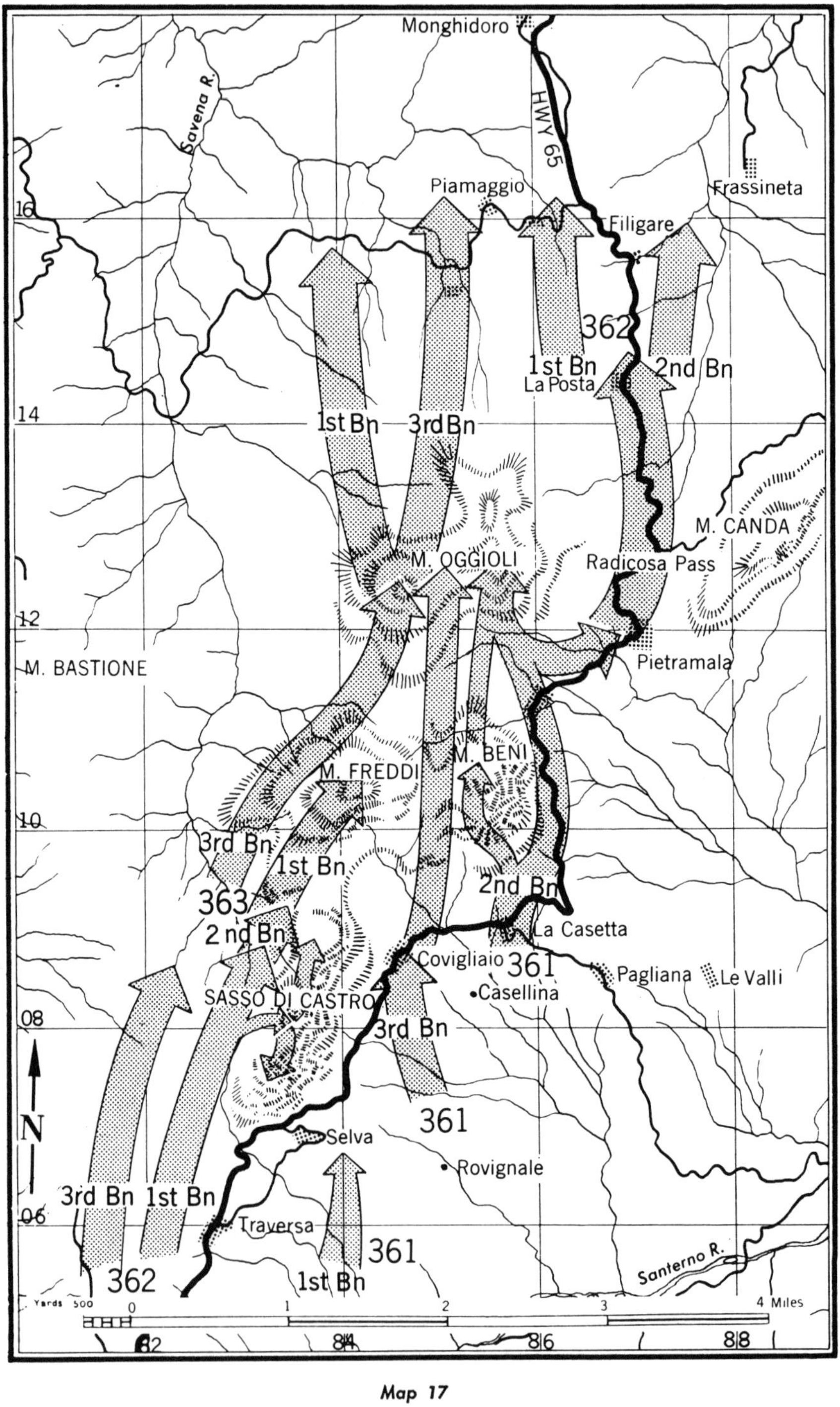

Monghidoro
HWY 65
Savena R.
Piamaggio
Frassineta
Filigare
362
1st Bn
2nd Bn
La Posta
1st Bn
3rd Bn
M. CANDA
M. OGGIOLI
Radicosa Pass
Pietramala
M. BASTIONE
M. BENI
M. FREDDI
3rd Bn
1st Bn
363
2nd Bn
2nd Bn
La Casetta
Covigliaio
361
Pagliana
Le Valli
Casellina
SASSO DI CASTRO
3rd Bn
361
N
Selva
Rovignale
3rd Bn
1st Bn
Traversa
361
Santerno R.
362
1st Bn
Yards 500
4 Miles

Map 17

an aggressive infantry. The natural defensive features of the terrain made the area stronger, in that respect, than the Gothic Line.

On 23 September General Livesay ordered the Division to attack north to Mt. Oggioli on the following day. The main effort was to be made west of Sasso di Castro. When the 362d Infantry, which was to attack on the left of the Divisional sector, had bypassed Sasso di Castro, it was to be prepared to assist the 34th Division in capturing Mt. Bastione. On the right, the 361st was to maintain pressure on the enemy, with special attention to Selva, Covigliaio, Pagliana, and Le Valli. Attached to both regiments were elements of the 755th Tank Battalion and the 804th Tank Destroyer Battalion. Maximum use was to be made of this armor, wherever possible, to reduce sniper and machine-gun fire.

The opening phases of the campaign for Bologna, 24 to 30 September, proved to be less difficult than had been anticipated. As the advance developed, it became apparent that the enemy had left strong delaying forces on Mt. Freddi and Mt. Oggioli, and that a strong defensive line of hasty fortifications was being set up at Monghidoro. Much of the difficulty experienced by the Division was due to the weather. Rain, cold, and fog hampered movement, and the visibility was rated "poor" or "limited" on eight of the last nine days of the month.

On 23 September the 361st Infantry sent out three patrols of platoon strength. After a half-hour artillery preparation, one platoon moved out at 1320 for Rovignale, one for Selva, and one for Casellina. By 1515 the patrol to Casellina had reached its objective without incident, and at 2330 Selva had been secured.

At 0530 on 24 September the attack jumped off. Enemy artillery and mortar fire was heavy throughout the day, especially from the vicinity of Sasso di Castro, Mt. Freddi, and Mt. Beni. Company K, 361st Infantry, reached Covigliaio by 1745. Throughout the rest of the day, however it was subjected to heavy artillery and mortar fire. It was found necessary to clean the town out house by house with grenades and rifle fire. Progress was slow because of the heavy mortar fire covering all roads in the town. After dark the company completed its mission, and the town was entirely cleared of enemy. In other parts of the 361st's sector progress was excellent, and by the end of the day positions were stabilized along a line through La Casetta and Sartone.

The 362d Infantry reported at 1605 that a patrol had cleared out an artillery observation post on Sasso di Castro and that the forward elements were on a line generally even with the 361st Infantry on the

Reserve battalion moves up Highway 65 south of Futa Pass

Typical October weather

right. Fighting was considerably hampered by the fog patches in the woods. During the night the 362d, except for the 3d Battalion, was relieved by the 363d, which had been brought forward from its assembly area in the vicinity of Villanova. The 362d's 3d Battalion remained on Hill 1302 and was attached to the 363d until 27 September.

Fighting the next day centered on the taking of Mt. Beni and Mt. Freddi. Sasso di Castro was bypassed by both regiments. The 361st Infantry advanced on Mt. Beni, with a platoon of Company G moving to the left of the mountain and Companies E and F to the right. Resistance was characterized by scattered pockets of fanatical enemy troops. By noon one platoon, followed by the remainder of Company G and supported by a tank, moved up the slope of the mountain and knocked out five machine-gun positions. Few prisoners were taken. Most of the crews fought to the bitter end and were eliminated by rifle fire and grenades. Company E and Company F consolidated their positions astraddle Highway 65 at the foot of the mountain. In the sector of the 363d Infantry, progress was slowed by heavy artillery fire from the vicinity of Mt. Bastione. Artillery and air bombing were directed on the position, and at 1500, 25 September, the attack was resumed. By 1815 Company G and Company A had reached the southern foot of Mt. Freddi.

The attack jumped off again at 0630 on 26 September, with the 363d Infantry advancing toward Mt. Freddi and the 361st attacking north toward Mt. Oggioli. Cold rain drenched the troops and hampered progress considerably. In the sector of the 361st Infantry, during the night of 25-26 September, the 361st's 1st Battalion relieved the 363d's 2d Battalion, which reverted to control of the 363d and moved into an assembly area in reserve.

Resistance in the 361st sector consisted mainly of well placed mortar and artillery concentrations, directed on the front-line troops from excellent observation posts on Mt. Canda and Mt. Oggioli. There was also heavy machine-gun fire from the vicinity of Pietramala on the exposed right flank of the Division. However, by 1030 elements of the 361st had reached the 11th Northing, a few hundred yards south of La Mazzetta, and were ordered to consolidate their positions and hold until the flanks had been brought up.

Mt. Freddi proved a more difficult objective. An estimated two battalions of enemy occupied the position, and heavy artillery fire again came from the vicinity of Mt. Bastione. Hill 1108, which Company A attacked from the east and Company C from the south was even more heavily defended. It was only after a touch-and-go battle lasting all afternoon and night of 25 September that the hill finally fell to a daring daybreak attack by Company A. Fourteen prisoners were taken, the remainder of the defenders having pulled back to fight at Mt. Freddi. The attack continued northeast toward Hill 1254, the nose of Mt. Freddi, with Company A on the right and Company C on the left.

At this time the 363d's as-yet-uncommitted 3d Battalion, which was to pass through the 1st Battalion when Mt. Freddi fell and attack Mt. Oggioli in conjunction with the 361st, committed Company K on the left of the 1st Battalion. It was to reduce Hill 1035, a western spur of Mt. Freddi, to enable the entire battalion to jump off for Mt. Oggioli at the earliest possible moment after the capture of Mt. Freddi.

Company B, 1st Battalion reserve, went around the west of Poggio Savena and cut between Company C and Company K to capture the saddle between Mt. Freddi and Hill 1035. The companies were now disposed from west to east thus: K, B, C, A. By 0630, 27 September, after a bitter attack in the rain, Companies A and B were on Mt. Freddi. Company K had taken Hill 1035 and, while mopping up and consolidation of positions were carried on in the 1st Battalion sector, led the 3d Battalion in the attack on Mt. Oggioli with the support of machine guns and mortars set up on Mt. Freddi.

On the right, the 361st Infantry pushed forward to capture La Mazzetta, where the 2d Battalion was again ordered to hold until the units on the flanks had been brought up. The attack was resumed at 0930, after a twenty-minute artillery preparation, and by nightfall, 27 September, Company E had secured the eastern base of Mt. Oggioli.

With both the 361st and 363d using the same trail and often intermingling, the capture of Mt. Oggioli was truly a joint operation. Both

Radicosa Pass

units took the rocky objective at the same time, the early afternoon of 28 September.

In the meantime, the 361st's 2d Battalion, after capturing Pietramala, was directed to continue the attack northward along Highway 65. After a brief halt, it swept over Radicosa Pass and by 1330 had completely occupied La Posta. The Germans, after a successful delaying stand, had withdrawn to the next defensive line at Monghidoro, and the only enemy encountered by the 2d in its drive to La Posta were stragglers and die-hard fanatics cut off from their own lines.

Artillery was extensively used in this period despite the generally poor visibility, which hampered observation. On 26 September, the clearest day during the latter part of the month, the artillery had a field day. Assisted by the 77th Field Artillery Group, Division Artillery fired twenty-one observed counterbattery missions, which silenced the enemy batteries. Five direct hits were observed. The next day the 348th Field Artillery Battalion had its heaviest day's firing since the preparatory phase, with a total of 1,693 rounds. This was the peak of the neutralization fires, for the phase of pursuit was reached on this day. In addition to the artillery firing, very heavy air bombardments were loosed, principally on Mt. Oggioli.

On 28 September, in accordance with II Corps' orders that all divisions should rotate units so as to have one regiment resting in preparation for early employment, General Livesay ordered the 362d Infantry

"Engaging the enemy with accurate rifle fire, Sergeant Ambrose wounded their leader."

to relieve the 361st. After a gruelling march through driving rain, the battalions closed in their assembly area just south of La Mazzetta. The movement north to the relief line was also complicated by weather. The bypass which the 316th Engineers had built around two large craters in Highway 65 just below La Mazzetta was washed out by rain-swollen streams, but by daybreak of 29 September a Bailey bridge was completed, and traffic again moved north on the highway.

After the relief had been accomplished, the 363d and 362d pushed off through a dense fog that slowed the advance markedly. Some machine-gun and mortar fire was encountered north of Filigare, but it was reduced, and the advance continued. Throughout the day of 29 September the regiments pushed ahead slowly but steadily against light resistance. The enemy had withdrawn to Monghidoro, where the Division made contact with them the following day.

Only limited advances were made against stiffening resistance on 30 September. Company E, attacking to the right of Highway 65 in the center of the 362d's 2d Battalion sector, became engaged in one of the severest actions of the day near La Costa. When the attacking platoon, of which Staff Sergeant Francis G. Ambrose was a member, encountered deadly enemy machine-gun fire, Sergeant Ambrose voluntarily crossed an open field alone toward the hostile weapon. Engaging the enemy with accurate rifle fire, he wounded their leader. He continued his advance, while the Germans made a concerted effort to kill him, wounded a second enemy soldier, and forced the remaining four to surrender. Later, when his platoon received heavy machine-gun fire during an enemy counterattack, Sergeant Ambrose again moved forward to eliminate the menace. He crawled a hundred yards to an advantageous position, killed three Germans, and silenced their machine gun in a brief fire fight. He was mortally wounded by a sniper's bullet while returning to his platoon. His daring, intrepidity, and sheer determination were vital factors in his platoon's success, and his inspiring actions will long be remembered by his comrades of Company E.

To the left of Highway 65, moving against the same type of resistance the 362d Infantry was meeting, the 363d pushed slowly forward and by nightfall had reached positions generally along the lateral road running west from Piamaggio.

Inclement weather seriously hampered operations along the entire Division front, washing out roads and bridges, miring the artillery, and impeding the infantry. Rain and fog made visibility poor and coordination most difficult, but the delaying positions on Mt. Freddi, Mt. Beni, and Mt. Oggioli had been reduced, and the attack on the next prepared enemy line was about to begin.

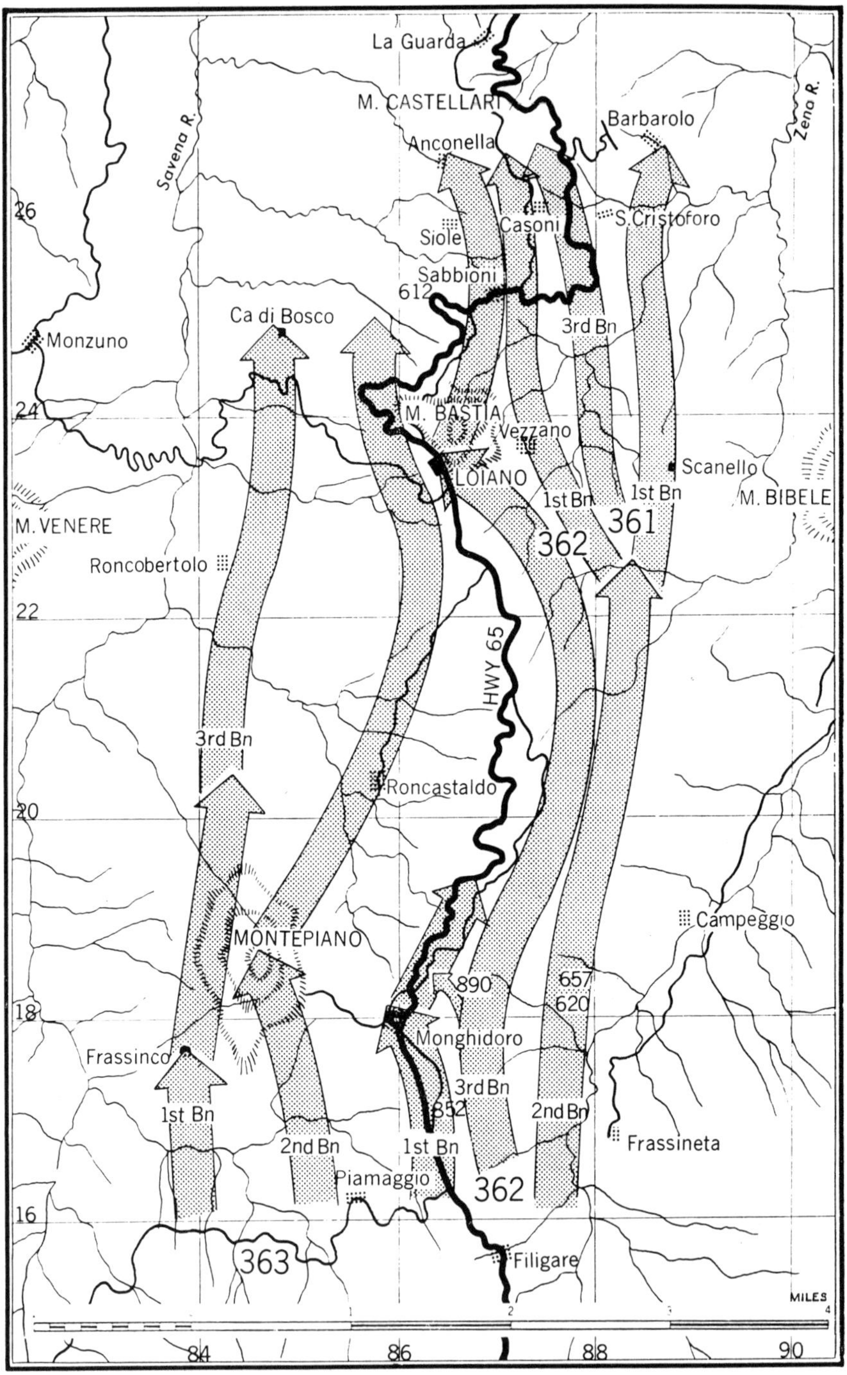

La Guarda
M. CASTELLARI
Barbarolo
Zena R.
Savena R.
Anconella
26
Siole
Casoni
S. Cristoforo
Sabbioni
612
Ca di Bosco
Monzuno
3rd Bn
24
M. BASTIA
Vezzano
Scanello
LOIANO
1st Bn
1st Bn
M. BIBELE
M. VENERE
361
362
Roncobertolo
22
HWY 65
3rd Bn
Roncastaldo
20
Campeggio
MONTEPIANO
890
657
620
18
Monghidoro
Frassinco
3rd Bn
852
2nd Bn
1st Bn
Frassineta
2nd Bn
1st Bn
Piamaggio
362
16
363
Filigare
MILES
84
86
88
90

Map 18

CHAPTER 10

THE ATTACK: MONTH OF OCTOBER

AT THE beginning of October, the 91st Division, having fought through the Gothic Line, was driving toward Bologna. Mt. Oggioli had been taken, and the Passo della Radicosa freed. At the end of the day's fighting on 30 September, the Division had two regiments on line: the 363d Infantry on the left, and the 362d on the right. Contact had been made with the main line of resistance before Monghidoro—a strong line of hasty defenses which the enemy had organized after the fall of Futa Pass. In anticipation of the powerful resistance expected along this line, General Livesay, on 30 September, ordered the 1st Battalion of the 362d to attack along Highway 65, capture Monghidoro, and then move to the dominating ground north of the town, from which it could successfully defend the position against possible counterattacks. The 3d Battalion, which had been in reserve, was ordered to attack through Company E in the center of the regimental zone and, after assisting the drive of the 1st Battalion, continue northward, eventually pinching the 1st out.

THE CAPTURE OF MONGHIDORO

At 0600, 1 October, the five battalions jumped off after an artillery preparation of 3,523 rounds. Opposition was intense, with very heavy fire of all calibers directed upon the attackers. With the help of Division Artillery, which fired 10,587 rounds during the day, limited gains were made. In the sector of the 362d Infantry, Company F secured Hill 585 and Company G secured Hill 620. Company E, in the center of the regimental zone, met and repulsed a severe counterattack.

When the enemy attacked Company E's position, Private First Class Jack Green, acting leader of a machine-gun squad, moved his machine gun to an exposed forward position and opened fire. His accurate bursts killed seven of the enemy and dispersed the remainder. When the other weapon in his section jammed, he turned his machine gun over to his assistant, made his way to the other gun, and put it in workable condition again. Then, observing several more Germans approaching from another direction, he opened fire and killed five of them. Two others of the enemy succeeded in approaching within hand-grenade range of him. Unable to fire his machine gun at them, he rolled back a few feet, picked up a rifle, and shot both of them. Later, when an adjacent squad lost its leader, Private First Class Green rallied the members of his squad and placed them in position to aid in re-

"Green rolled back a few feet, picked up a rifle, and shot both of them."

pelling the attack. At this time, he received a painful bullet wound in his arm, but he refused to be evacuated and continued to direct his men and fire his weapon at the attacking enemy. Still later in the engagement, a platoon of an adjacent company became disorganized, and several of its members wandered about exposed to direct enemy fire. Despite his wound, Private First Class Green took control of these men and assigned them to positions which strengthened the line. By his aggressive action and leadership, he was instrumental in turning back the enemy counterthrust and aided materially in stablizing his company's line.

Following Private First Class Green's heroic action, Companies I and K attacked through Company E and reached positions approximately 600 yards north before they were stopped. By 1500 Company B had captured Hill 852, east of Highway 65, the dominating terrain feature south of Monghidoro. The artillery was especially helpful in the fierce fight for the Company B objective. In support of the assault on Hill 852, the 346th Field Artillery Battalion fired more than 500 rounds. No sooner had the hill been occupied than a strong enemy counterattack was launched to regain it. The 346th Field Artillery Battalion, with the reinforcing fires of the 916th Field Artillery Battalion, fired more than 1,500 rounds into the counterattack, which wavered and then distintegrated.

Hill 852 had proved to be, with its excellent and thorough organization of the ground, one of the strongest positions met by the 362d in the entire Italian campaign. Company B took the hill once, only to be forced off by bypassed enemy paratroopers firing from dug-in defenses. The German desire to hold 852 was indicated by the fact that after one of the two defending companies had lost its commander, three other commanders were in turn sent to it. Company E and Company I helped the 1st Battalion by clearing out the eastern defensive positions during the fight for the crest of the hill.

Meeting the same kind of strong opposition, the line of the 363d Infantry worked forward slowly on the morning of 1 October. Company A, attacking on the left flank of the regiment, reached positions just north of Valgattara, where it received a severe counterattack. Supplementing its small arms and 60mm mortar fire with artillery and Cannon Company fire, it successfully repulsed the enemy.

Company F, attacking on the right flank of the 363d, saw the severest action of the day. Private Howard E. Weaver, a replacement with just four days of combat experience, was acting as first scout of the 2d Platoon of Company F. This platoon had the mission of clearing a way for the advance of the company over the strongly held enemy ridge, sloping up toward Monghidoro, that had been holding up the advance of the battalion. As Private Weaver led the advance, a heavy concentration of enemy machine-gun fire barely missed him and killed his squad leader behind him. The platoon, composed mainly of replacements, became disorganized, and the attack was slowed. Private Weaver assumed command of his squad, and after reorganizing it, led it up the hill until fire from an enemy machine gun was encountered. Exposing himself in order to draw fire from the gun, Private Weaver located the emplacement and crawled to within 35 yards of it. Then he threw a grenade, killing one of the German gunners and wounding another, whom he took prisoner.

After climbing up a hundred yards, the squad was again subjected to intense machine-gun fire from a well concealed emplacement farther up the slope. Private Weaver located this weapon, too, by exposing himself to its fire. Crawling to within 25 yards of it, he hurled his last two grenades. He captured the two Germans who were still alive when he reached the position. With this second pillbox eliminated, Private Weaver signalled his squad forward and continued the assault up the hill.

Troops move through Monghidoro after its capture by the 362d Infantry

"After killing a sniper hiding in a tree, Weaver motioned his men on."

Observing a covered position for his squad, he led it to the shelter of a steep bank, climbed over the top, and reconnoitered for a route of approach to the objective. After killing a sniper hiding in a tree, he motioned his men on.

Again leaving his squad in a covered position, he advanced toward a house in the face of sniper fire, entered the building alone, and captured two enemy snipers. His courageous performance, inspiring leadership, and willingness to assume responsibility at a critical time made it possible for his company to take its objective. A short time later he received a field promotion to second lieutenant for his continued outstanding combat service.

At 1300 the 363d Infantry overran the enemy's main line of resistance in its sector, and by nightfall the regiment's line was stabilized along the 17.3th Northing. During the night, patrols seeking to determine whether the enemy strongpoints could be bypassed met stiff resistance everywhere. Artillery fire was light, however, and the night was comparatively quiet.

At 0600, 2 October, all units of the Division launched the attack which achieved the success earned by the hard fighting of the day before. Monghidoro fell. Beautifully coordinated, all battalions con-

Germans used the church steeple in Loiano as an artillery OP

tributed to the victory. On the right flank, the 2d Battalion of the 362d Infantry took Hill 620 and Hill 657. The 3d Battalion, in the center, flanked Monghidoro and occupied Hill 778 and strategic Hill 890, five hundred yards northeast of the town, one platoon of Company I having captured two tanks and two self-propelled guns. The 1st Battalion, advancing up Highway 65, was reinforced at 1145 by a platoon of tanks. At 1315 another platoon of tanks followed the first into town. By 1415 Company A was cleaning out Valle, and elements of Company C were in Monghidoro. House by house, the town was cleared of enemy.

Just north of the town the 1st Platoon of the 91st Reconnaissance Troop found that a machine gun was holding up the progress of elements of Company C. Several shots from the armored cars silenced the machine guns, and two of the enemy gunners were captured. In the meantime, movement had been observed in a long house nearby. Several shots fired into the building produced no results, but a few more shots brought out 29 Germans with their arms raised in surrender. By 1845 Company C, reinforced by two platoons of tank destroyers and nine tanks, was in complete possession of the town and had taken 103 prisoners of war. The artillery prepared defensive fires to protect the prize, and the Cannon Company moved to positions just south of the town.

On the left, the gains of the 363d Infantry had been no less gratifying. The 1st Battalion captured Frassinco and Acqua Calda. The 2d

Battalion, with Companies F and G in the lead, advanced steadily. At noon Company E was directed to cut the lateral road running northwest from Monghidoro to assist the 362d Infantry and to isolate Montepiano. At the same time Companies F and G pressed forward in a coordinated attack to capture Montepiano.

A very strong counterattack by 200 enemy was launched at 1400 along the lateral road from Monghidoro. Visibility was poor, but the artillery forward observer adjusted fire as best he could. A fortunate break in the fog exposed the 200 Germans, and it was estimated that, in a few seconds, 190 of them were killed.

Advancing steadily through the foggy darkness, the companies occupied the crest of Montepiano. By the end of the day the enemy's Monghidoro–Montepiano defenses had been completely overrun. It was from the heights of Monghidoro that the battle-scarred troops of the Division got their first view of the broad Po valley and the snow-covered Alps beyond. That first view, on a clear day, was a thrill that defies description.

General Keyes, Commanding General, II Corps, expressed his pleasure at the swift success in a telegram to General Livesay.

CONGRATULATIONS UPON THE CAPTURE OF MONTEPIANO AND MONGHIDORO [he said]. THE CONTINUED DRIVE OF THE 91ST DIVISION AGAINST A STUBBORN ENEMY AND DESPITE THE ADVERSE ELEMENTS IS A TRIBUTE TO YOUR FINE DIVISION.

THE LOIANO LINE

The Monghidoro–Montepiano line having fallen, the enemy fell back to establish his next line of defense—the Loiano Line. Anchored on Mt. Venere in the sector of the 34th Division to the west, the line swung in a huge arc toward Mt. Bastia, just north of Loiano on Highway 65, and, including Loiano, continued eastward to Mt. Bibele in the sector of the 85th Division. Well prepared machine-gun, antitank, and mortar positions had been dug by forced Italian labor.

Resistance was slight on 3 October. The 1st Battalion, 362d Infantry, remained in Monghidoro for rest and rehabilitation. The 2d and 3d Battalions moved forward rapidly, until their advance was halted along the 22.2d Northing at 1700 to allow them to prepare for a coordinated attack at dawn.

In the 363d's sector sporadic resistance was encountered, but the advance was not seriously delayed at any time. At 1445 the 3d Battalion, with Company C attached, relieved the 1st along the 19.5th Northing and continued the advance. By the end of the day the 2d and 3d Battalions had stabilized their lines along the 22.1d Northing after an advance of over four kilometers.

In ground gained, 3 October was the most successful day of the month for the Division. Its rapid advance brought the infantry up to the Loiano position, where very stiff opposition was encountered.

Continuing the attack on 4 October, the 363d, on the left, made small advances in the face of heavy small-arms, machine-gun, and mortar fire. Machine guns, firing from hastily prepared positions under houses at Roncobertolo and Ca' di Parisoni, were destroyed by antitank guns.

On the right, small-arms and mortar fire checked the 362d Infantry's advance toward Loiano. At 1500 tanks were sent into the town, but when the two leading tanks were destroyed, the remainder were withdrawn, and artillery fire was laid on the area. During the remainder of the day, 4,018 rounds were fired into Loiano and Mt. Bastia, the dominating peak just north of the town. At 1920 Colonel Cotton reported: "Seems to be lots of stuff in Loiano. Have put artillery on houses and knocked them down, but Jerry still comes out."

At 1500 General Livesay issued Field Order No. 22, outlining the tactics of the next advance. The 361st Infantry was to pass through elements of the 362d Infantry in zone and attack, initially with one battalion, on the right of the Division sector. Successive objectives were Scanello, Castiglioncello, and San Cristoforo. Strong contact was to be maintained with the 85th Division on the right. The 362d Infantry, in the center, would continue the attack to the north, gradually spreading west to pinch out the 363d Infantry. Its successive objectives were: Mt. Bastia, Hill 612, Siole, Hill 506, and Hill 518, and, on the right, Bellaria, Hill 705, Hill 708, and Mt. Castellari. After seizing limited objectives, the 363d Infantry would revert to Division reserve.

The next morning, 5 October, the artillery continued to give powerful support by laying down a thousand-round preparation for the attack at 0600. Twenty-five minutes later, under tremendous artillery preparation fired by the 346th Field Artillery Battalion, the 3d Battalion of the 362d Infantry attacked Loiano. Company L fought its way into the southern outskirts of the town. One enemy tank was destroyed, and the entire area was methodically cleared of all resistance by savage house-to-house fighting. With the help of the 916th Field Artillery Battalion, firing 204 rounds between 1510 and 1525, the company beat off a serious enemy counterattack which threatened to retake Loiano. With Company K, Company L then continued the advance in the face of heavy enemy fire directed from Mt. Bastia, and by 2100 elements of both companies were securely dug in on the mountain. Meanwhile, the 1st Battalion, moving in the 3d Battalion zone, took Vezzano.

On the flanks the Loiano Line was also broken. Rain and limited visibility hampered the advance, but did not halt it. The 1st Battalion, 361st Infantry, on the right, attacked at 0600 and initially met no resistance. By 0835 Company A had occupied Scanello, and Company C was deployed to its left, 200 yards south of Ca' di Ricci. The advance continued against gradually stiffening resistance. Company L was committed between the 1st Battalion and the 362d Infantry, but difficulty was experienced maneuvering in the narrow sector. At the end of the day, however, the 3d Battalion was deployed on the left of the regimental sector along the 23d Northing, while the 1st Battalion had advanced to positions along a 600-yard front extending through Ca' di Ricci.

To the west of Loiano, the 363d Infantry advanced slowly against stiffer resistance than was met anywhere else during the day. Machine-gun fire was received from Bibulano, and small-arms and self-propelled-gun fire from Mt. Bastia. At 0825, 6 October, the 2d Battalion (minus Company E) assembled in the vicinity of Loiano. The 3d Battalion, after moving forward enough to outpost Company E at Ca' di Bosco with one platoon in Olive, assembled at Bibulano. The 362d then assumed command of the left half of the Division sector.

The fighting during the period 5-9 October was of a peculiar nature because of the extremely limited visibility. Rain and fog, while they did not stop the advance, made the fighting confused and difficult. The enemy, pushed out of his Loiano Line, did not fall back to his next defense line, but made use of the terrain for shrewd delaying tactics. Not so mountainous nor so high as the ridgeline of the Apennines, the countryside between Loiano and Livergnano is hilly and open. The gradually descending northern slope of the Apennines is broken by rough spurs running generally north and south and by hills of dominating height which command the ravines and draws. In the 91st's zone Mt. Castellari was the dominating feature. For the enemy, the terrain afforded unlimited opportunities for delaying positions and elastic defense. For the 91st, attacking north, the mountains and valleys would have been difficult terrain to advance over even in normal circumstances. Made slippery and muddy by the autumn rains, they challenged the best powers of the troops.

Companies maintained contact satisfactorily by radio, and the advance was at all times well coordinated, but enemy positions were frequently bypassed or discovered by accident, and fire fights flared in fog-isolated areas across the entire front. Further, the limited visibility hampered the artillery support of infantry units. Between 5 and 9 October, not a single air observation mission was flown. Information received from shell reports, partisans, and previous photo intelligence formed the basis for most of the firing. Registrations observed from the air, the most certain assurance of accurate firing, were impossible. Yet during the five days, 18,298 rounds of artillery were fired with notable success.

On 6 October, the 361st Infantry advanced through rain and fog against little resistance. Elements of the 1st and 3d Battalions made their way in the fog through Castiglioncello, Burzano, and Ca' dei Salmi. Company L was unable to advance because of heavy frontal fire from Casoni. In an attack, coordinated by the commanding officers of the 361st and 362d Infantry Regiments, Company L moved to the west behind the 362d's Company B and attacked Casoni from the west flank. Company I advanced up Highway 65 to reinforce Company L, and by 1815 Company B, 362d Infantry, and Company L, 361st Infantry, had taken Casoni. Company K then set up outpost positions at Scope. Throughout the night heavy mortar fire fell in the regimental sector.

On the left, Company K, 362d Infantry, occupied Sabbioni on Highway 65, and Company L reached Hill 612, to the left of Highway 65. Both companies moved ahead, and at 1050 Company K reported an estimated company of enemy at Anconella. Progress was halted in mid-afternoon along the draw 200 yards south of Siole. With patrols working up Hill 705, the companies moved forward, and by 0100, 7 October, two platoons of Company A had reached the crest of the hill. At 0500 the same day Company L moved into Anconella.

Both battalions of the 362d were in position and ready for the attack on the morning of 7 October. The 3d Battalion was in a column of companies, with Company L in the lead at Anconella, Company I ready to move up the left flank, and Company K in reserve. Company A was on the north slope of Hill 705; Company B was in Casoni; and Company C was between Companies A and B. By taking advantage of the dense fog and ordering the artillery not to fire after 0400, the battalions achieved some element of surprise when they jumped off, without preparation, at 0800.

Resistance in the sector proved heavy, however. Small-arms, mortar, and some artillery fire, especially from Mt. Castellari, made progress slow. At 1515, three hundred rounds of artillery fire fell on the 1st Battalion, an indication of the sort of opposition that characterized the day's fighting. Company B, moving out from Casoni, reached the foot of Mt. Castellari, where small-arms fire and machine guns located in concrete emplacements checked the advance.

In the sector of the 361st Infantry weather and terrain were the principal delaying factors. Numerous isolated fire fights were joined in the fog, and by mid-morning the 3d Battalion reported that it had used up its ammunition and was bringing more forward. San Cristoforo fell to the 1st Battalion in the afternoon, and Company A continued to push on under heavy enemy fire to occupy Hill 303 at 1730. The remainder of the battalion made its way slowly over very rough terrain. The deep slopes of the ravines, slippery and muddy because of the recent rain, forced the infantrymen to crawl forward on their hands and knees. Tanks rendered effective fire support by shelling Trebbo, and artillery fired with good effect on enemy weapons and personnel.

When, at 1900, prisoners reported that the heavy artillery had so disorganized the enemy that they were retreating, both battalions were ordered to advance throughout the night. Company I fought its way forward to Barborolo, while the 1st Battalion attacked Trebbo. Trebbo was cleared by 0720, 7 October, with eight prisoners taken and the rest of the garrison killed. Immediately Company C proceeded to Valle in an effort to cut off Poggiolo.

During the night heavy artillery concentrations fell in the vicinity of Casoni. At times three to four rounds per minute fell in the area. Company I reported at 0810, 8 October, that it had captured Barbarolo. In accordance with the order of the regimental commander, the 2d Battalion moved forward to pinch out the 1st. Company A was then given the mission of protecting the right flank of the regiment and the Division.

In the sector of the 362d Infantry, rain, fog, and heavy enemy resistance handicapped the advance toward the crest of Mt. Castellari on 8 October. Enemy fire from the mountain swept the whole area. The 2d Battalion went in on the left of the 3d and reached the 27th Northing by 1000. After fighting all morning, forward elements of the 1st Battalion reached the 26.9th Northing, within 75 yards of the crest, where the advance was halted in order to reorganize for the continuation of the drive.

At 1500, in a coordinated attack, the 2d Battalion moved westward to seize Farne, while Companies A and B started up Mt. Castellari. Company B, while still probing for the crest of Mt. Castellari, received a counterattack of between 50 and 75 enemy at 1800. This had been anticipated, for radio intercepts had picked up earlier in the day an enemy order directing a company to this spot. Due to the position of Company B so near the crest of the mountain, no supporting fire was possible except that of the 60mm-mortar platoon, but Company B was ready and beat off the counterattack. The enemy withdrew to a draw a short distance to their rear. A patrol of Company B managed to work forward close enough to overhear the preparation for a second counterattack, and as soon as the news was relayed back, the 60mm mortars opened up again and thoroughly dispersed the enemy. More than 600 rounds of 60mm-mortar ammunition were fired against the counterattack. Company B then moved ahead and occupied Mt. Castellari's crest, and by 0100, 9 October, had cleaned out all ground within reach and had ladders and ropes up to aid in searching the cliffs. At 0215 the objective had been completely occupied, but many small groups of enemy had been bypassed and had to be dug out by reserve elements. Company A sent out a patrol to La Guarda in advance of the remainder of the company.

At 0430, 9 October, the Division resumed the attack. Early that morning the command post of the 3d Battalion, 362d Infantry, between Sabbioni and Casoni, was surprised when a strong enemy counterattack of 47 Germans materialized out of the fog. The battalion commander, Lieutenant Colonel Thomson, directed mortar fire by sound, as the fog had closed in, and brought the bursts of the 81mm shells closer and closer around his CP. The six men of an enemy light-machine-gun section came into the CP to surrender. Shortly thereafter a platoon of Company I arrived, and all of the attacking enemy were accounted for except two men and the officer commanding. Prisoners stated that they had been directed to seize "that pink house, as it would make a good OP."

At 1115, 9 October, General Livesay directed the 362d Infantry to occupy Le Fosse, on the 29.9th Northing just south of the Savenna River, and then revert to Division reserve. In order to carry out the new instructions, the regimental plans were changed. Company I occupied Farne and then proceeded to La Valle; Company L occupied La Guarda; and Company K secured Hill 513, completing the capture of Mt. Castellari, which could have been so important to the enemy. The advance northward continued despite strong and continuous small-arms, machine-gun, and mortar fire. By 2245 Company E had reached the 29.1th Northing and Company F had moved to Hill 518.

Advances in the sector of the 361st Infantry on 8 October were extensive. The enemy attempted in vain to hold up the regiment with 170mm artillery fire from the northwest and 120mm mortar fire from the northeast. Company K searched every building in Barbarolo, while Company I moved on the left to occupy Vaiarano. Company L continued clearing Trebbo. At 1305 Company I was reported at Prato di Magnano. The 3d Battalion immediately took advantage of its remarkable advance to the southern escarpment, just south of the 29th Northing. Company L moved westward to Cant, while Company I followed the rim of the southern escarpment to try to cut Highway 65 south of Livergnano.

At Cant Company L found that Hill 708 had not yet been completely cleared of hostile forces and that the town was being subjected to heavy machine-gun fire from the high ground. The 2d Platoon led the attack —which was made at night—followed closely by the 1st Platoon, with the 3d Platoon in support. Although enemy mortar fire caused eight casualties in the 2d Platoon, sniper and machine-gun fire was very light, and the platoon was able to occupy Cant. The 1st Platoon then occupied the high ground to the rear of the town. An enemy counterattack was beaten off by the 1st and 3d platoons, but in the fog the enemy succeeded in cutting between them, and the 2d Platoon and all troops in Cant were captured.

Company I, making its way carefully through the fog and darkness, succeeded in moving behind the enemy positions and cut Highway 65 at La Fortuna, 2,000 yards behind the enemy lines. Just east of the road, Company K followed in close support. Both companies engaged in continuous bitter fighting throughout the night. In the foggy darkness several large groups of enemy withdrew down the highway, and many were killed or captured. In the impenetrable darkness First Sergeant Myers of Company I lured many enemy soldiers into a trap by barking orders in excellent German.

During the night of 8-9 October the 361st Infantry occupied the following positions: Company I, La Fortuna; Company K, just east of Highway 65 on Hill 470; Company B, Prato di Magnano; Company C, east of Company B, in the vicinity of Le Trece; Company L, Cant; and Company A, Hill 409. The 2d Battalion was in position in the vicinity of Vaiarano.

Breaching the Caesar Line

The 361st Infantry had arrived before the most formidable natural barrier between the Santerno River and the Po Valley. It is a sheer, rocky escarpment, rising at some points over six hundred meters in

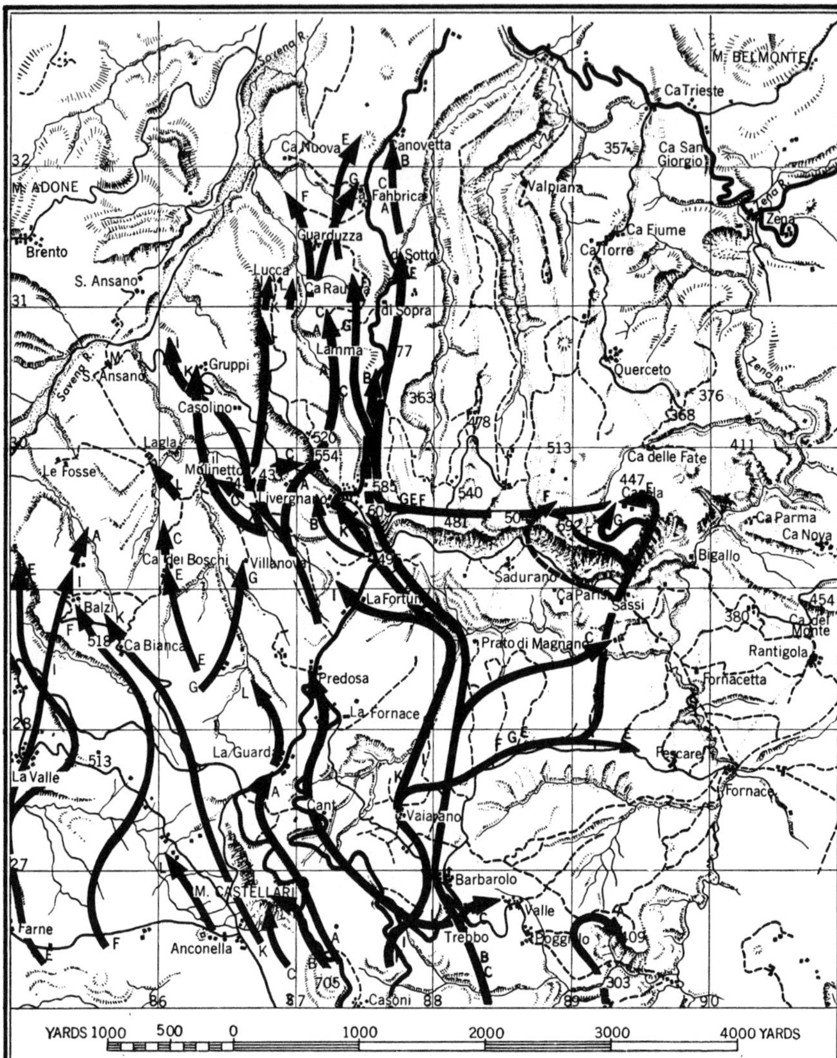

THE CAESAR LINE

Color Legend

- 361 INFANTRY ●
- 362 INFANTRY ●

Map 19

height. In places, especially in the upper half of the cliff, it is a perpendicular rock wall. From the rim the enemy commanded every approach from the south. Rising above the rim was a lateral series of hills, each one a prepared strongpoint from which the high plateau could be covered with machine-gun and mortar fire. Only two breaks by which the plateau could be reached existed. One, a foot path, lay just north of Bigallo to the east, and the other was a cut at Livergnano through which Highway 65 ran, curving to the northeast behind the protecting escarpment.

As the regiment faced this escarpment, it was considerably in advance of its adjacent units. It was exposed on the right to fire directed from Santa Maria di Zena, a fortress-like Italian church situated on top of the 638-meter-high Mt. delle Formiche. On the left, the 362d Infantry had reached La Guarda and was moving down the northern spur of Mt. Castellari. In the fog and rain the 361st had surprised and overrun many enemy positions and had bypassed many others. Thus it had become a spearhead, thrust forward through the enemy lines on its flanks. Its campaign was being fought by companies instead of battalions because of the reduced visibility and the nature of the terrain—in which most avenues of approach ran north and south, making it difficult to maintain lateral communications. The 1st Battalion at this time was using rifle companies K, C, and B, operating on the left of the regimental sector along Highway 65. The 2d Battalion, consisting of companies E, F, and G, occupied the right of the sector. In the 3d Battalion—consisting of companies A, I, and L—each unit was carrying out independent missions. Company A was on Hill 409, more than a mile behind the point of farthest advance, protecting the right flank of the Division. Company L, strengthening its hold on Cant, was also more than a mile to the rear of Company I, which was maintaining its roadblock across the highway at La Fortuna.

There was no choice in taking the rocky escarpment. The 2d Battalion was ordered to move eastward, cut north of Bigallo, make its way up the rim rock, and then move westward to seize in succession Hills 592, 504, and 481. It was then to fight its way back to Highway 65. On the left, the 1st Battalion was ordered to move through Livergnano, neutralizing its two sentinels, hills 544 and 603, and continue west to pinch out the 362d Infantry. For both battalions the missions were to mean five days of heart-breaking, bitter fighting and difficulties which were well-nigh insurmountable.

At dawn of 9 October, Company I and Company K, on either side of Highway 65, were heavily counterattacked by an enemy force which

This picture of the eastern edge of Livergnano was taken a few hours after its capture. No

smoke and dust caused by German artillery concentration falling in the center of the town.

included tanks. The fire fight was desperate, but the enemy was forced back with many killed and captured. At 1040 General Livesay directed the 3d Battalion to hold its positions along the southern escarpment until its left flank was more secure. One platoon, however, was to make its way into Livergnano.

By 1500 the 3d Platoon of Company K had reached the southern outskirts of the town, where it was met by withering bursts of machine-gun and sniper fire and was driven to the cover of a large four-story building. In the hail of machine-gun bullets about ten men became separated from the rest of the platoon and took refuge in a pigsty under a house nearby. It was impossible for the 1st and 2d platoons to reach the town in daylight because of the deadly sniper fire coming from Hill 603. About 1900 they moved up to Hill 495, and from this point the 2d Platoon was guided to the 3d Platoon command post. Buildings in the vicinity were searched, and local security was posted. When the 1st Platoon and the Weapons Platoon moved into town, they were also struck by intense fire from machine guns and mortars on Hill 603 and in the town. Herded by gunfire, they took cover in the nearest shelter—the large building occupied by the 3d Platoon.

Just before daylight on 10 October, the enemy attacked with rifles, machine guns, and hand grenades, but the company stood them off successfully and they withdrew to bring up heavier weapons. At about 1000 two enemy tanks and many mortars began to fire point-blank at Company K's building. The enemy changed his positions methodically so that every part of the building was damaged and numerous casualties were inflicted on Company K. Finally, the building began to fall apart under the steady pounding. Then suddenly, the firing ceased, and the enemy assaulted, killing or capturing practically all of the company. The only ones saved were those who had hidden in the pigsty. After a day of tense waiting, these men escaped back to La Fortuna. A new Company K was reorganized that night and with high morale, struck out on a new mission the following morning.

Meanwhile, the 2d Battalion, moving laterally across the regimental front from the vicinity of Vaiarano, had reached Hill 384, where it was momentarily halted while a platoon of Company E surprised and captured ten Germans at Pescare. The Company E platoon reported a large number of enemy in a draw to the east, near Fornace and La Casona. The platoon remained in position around Pescare as flank protection, while the remainder of the battalion continued on forward to the vicinity of Sassi. The advance was again halted here while reconnaissance patrols were sent out to look for possible routes over the escarpment. After several hours of reconnaissance, it was determined that the only route lay through the cut between Sassi and Bigallo. However, Lieutenant Jack Taylor of Company G, leader of one of the patrols, commented, "Where Ah come from in Texas, they drive hosses into places like that. It looks mighty like a trap to me. Ah don't like it."

There was no trail leading up to the plateau at this point, but it was possible by scaling and climbing to reach the top. Riflemen slung their rifles over their shoulders and crawled up by their fingers and toes. The machine gunners disassembled their weapons, and each squad member carried parts in his pockets or pack. At one point on the way up, Companies E and G found that they had to cross a narrow ledge which the enemy had zeroed in. By running a few men across at a time, they cleared the obstacle and made their way forward.

By the time Company E reached the rim of the escarpment the fog had lifted, and the visibility was excellent. The two scouts of the leading platoon were allowed to pass over the rim of the escarpment unmolested, but as the next four men rushed over, they were met by a withering blast of machine-gun fire from the vicinity of Hill 592 to the left and instantly killed. Almost simultaneously enemy machine-gun and sniper fire opened up on the two scouts. One was killed instantly; the other managed to escape for the moment by jumping into a ditch, but was killed later when he attempted to move back to the rest of his platoon. The leader of the leading platoon of Company E immediately began to deploy his men on a firing line along the ridge.

Toward evening, after an afternoon of almost superhuman effort, two platoons of Company E were in position on the top of the escarpment overlooking Casola, a scant 300 yards to the northwest. Company G, working its way forward in much the same manner, had moved into position to the left of Company E. Taking advantage of the concealment of growing darkness, the enemy counterattacked the position. While the fight was in progress, the Company E platoon, which had been left near Pescare, worked its way up the escarpment to a point just in rear of Companies E and G. One squad of this platoon was sent to the right flank of Company E to assist in beating off the counterattack. At the same time, two heavy counterthrusts were directed by the enemy at the left flank of Company G and at the center of the line between the two companies. The remaining two squads of the E Company platoon were sent to bolster these positions. All enemy attempts to dislodge the companies proved unsuccessful.

So, at 2000, 9 October, companies E and G were on top of the escarpment just south of Casola, and Company F was at the escarpment's base near Sassi. After a day of tortuous movement and savage fighting, the companies found themselves at the bottom of a "tilted saucer" with high ground, occupied by the enemy, completely surrounding them.

The next morning, Companies E and G were again attacked by the enemy in an attempt to push them off the escarpment. Company F, moving forward to support them, found its movement considerably hampered by heavy machine-gun fire from the vicinity of Pescare, which had been reoccupied by the enemy.

By 1000 they had only managed to reach Bigallo. Throughout the day the fighting continued, but the enemy never succeeded in driving Companies E and G off the escarpment. The terrain was reported by the companies that had fought over hills 844 and 856 in the Gothic Line to be the ruggedest yet encountered. The men were under constant fire of every sort—small-arms, machine-gun, mortar, artillery, and self-propelled-gun—from prepared enemy positions on the plateau, from high points on Hill 592 to the left, and from Mt. delle Formiche to the right rear.

While these determined efforts were being made at either end of the escarpment, Company A was fighting up Hill 409 against unexpectedly strong resistance. By 1400 the position had been occupied in force, and the company took up its mission of protecting the right flank. Company L at Cant remained in position ready to move forward with elements of the 362d Infantry. As these elements moved up the west flank, Company I continued to maintain the roadblock at La Fortuna, despite repeated enemy attacks.

At 1700, 9 October, Division Field Order No. 23 was issued by General Livesay. The 363d Infantry was to be committed on the right of the Division sector. It was to attack north, seize Ca' Trieste, Hill 377, and Hill 357, maintain close contact with the 85th Infantry Division, on the right, and assist in the latter's advance. The 361st Infantry was to continue its attack to the northwest, pinching out the 362d and protecting the Division's left flank north of the 31st Northing. The objectives of the 361st were Mt. Adone and Mt. dei Frati. After the 362d Infantry had seized Le Fosse, just south of the 30th Northing, it was to revert to Division reserve. Commanders were cautioned to avoid, whenever possible, obvious approaches to enemy positions and to make every effort to locate possible targets for the artillery.

On 10 October, the 362d Infantry, on the left of the Division sector, continued its push toward Le Fosse. The 1st Battalion had gone into reserve; the 3d held in position; and the 2d, with Company I attached, pressed the attack. Progress throughout the day was slow, and it was necessary to maneuver for positions. By noon Company F had reached Hill 518 near Ca' Bianca, and elements of Company G had reached Villanova. The attack was halted during the night, but at 0600, 11 October, the companies renewed their efforts. There was little resistance. As Company I moved through i Balzi, it captured 50 prisoners. Company E, to the east, found Ca' dei Boschi a very heavily defended point, and a running fire fight kept up there all day.

During the night of 11-12 October, since the 362d's 2d Battalion had not reached Le Fosse, the 1st Battalion relieved it. At 0700, 12 October, the attack jumped off again, and Company C, which had replaced Company E before Ca' dei Boschi, seized the town and captured twenty prisoners.

The enemy defense stiffened throughout the day and became the same fanatical resistance that the other two regiments were meeting to the east, with heavy small-arms, mortar, and artillery fire. The enemy remained in his positions until he was overrun. The fighting was extremely bitter, and the progress was slow. Many enemy dead were counted, and many prisoners were taken after the German positions had been assaulted and captured. In one engagement just south of Ca' dei Boschi, Company G captured approximately 40 prisoners after vicious hand-to-hand fighting.

At 2400, 12 October, control of this sector passed to the 34th Division, and the 1st Battalion, 362d Infantry, was relieved by the 2d Battalion of the 135th Infantry. The 362d, reverting to Division reserve, then assembled in the vicinity of Anconella for rest and rehabilitation.

At Livergnano, when the desperate plight of Company K became known, on the morning of 10 October, every resource was mustered to neutralize the enemy's advantage. The 362d Infantry pressed forward aggressively on the left. Company G, the left company of the 2d Battalion, 361st Infantry, although already engaged in a desperate fire fight near Casola, was ordered to try to cut its way to Highway 65 north of Livergnano. Company C maneuvered into position to attack the town from the southwest. Company B was to push at least one platoon up Highway 65 into Livergnano. The artillery poured shells into the northern edge of the town. The 348th Field Artillery Battalion and the 916th Field Artillery Battalion each fired 300 rounds. After considerable difficulty, caused by fog in the vicinity of the airstrip, an artillery liaison plane took off and throughout the afternoon directed the firing. Three bombing missions were also flown, with good success. But on the ground the enemy held grimly to his positions, and the infantry advanced only by the bitterest sort of fighting.

At 2100 Company C reported one platoon in the vicinity of Il Molinetto and one platoon just to the east on Hill 433. Company B was on Hill 495, along Highway 65 just to north of the 29th Northing. Company I was at La Fortuna with Company A, which had swung northwest from Hill 409. One platoon of Company L was at Predosa, and one platoon was just behind at La Fornace.

On the escarpment, fighting during the day had been equally bitter. Although the Germans attacked again and again, the companies did not give ground. At 2100, 10 October, they had dug in securely and had left no gap in their line along the rim of the escarpment. Company F was in position to the rear of companies E and G along the cut in the escarpment. The supply and evacuation problems had become extremely difficult, however. It took twelve to sixteen hours to bring a wounded man down over the escarpment by hand-carrying, which was necessary since litters were too bulky for the width of the trail. Communications were consistently disrupted by enemy fire into the draw. It took an individual runner four hours to negotiate a one-way trip to the battalion command post.

Farther to the east, the 363d Infantry attacked in accordance with Field Order No. 23. The 1st Battalion, attempting to reach Bigallo, was slowed by the treacherous terrain and enemy opposition and at nightfall had advanced only to Pescare. Company A, continuing its advance during the night, established a line from Sassi to Ca' Parisi. The 2d Battalion, on the extreme right flank, contacted the 338th Infantry of the 85th Division at Fornace and then continued to move northward. Company G made its way to Fornacetta, and Company E reached Rantigola. During the night of 10-11 October, Company E attempted to continue its advance from Rantigola, but 400 yards north of the town it was met by heavy mortar and machine-gun concentrations, necessitating a slight withdrawal to reorganize. Company E then sent patrols to Bigallo.

On 11 October, the advance continued very slowly through rain and mud against a very determined enemy entrenched in excellent positions. The 363d's 1st Battalion moved forward from Pescare to Sassi, while Company F fought its way to Bigallo. Company E, attempting to advance up the spur of the escarpment north of Rantigola, changed its plan of attack when one man of the leading squad was met by a hail of machine-gun fire and instantly killed as he attempted to move over the top. It was decided to send the company around the obstacle through Ca' del Monte and Ca' Nova to Ca' Parma. By 2100 it had reached Ca' del Monte.

In the meantime, the situation of the 2d Battalion, 361st Infantry, on top of the escarpment had become increasingly grave. After it was reported that Companies G and E were being decimated by enemy fire from the east, General Livesay ordered that elements of the 363d Infantry must get to the top of the cliff during the night to help relieve the desperate position of the 361st's 2d Battalion. In compliance with this order, Companies A and C were ordered to scale the cliff. By 0645, 12 October, both had reached the top, were in position to the right of companies E and G, 361st Infantry, and were engaged in a fire fight. The 363d's Company E advanced from Ca' del Monte and shortly after daylight seized Ca' Nova. Farther to the left, the 3d Battalion, 363d Infantry, was committed to attack Hill 504 up over the escarpment north of Sadurano and to continue the attack to the northeast and seize Hill 513. At 2330 the battalion was reported at Barbarolo, its vehicles mired down in a sea of mud. Yet by 0630, 12 October, one platoon of Company K had reached Hill 504, and another platoon was on the way. Company L was at Sadurano, ready to follow Company K up the rocky wall.

While the 363d was maneuvering in the center and on the right, the 361st, after initial attempts to advance, was ordered by General Livesay to hold its positions until the units on its flanks came abreast.

On the morning of 11 October the 2d Battalion of the 362d Infantry was attacking i Balzi and Ca' dei Boschi and making slow progress against very heavy resistance. On the right the 363d was engaged in bitter fire fights throughout its zone of action immediately after it was committed. Farther east, in the sector of the 85th Division, the 338th Infantry was pushing slowly toward Santa Maria di Zena, the German strongpoint on Mt. delle Formiche from which the enemy was directing artillery east and southeast to Highway 65.

At 0735 one platoon of Company B, 361st Infantry, attempted to advance directly upon Hill 603, while another platoon flanked the town from the west. Both platoons were driven back by small-arms fire from the enemy holding the caves in the cliffside overlooking Livergnano. It was estimated that in the northern part of the town, lying on the reverse slope of Hill 603, there were 200 German riflemen reinforced by three tanks and a platoon of mortars. In their protected positions they constituted a very formidable force. On top of the escarpment, companies E and G were being heavily pounded by heavy

Our artillery blasts a German OP on Mount delle Formiche

fire from self-propelled guns due east of Ca' Nova and by a murderous volume of heavy-mortar fire from the well protected ravines to the north and northeast.

Company F, guarding the cut north of Bigallo, was trying to alleviate the desperate supply problem. Mules were used to bring supplies from Trebbo to Sassi, and from there they were carried by hand up the escarpment to the company.

At 1350, Lieutenant Colonel Howard W. Reynolds, commander of the 1st Battalion, 361st Infantry, was directed to prepare companies A and C for a night attack in an effort to cut Highway 65 north of Livergnano. Plans called for Company C to lead out at 2400, following an artillery preparation. The attack jumped off on time, and by 0800, 12 October, one platoon of Company C was at Il Molinetto, and another was on Hill 433, just to the east. Company A had a platoon and a half on Hill 554, the western guardian of Highway 65.

In addition to the reinforcement of the 361st Infantry's positions by the 363d and the notable advance of the 361st's 1st Battalion, the general situation at dawn, 12 October, was better for other reasons. At 1055, 11 October, General Livesay had arranged for all possible air support to be given the Division. Nine bombing missions were flown with great success during the day, some of the targets being only 350 yards ahead of the front-line positions. The artillery fired missions of neutralization or interdiction all day, while the tanks and tank destroyers pounded the caves of the cliff wall hour after hour. The advance of the 338th Infantry, on the Division's right flank, also strengthened the position materially. At 1900 it was reported that elements of that regiment had occupied the crest of Mt. delle Formiche and seized Santa Maria di Zena, which had been used as an enemy strongpoint and observation post.

Although the situation on the right had, by the added weight of the 363d Infantry, been made somewhat more secure, the situation around Livergnano was still touch-and-go. Most of the activity on 12 October centered in the effort of the 1st Battalion, 361st Infantry, to take Livergnano and the heights dominating it. The day also brought a concerted effort on the part of the Division Artillery and air support to knock out gun positions, emplacements, ammunition dumps, and enemy strongpoints to permit the infantry to advance.

In the sector of the 361st's 1st Battalion, each company was assigned a specific mission. Company A was to make the main effort against Hill 554, with Company C assisting by maintaining strong pressure north and northwest from Il Molinetto on the left flank. To the east

Looking down on Livergnano from Hill 603, with the Church on Hill 544 in the background. Most of the buildings seen here were leveled by German artillery before the winter was over.

of Highway 65, Company B was to assault Hill 603. The two hills proved to be difficult obstacles indeed.

On the left, Company I moved westward toward Casolino under observation of the enemy on Hill 544 and in the buildings on the western outskirts of Livergnano. Tank-destroyer fire was employed to demolish these buildings, and the company made fair progress under its cover.

On Hill 554 Company C fought bravely, but failed to achieve its mission. By 1230, 12 October, one platoon had reached the crest of the hill, knocked out two machine-gun positions, captured an officer, and killed ten German soldiers. Enemy fire drove them to take cover in a building, however. Then the enemy brought up a tank, fired high-explosive shells at point-blank range, and registered 120mm mortar shells on the building, and the platoon was forced to withdraw down the hill and rejoin the rest of the company at the base.

On Hill 603 Company B drove hard to capture the crest. In the face of fire from caves in the side of the cliff, progress was slow. But at 1330 under cover of direct artillery fire from the tank destroyers, a squad closed in on the cave positions with hand grenades. Fourteen prisoners were taken. Advancing from the caves to the top was a bitter struggle, however. The enemy dropped hand grenades on the attackers from positions almost directly overhead, and machine-gun fire from Hill 481 raked them from the right. Although they had not taken the crest by nightfall, they had made notable progress after the most difficult fighting.

During the night the missions of Companies A and C were exchanged. Company A was to resume the attack on Hill 544, and Company B was to remain in its position prepared to move at daybreak.

On the right flank of the 361st Infantry, companies E and G were still unable to move. Caught on low ground, they held firm in their positions despite the continuous pounding by artillery and mortar fire from the north and east and by machine-gun fire from Hill 592. From their observation posts on the surrounding hills, the enemy could see the foxholes of almost all the men. Company F tried to relieve the positions by sending a rifle platoon, reinforced by a platoon of heavy machine guns from Company H, to attempt to maneuver around the southwest side of Hill 592. After going up over the saddle between Hill 592 and Hill 504, they were to attack Hill 592 from the west. The move was accomplished, and the platoon very nearly reached the crest of Hill 592 before it met opposition. In the ensuing fire fight, the platoon of heavy machine guns was surrounded and captured, and the infantry platoon, after sustaining heavy casualties, was forced to return to Sassi.

In the sector of the 363d Infantry, the companies maneuvered for better positions. Company K remained on Hill 504, unable to advance because of the heavy small-arms fire. Company L got over the escarpment and occupied Hill 481, ready to attack north the following day. In the sector of the 1st Battalion on the right, Company A and Company C were held up all day, but after dark the battalion reorganized, and Company B moved to Hill 411, where it remained in position for the next day's attack. The importance of these two moves to advantageous positions for the attack at 0600, 13 October, cannot be overestimated, for in them lies the explanation of much of the next day's success. To the right of the 1st Battalion, at the foot of the cliff, the 2d Battalion reorganized at Ca' Parma. The battalion had been given the mission of advancing north and east along the Division's right boundary. It was to attempt a crossing of the Zena River at Zenarella with a view to securing Hill 411. A patrol sent to Zenarella at 2120 reported that the enemy had fled the town and that Company B had moved onto Hill 411.

While the infantry maneuvered to secure better positions, the artillery and air support blasted the enemy strongpoints. The artillery fired 8,400 rounds of all types, most of them in an arc about Livergnano in support of the 1st Battalion, 361st Infantry. Twenty-five TOTs[1] were fired on every sort of target—strongpoints, counterattack formations, troop concentrations, and supply dumps. The artillery's power was augmented by the firing of tanks and tank destroyers. Company A, 804th Tank Destroyer Battalion, expended 324 rounds of direct fire in support of the 361st Infantry, knocking out two mortars and one machine gun and prompting the surrender of 28 prisoners, who admitted as they came out of the caves above Livergnano that they could not stand direct self-propelled-gun fire. Company B, 804th Tank Destroyer Battalion, fired 1,296 rounds in support of the 363d Infantry with excellent results.

In the air, 222 medium bombers attacked bridges, a munitions factory, a bivouac area, and supply, also with excellent results. Fighter-bombers continued their close-in support of the II Corps, flying 250 sorties. Strongpoints, troop concentrations, and gun areas were the day's targets in front of the fighting troops. The air attack started

[1]*Time on Target.* This is a type of artillery concentration so fired that all shells arrive at the target simultaneously. No warning of any kind is allowed the victims.

many fires and explosions, which aided materially in reducing enemy resistance.

In the attack at 0600, 13 October, the artillery laid down a tremendous concentration—2,120 rounds in sixteen minutes. Better progress was noted everywhere during the day, and it appeared that the Germans had at last begun to give way under the steady pounding from our artillery, planes, and infantry. On the right, the 1st and 2d Battalions, 363d Infantry, made good gains. Although progress was slow during the morning because of the flanking fire from the vicinity of Casola, resistance lessened in the afternoon, and satisfactory gains were made.

By evening, in the 1st Battalion's sector, one platoon of Company B was on Hill 368, and one platoon of Company C had occupied Hill 376. Companies A and B continued their advance during the night, and at 0740, 14 October, they had taken their objective, Querceto.

In the sector of the 3d Battalion, 363d Infantry, Companies I and L led the advance against very heavy resistance. Pushing as rapidly as possible over the rugged terrain, they bypassed Hill 592 and swung slightly to the west. By nightfall Company L was on Hill 540, and Company I was on Hill 510 north of Hill 592. Company K was employed to clean out bypassed enemy strongholds.

Notable gains were also scored by the 361st Infantry. Casola, where Companies E and G had been held up for three and a half days, was captured by the 2d Battalion; Hill 603 was secured; and on the left flank, Casolina was neutralized. Only failure to occupy Hill 554 marred the day's record, and here the enemy was forced to withdraw during the night of 13-14 October.

Led by an Italian *paesano* over a heretofore unknown path leading over the escarpment north of Sassi, Company F attacked Hill 592 with great stealth and suddenness, seized it, and captured 26 prisoners. At the same time companies E and G attacked Casola. The enemy, occupying strong positions in underground dugouts, manned machine guns with deadly accuracy. Company E established a base of fire, while Company G flanked the three houses that comprised the town. Using bazookas and antitank rifle grenades to reduce the emplacements, Company G overran the enemy positions and captured 57 German enlisted men and an officer. It is indicative of the complete preparations the enemy had made to hold the escarpment line that ample supplies of bread, butter, meat, and wine were found in the buildings. They were a welcome prize to the men of Companies E and G, whose supplies had been reduced to a minimum during the

Powder River doughboys will never forget Purple Heart Corner in Livergnano. The house in the upper left corner, on Hill 585, was used as a Division OP.

past few days. Subterranean storage rooms were found to be well stocked with all types of ammunition and weapons. The 2d Battalion immediately began a reorganization in order to move westward to Hill 481, then around Hill 603 to cut Highway 65 one kilometer north of Livergnano.

At Livergnano Company B renewed its attack on Hill 603. Jumping off at 0600, 13 October, the company moved halfway up the hill without meeting enemy resistance. With the protecting fires of the artillery ringing the hill, the company reached the crest by 1145. During the afternoon it was subjected to a heavy concentration of artillery and mortar shells from the north and northwest, but held the position.

On Hill 554, the attack of Company A was not so successful. In a repetition of Company C's experience the day before, Company A reached the ridge of the hill after advancing through heavy mortar and machine-gun fire and knocking out with rocket launchers a machine gun which swept the only avenue of approach to the objective. When they had occupied the ridge, machine-gun and small-arms fire forced them to take cover in the building, as Company C did the previous day. Two self-propelled guns began firing point-blank into the building, and the men were forced to withdraw to the reverse slope of the ridge.

On the left, the 3d Battalion, 361st Infantry, secured the left flank of the regiment by neutralizing Lagla. The 3d was alerted for the movement at 0015. Companies I and L moved westward toward Lagla and by 1525 Company I had captured the town, while Company L was entrenched to the east on Hill 345.

Thus, at the end of the day, the lines had been straightened and the flanks secured. With Lagla on the left and Querceta on the right in the Division's hands, and with the enemy strongpoints, Hill 603 and Casola, reduced, the enemy line, referred to by many of the prisoners questioned by the 361st Infantry as the Caesar Line, had been overrun, and the escarpment had been surmounted. Enemy casualties had been heavy, and many prisoners had been taken, 225 on 12 and 13 October alone. Enemy pockets of resistance still remained, and the next day's activities, most of which were undertaken by the 2d Battalion, 361st Infantry, were designed to mop these up.

At Lagla, Company L moved abreast of Company I, and together the two companies held the left flank until the 361st's 2d Battalion had swung into its position along Highway 65. At Livergnano, Company B worked forward on the spur called Hill 585, north of Hill 603. Here very careful observation revealed that the enemy had withdrawn from the town and from Hill 554. A patrol of Company K was sent into the town to investigate, and at 1645 it confirmed the report.

The following is a story written by Mr. Jack Bell,[2] Associated Press War correspondent covering the Fifth Army front:

WITH THE FIFTH ARMY IN NORTHERN ITALY, October 25, 1944—There is a little Italian town high in the Apennines a few miles south of the big industrial city of Bologna, a town called Livergnano.

Perhaps I should say "there was a town"—for Livergnano isn't any longer. Even the old church on the hill has only a steeple now. The homes chiseled out of solid stone into a high cliff are still there, though their furnishings are smashed. The caves under the cliff at the end of town opposite the church remain, made even deeper by tank guns which were wheeled up and fired point-blank at twenty feet. The buildings are gone except half a dozen ground-floor rooms which are virtual caves.

"We took Livergnano today."

---

[2]Bell lost an arm while serving as a machine gunner with the 363d Infantry in World War I and was awarded the DSC.

Artillery fire reduced Liverano to a mass of rubble

The statement made by a Public Relations Officer at Fifth Army Headquarters the day I arrived there didn't mean much to me and apparently little more to him.

I remember asking, "Was it an extensive battle? Is it important?"

"Oh, just another town on Route 65," he replied. "Held us up for three or four days."

Yes, it had held us up for a few days. It *was* important and the battle of Livergnano was perhaps the most desperate and costly of the entire Italian summer and autumn offensive. American dead still lie buried under crumbled Livergnano and men still grow quiet and tense when they speak of Livergnano. Men are still under daily German artillery fire in Livergnano.

Ask the men of Company K if Livergnano was important. But I should not have said that; for the men of K Company—eighty-one of them—are German prisoners; twelve escaped, the rest died there. Ask the men of I Company who took La Fortuna, a few hundred yards short of Livergnano. They were shot to pieces for three days while other companies fought fiercely up the road. Ask the men of A Company who fought their way into a church, just short of the town, found the basement full of Jerries and the units threw hand grenades at each other until a German tank gun rolled up, literally blew the Americans out of the church into machine-gun crossfire. Ask the men of B and C Companies who fought blindly but with unbelievable courage against the German Army, stronger in number, with better observation and greater fire power.

Ask any of them who survived. They'll shake their heads and tell you they don't know how they did it. Colonel Rudolph W. Broedlow, regimental commander, sits staring into space, then turns and says, "I guess the boys just had so much guts the Krauts couldn't understand."

"Occupy Livergnano at all costs" was the command, fashioned way up among the high brass somewhere. The high command told the general, who told the colonel, who told the battalion commander, who told his company commander.

On September 8, I Company crawled up out of the woods and threw a roadblock at La Fortuna, which is only a house alongside of the road. This was to keep the Jerries from coming south on Route 65, if they didn't cut I Company to pieces too fast. That night K Company drew the assignment to enter Livergnano. The German strategy never was better. It was one of those things. Everything the Boche tried worked. K Company couldn't do anything effective.

If you could drive up the highway into Livergnano you could see why it happened. How the 1st Platoon of K Company, moving at night and at the edge of the city, dodged into the big house there when a machine gun suddenly cut loose, and sighed with relief. They listened. All was quiet. Guess he was just shooting for the hell of it, they decided.

The 2d Platoon ran into fire from above and broke down the hill. Guns down there spoke, and the Americans turned back into the convenient house at the edge of town. The 3d Platoon, with no knowledge of the advance, was neatly maneuvered into—yes—that same house.

A whole company of Americans in one house. They had no antitank guns, no place to use mortars. The Germans had herded them into that house and some time after midnight brought up a tank gun—maybe two—and started point-blank fire. An 88 shell has tremendous power. Steadily the building crumbled.

This picture of Captain (then First Lieutenant) Chetlain Sigmen and members of ill-fated Company K, 361st Infantry, ironically enough was one of the very few group pictures taken of Division units en route overseas.

There was a sergeant, Dave Covington, a little guy from Kenosha, Wisconsin, and another sergeant, Luke Owens, a wiry little fellow from Griffin, Georgia. They saw what was happening, and as the last platoon broke for the house they dashed for the next house. The Germans had figured even that possibility and had a machine gun zeroed on the door. The first man who tried to go through that door was cut down—instantly killed. Sergeant Owens dashed to the side, the others followed. They found a hole, crawled in and were deep in a pigsty under the house.

B Company, supposed to come up and support K Company, was pinned down; their captain, who had gone on a scouting patrol, was pinned in the house with K Company. A Company was pinned in the church below. C Company was under terrific fire back along the road at La Fortuna.

Sergeant Troy Pons, Silver Springs, Florida, was one of the men who had followed Sergeant Owens into the pigsty. Sometime during the night, when a Jerry who evidently didn't know the men were down there came along, Sergeant Pons picked him off. Then Sergeant Pons told his mates he was going to find the rest of the company. He hasn't been seen since, probably being a prisoner . . .

For when daylight came, the Jerries methodically shelled that house down around the heads of K Company. Captain Chetlain Sigmen, Seattle, Washington, talking with the battalion headquarters by radiophone, said that he had been asked to surrender. He was talking to his commanding officer when the voice stopped.

Back at American observation posts they watched the Americans filing out of the shattered house. Colonel Broedlow, hoping a desperate measure might free them, ordered a terrific artillery barrage. But the Germans didn't run, leaving their prisoners behind. They got away, somehow, taking eighty-one Americans with them.

All that was left of K Company, the ten lads who had been led down into the pig pen by Sergeant Owens, never knew what had happened. They lay all day listening to the shelling, heard German voices and moving vehicles. Late at night they escaped.

So! The Germans are disorganized, are they? They're short on artillery—yes? They've lost morale? Don't talk thus before any United Nations soldier in Italy.

Sadly the Americans viewed the scene from which their comrades had been taken as prisoners. Grimly they set their teeth and went back to work. They called for bombers, and for two days Spitfires dived and blew the town apart.

And there was a guy from Bakersfield, California—Sergeant John Sumpter. Back in the States he had been a typical "eight-ball" soldier. He always was in trouble, always punching someone in the nose, always out of step. Then the outfit came over here, where the chips were down. The battalion eight-ball took things in his stride. He became a "hell of a soldier, afraid of nothing, and a born leader." It wasn't long after the outfit got here that he moved up the scale.

So, when Company B went back into action, determined to take Livergnano, tough Sergeant Sumpter led the 1st Squad for the battle to take the high cliff overlooking the city. Machine gunners were in the caves at the foot of the cliff and opened fire. Sergeant Sumpter drew his men around the bend in the road twenty-five yards away and called for tank guns into the cave. The Jerries poured out and were disarmed.

Then Sergeant Sumpter went to the reverse side of the hill, shouted gruffly, "Come on, you guys," and started up the hill. Every German mortar in Italy poured shells on the hillside. Machine guns cracked. Tank guns thundered. The sergeant and his platoon went on, shooting with tommy guns, throwing grenades, evading enemy shell holes. On and on they walked—and the Germans up there threw up their hands in despair.

"You must be the élite American troops," an English-speaking German prisoner said. "It seems impossible for you to come through that mortar fire."

"I don't know what élite means," growled Sergeant Sumpter. "But if you mean good, that's right . . . B Company, 361st Infantry, buddy!"

Up came the remainder of B Company. They set up observation. They directed American tank fire. The other companies, A and C, moved in almost unmolested—because this time the Americans had the upper hand to direct their tanks and artillery fire. Remnants of the companies cut to pieces three days before went in with grim determination and gun-spouting death. Even the remnants of the ill-fated K Company, which had been reorganized, led replacements into the general triumph.

They went into a city battered by American big guns, by dive bombers—and after we got there, by German artillery. They had carried out the general's order to take Livergnano "at all costs."

Yes, the PRO was right. We took Livergnano today.

On the right, in the sector of the 363d Infantry, Companies A and B, having reached Querceto at 0740, 14 October, consolidated their positions and maintained contact with the 335th Infantry on the right and the 3d Battalion on the left. Company I advanced northward under moderately heavy mortar and artillery fire, while Company L mopped up Hill 387 and then advanced to Hill 384. The 2d Battalion, in reserve, moved to an assembly area in the vicinity of Ca' delle Fate. Leaving Sassi at 1930, the battalion pulled in to its new area at 0050, 15 October—a significantly short time for the trip in contrast to the bitterly slow advance of the 1st and 3d Battalions over the escarpment just a few days before.

The main effort of 14 October was made by the 2d Battalion, 361st Infantry. Moving westward from Casola, they fought their way across the rough terrain of the top of the escarpment under constant heavy artillery and mortar barrages. By 1800, however, Companies E, F, and G were on Hill 603, and two hours later all companies were in their planned positions. Company G, skirting the west slope of Hill 603, threw a roadblock across Highway 65 one kilometer north of Livergnano. Company E, skirting the east slope of Hill 603, moved to a point 200 yards short of Casa di Achele, abreast of Company G on the left. Company F, in close support, took up positions on Hill 585. The artillery laid down a screen, behind which the companies were able to dig in. Throughout the night the forward elements of Companies E and G could hear the shouts of German soldiers and much motor movement up and down Highway 65.

Another protective screen, this time of sound, was laid down by the artillery during the night of 14-15 October. Just south of Livergnano, where Highway 65 is built on a ledge on the side of Hill 603, the retreating enemy had blown a huge crater in the road. The crater was 75 feet long, sloping between 45 and 60 degrees, and effectively blocked the movement of all armor and trucks into the town to support the infantry. On the night of 13-14 October reconnaissance by the 316th Engineer Battalion determined that hand labor could not fill the crater. Yet the site was so close to the front line that the noise of a bulldozer would surely bring down artillery and mortar fire.

The next night, after a tankdozer proved incapable of doing the job, a regular bulldozer with an angled blade was brought up. Then, under the personal direction of Lieutenant Colonel William C. Holly, Commanding Officer of the 316th Engineer Battalion, and under cover of artillery fire, the dozer operator, Technician Fifth Grade Archie Childers of Company A, working swiftly and expertly, dug dirt from the side of the hill and pushed it into the hole. He stopped once when he felt the dozer blade strike something hard and metallic and got down on his hands and knees to look for the object in the darkness. Suddenly his hand grasped the familar handle of a German Teller mine. Without hesitation, he picked it up and threw it off the road. In less than fifteen minutes, he made the crater passable one way. Immediately, tanks, tank destroyers, and antitank guns, which had followed the engineer crew to the site and stood by in darkness, moved forward into Livergnano and prepared to support the drive of the 361st Infantry northward.

At 1500, 14 October, Field Order No. 24 was issued, outlining the tactics for the final push to Bologna, which was to begin on 15 October. The 363d Infantry was to continue its drive on the right of the Division sector, maintaining close contact with the 34th Division, which had relieved the 85th Division on the right. The 361st Infantry was to press on north in the center, up Highway 65. The 362d Infantry, which had had a brief period in Division reserve, was ordered to pass through elements of the 135th Infantry on the left flank of the Division and, making the main effort on the left, bypass Mt.

Adone. Then, in a coordinated effort with the 1st Armored Division, it was to seize Mt. Adone from the northeast.

Thus, operating in a very narrow sector for the first time, the Powder River Division would proceed along Highway 65 in the shadow of two important hill masses, Mt. Belmonte to the east and Mt. Adone to the west. The flank regiments, coordinating with the flanking divisions, were to assist in the capture of the mountains. While the final preparations for this concerted attack were being made on 15 October, the 361st and 363d Infantry Regiments pushed on.

In the sector of the 363d, the 1st and 3d battalions moved forward from Querceto, while the 2d Battalion, in reserve at Ca' del Monte, outposted the Division's right flank until elements of the 34th Division could push forward to protect it from fire from Mt. della Vigne. Gains were small all across the line, and at the end of the day the lines were stabilized on the 31st Northing. The next morning, 16 October, at 0500, the attack jumped off without artillery preparation in an effort to achieve surprise. So successful was the plan that at 1000 the companies were ordered to consolidate their positions and hold until the flank units pulled abreast. Company B halted at the road junction just north of Ca' Fiume; Company A, at Cassa Torre; and Company K, on the 31.6st Northing, one kilometer east of Highway 65.

No further attempt was made by the 363d to advance on 16 October, but extensive artillery, cannon, and mortar fires were laid on the draws and gullies immediately to the front to forestall counterattacks and soften the enemy for the resumption of the advance. In the afternoon arrangements were completed with the 133d Infantry to assist in the capture of Mt. Belmonte.

The night attack jumped off at 2000, the 1st Battalion coordinating its effort with the 133d Infantry, 34th Division, on the right. Very little advance was made, however, because no definite location could be reported for the regiment on the right, and there was consequent danger of firing on friendly troops in the dark. At dawn, 17 October, it was found that Mt. Belmonte had not been secured during the night, and consequently the 133d Infantry had not moved forward far enough to relieve the presssure on the right of the 363d Infantry. The 361st Infantry had made only slight gains up Highway 65, so the 363d remained well out in front and exposed on both its flanks.

Despite this exposed position, both battalions drove forward. Company B made a very substantial advance to the 32.1d Northing in the vicinity of Ca' San Giorgio, and Company K moved to Valpiana. During the night every attempt was made to bring forward antitank guns and tank destroyers to give closer support to the infantry. Antitank prime movers were unsuccessful in moving the guns, so oxen were employed, but were unable to do the job because of the soft footing. The engineers sent two bulldozers to assist in preparing a passable road, but it was not until after daylight that one antitank gun was moved over the improvised trail and put in position. The tank destroyers moved forward 500 to 1000 yards to better positions, but they were so imbedded in the mud that they could not advance with the troops.

With the reinforcing fires of this armor, however, the 1st and 3d Battalions, 363d Infantry, attacked again on 18 October. The 3d made little or no gain in the face of heavy enemy flanking fire from the northwest, but the 1st advanced toward Ca' Trieste and the lateral road running between Zena and Zula. Heavy casualties were sustained in the sector as a result of enemy firing from Mt. Belmonte, which was directly on the right and still had not been secured. In view of the situation, the Division Commander directed the 363d to consolidate its positions and hold its gains until the adjacent units came abreast.

In the meantime, the 361st Infantry continued its drive north along Highway 65. The attack jumped off at 0530, 15 October, with the 3d Battalion on the left and the 1st Battalion along the highway. The 2d Battalion remained in position. The 1st, led by Company A, made excellent progress. Passing beyond its objective, Hill 520, at 0600, Company A continued its advance and was north of Lamma by 0930. At the same time, in the 3d Battalion sector, Company I had secured Casolino; Company K was east of Il Molinetto; and the leading unit, Company L, was just south of Lucca. Throughout the afternoon the entire regimental zone was swept by the heaviest artillery and mortar concentrations yet encountered by the 361st.

Despite the heavy concentrations of fire, Company A continued to probe through the heavy fog towards its objective, Guarduzza. Just as it had reached a position between Ca' Raulina and Guarduzza, the protective cover of the fog lifted, and the company discovered itself in the midst of strong enemy positions. It was a target for enemy fire coming from every direction. Machine guns to the rear harassed every movement, and self-propelled-gun fire raked them from the direction of Castelazzo. In the face of this superior fire power, Company A was forced to withdraw to reorganize for continuance of the attack.

The withdrawal was covered by Captain Charles E. Mohler, company commander, Sergeant John E. Walraven, and one other enlisted man. These three successfully beat off an enemy counterattack, although Captain Mohler was seriously wounded early in the action. Sergeant Walraven succeeded in killing the crews of an enemy mortar and machine gun before they could set up their guns. The sudden loss of their heavy weapons disorganized the Germans, and the counterattack was broken. Hiding in the underbrush, the three men awaited the cover of darkness to rejoin their company. In a heavy mortar barrage laid on their position, Captain Mohler was killed, and both enlisted men were wounded. Sergeant Walraven, however, succeeded in carrying his comrade back to safety.

The next day, the 362d Infantry attacked on the left through the 3d Battalion, 361st Infantry, with Lucca as its objective. Despite heavy resistance, by 0800, 16 October, one platoon was within a hundred yards of the town. To facilitate the advance of the 362d, the 361st's Company I continued its advance to positions above the line from Mt. S. Ansano to Gruppi, and Company K pressed its attack toward Gruppi. The advance inched forward slowly in the face of very heavy machine-gun and small-arms fire. On the right of Highway 65, Company B advanced to Hill 477, where it established a roadblock across the highway.

At 2200 the 2d Battalion, 361st Infantry, attacked through the 1st Battalion, and by 0415 the next morning Company E had reached positions in the vicinity of Disotta, to the right of Highway 65 along the 31.5st Northing. Company F, to the left of Highway 65, was on Hill 422, while Company G was echeloned to the left toward Guarduzza.

Company E had to fight for its line of departure in the dark, early-morning hours, and in savage hand-to-hand combat, killed thirteen Germans and captured seven others. Almost immediately, the enemy unleashed terrific artillery and mortar concentrations on the Company E position. At daylight he brought up two self-propelled guns, supported by four Mark IV tanks, and deployed them along Highway 65 north of Canovetta. One of the guns succeeded in reaching a point within 75 yards of Company E's left flank, where it began systematically to blow down the company's observation post, a house at Disotta, by firing at it at point-blank range. Artillery and bazooka fire succeeded in driving off this armored threat, however. The extreme accuracy with which the 916th Field Artillery Battalion laid down protective fire for Company E averted what might have been a disaster. High-explosive shells were dropped within a hundred yards of the Company E positions without causing a single casualty among our troops, and the enemy attempt to drive a wedge through the 2d Battalion failed.

In the meantime, friendly tanks and tank destroyers were brought up, but the poor visibility caused by the rain and fog reduced their effectiveness. Two tanks of Company C, 755th Tank Battalion, in a bold and heroic attempt to relieve the situation, moved ahead of the infantry along Highway 65. The wily Germans held their fire until the tanks reached La Fahbrica and then sent armor-piercing shells screaming into them, reducing them to a mass of flaming wreckage.

To the left of Highway 65, Company G was checked in its attempted maneuver around the left flank of the battalion when its lead platoon was subjected to direct fire from an enemy tank across the Savena River in the vicinity of Ca' Boschi. One Company G platoon was caught on the open hillside in the vicinity of Guarduzza and dashed to the cover of a culvert. The enemy tank began systematically to annihilate the platoon by placing direct fire at either end of the culvert. In the face of such resistance, very small gains were registered by the 2d Battalion.

On the left, the 3d Battalion, 361st Infantry, drove forward over difficult terrain and in the face of stiff resistance to dominating positions 300 yards southeast of Gruppi. At the end of the day on 17 October all companies consolidated their positions and prepared for a possible counterattack down Highway 65 by enemy armored elements.

The attack of the 362d Infantry on Lucca continued during 17 October with Company I in the lead. Despite heavy small-arms, mortar, and artillery fire, augmented by direct fire from enemy tanks, the company broke into Lucca at 1400. Company G then moved into the town to reinforce Company I. During the night of 17-18 October, the 362d assumed command of the sector of the 1st and 2d Battalions, 361st Infantry, while the 361st's 3d Battalion continued to screen the left flank of the 362d.

The 362d's 1st and 2d Battalions resumed the attack at 0530, 18 October. Their mission was to sweep north along Highway 65, with the 1st Battalion driving for the key enemy positions at Canovetta while the 2d Battalion worked along the ridges west of Highway 65 to the high ground southwest of Canovetta. They advanced slowly against strong hostile resistance, and at 0930 Company F was reported on a ridge approximately 800 yards southwest of Canovetta, and Company E was on the 32.1d Northing east of Highway 65. Enemy resistance during the rest of the day was so intense that it was impossible for the battalions to move. By attacking at 0130, 19 October, the battalions bettered their positions. The 2d achieved its objective, occupying the high ground southwest of Canovetta and immediately east of Ca'

Nuova. The 1st seized its objective immediately south of the town at 0700.

The next three days were spent in maneuvering for positions in the face of heavy enemy fire of every type. At 1200, 21 October, after a bitter see-saw battle, a patrol entered Canovetta. Plans were immediately laid for the relief of the 2d Battalion by the 3d and for coordinating an attack by the 1st and 3d Battalions. However, since there were indications that the enemy might counterattack in great force, at 1650, 21 October, General Livesay directed the 362d to consolidate its positions and organize for defense.

At 1600, 22 October, for the first time since the Division had entered combat, General Livesay ordered it to revert from the offensive and take immediate steps toward the organization and defense of a main line of resistance. Three important considerations dictated the move. According to information received from the II Corps, two regiments of the 85th Division and the 88th Division were going off to the northeast, and the Corps commander did not believe that with the loss of the two regiments in the west zone, there was enough strength to move in west of the Corps' sector. A second reason for the shift may be found in a discussion of enemy capabilities found in intelligence reports of 25 October:

> Fearing the threat of II Corps advance, more than any other on the Italian front, the enemy had assembled in front of the II Corps elements of eleven divisions with an estimated strength of 13,000 men. Recently additional troops have appeared on this front—the 29th Panzer Division and the 90th Panzer Division from the Adriatic sector. A battalion of the 94th Panzer Division has been shifted from the west sector, and the 16th SS Division is moving east.

In the light of this enemy strength, a counterattack in great force from the northeast was considered very possible.

The intelligence report continues:

> The enemy has lost some key features in this sector, and if a stubborn defense is to be maintained for any length of time, he will probably attempt to regain Mt. Belmonte and Mt. Castellari. The object of this attack might well be to regain Mt. Castellari, which dominates Highway 65 and which prisoners state the enemy was very reluctant to lose. Loss of the high ground would make it difficult to maintain our advance positions east of Highway 65.

The third consideration which was weighed in reaching the decision was the artillery situation. On 19 October, the Division's ammunition allotment was drastically curtailed, and the greatest care had to be exercised in the selection and firing of targets. In the last week of the month an allocation of 7,000 rounds of high explosive and 700 rounds of white phosphorus was received, allowing the light battalions to fire but an average of 363 rounds per day per battalion. The last week's firing did not even equal the fires of one eight-hour period of the second day of the month. Of the last fifteen days of the month, on all but four more than 1,000 rounds of artillery of all calibers were received from the enemy, principally along Highway 65 and in the vicinity of Livergnano. On 20 October, 4,700 rounds of mortar and artillery fire were received, with 900 rounds falling in two hours in the area between Disotta and Canovetta along Highway 65. In the light of these circumstances, the decision is understandable.

During the remainder of the month the Division devoted itself to strengthening and defending the main line of resistance. Shaped roughly like the letter S, this line ran from Mt. Belmonte, on the east in the sector of the 34th Division, westward across Highway 65, approximately two hundred yards south of Canovetta, and then dropped southwest along the Savena River to just south of the 30th Northing, where it straightened out. The line was occupied by two battalions of the 361st Infantry on the left and two battalions of the 133d Infantry, 34th Division, on the right.

In accordance with a defense plan coordinated at Division Headquarters, defenses in the 91st's sector were set up along the main line of resistance by the infantry battalions assisted by the 316th Engineer Battalion. Sixty-eight truckloads of wire and other obstacles were obtained for the purpose, most of them coming from the port of Leghorn, which had been captured in July by Task Force Williamson. A total of 160,000 yards of barbed wire was procured, along with 106,000 sandbags, 9,000 yards of wire concertinas, 2,500 antitank and antipersonnel mines, and 8 tons of demolitions.

To maintain contact with the enemy, vigorous patrolling was carried on each night across the entire front. Patrols were sent to Fazzano, to the area north of Lucca, to Mt. S. Ansano, to Ca' Tianello, Ca' Nuova, Ca' Selvamaggiore, and Canovetta. Thus, constant contact was maintained with the enemy. To provide direct observation on enemy positions, two observation posts were established—one at Villa le Croci and another at Lucca—at 2125, 23 October. Manned by the 91st Reconnaissance Troop, these posts were invaluable in directing artillery fire and keeping a close watch on enemy activity.

Throughout the period, the forward positions were made difficult by heavy artillery shelling and by wretched conditions caused by the fall rains. Although every effort was made to bring hot meals to the men under the cover of darkness, supply trains were frequently targets for heavy artillery and mortar concentrations. Movements in the area during daylight were necessarily kept at an absolute minimum. On 27 and 28 October, the rain had turned the area into a sea of mud, and all foxholes and gun positions had to be revetted with sandbags. Roads became rivers of mud and water, and the engineers worked early and late to keep them passable. The battle-seasoned Powder River men bore the brunt of these unbelievable hardships with typical GI humor. They cursed the Italian mud and weather, the cold rations, the stubborn enemy, and their local draft boards back in the good old USA.

Early in the morning of 24 October the 362d Infantry had been entirely relieved and had closed in the area around Anconella for three days of badly needed rest and rehabilitation. The days and nights spent under the intense fire of the 120mm mortars and enemy artillery on Highway 65 had tried the men severely. Casualties had not been too severe, however, and the men had learned all the tricks of digging in and acquired a sort of sixth sense as to where the Germans would probably fire. The enemy, it seemed, concentrated his fire where he would have placed his men, but our troops used the ground in a somewhat different manner and thus escaped the direct effect of many enemy concentrations. The cumulative effect of the fire, however, was to tire the men to the utmost, and, while some of the area to be used for rest was in view of the German lines and in no place was it more than 4,500 yards from them, the prospect of three hot meals a day and comparative freedom from shelling gave a great feeling of relief and allowed a certain degree of relaxation.

The rest, however, was not to be; for just before noon of 24 October telephonic orders came for an advance party to go to Headquarters, 88th Division, for orders. At 1415 representatives of the 362d RCT met at La Posta and proceeded to the 88th Division CP at Villa di Sassonero, 5,500 yards behind Mt. Grande. The 88th had just captured the crest of the mountain and had started down the forward slopes into the plains of the Po Valley, and at this time advance elements were only 8,000 yards from Highway 9, which runs in a straight line from Rimini, on the Adriatic, to Bologna, on the flat plains of the Po Valley. In order to hold onto Mt. Grande and at the same time continue to advance, the 88th needed another combat team, and the 362d, though in need of rest, was the only one available deemed capable of the effort.

After a trip which took four hours to cover the thirteen miles of muddy roads, the 362d's commander and representatives of the 346th

Field Artillery; Company C, 316th Engineers; and Company B, 316th Medical Battalion reported to General Kendall, 88th Division commander, and received the situations and orders.

The 362d's mission was to relieve elements of the 88th Division on the night of 26-27 October and attack early on 27 October. The attack was to head east, up and down the ridges on the north bank of the Sillaro, and protect the Division's right in its drive down the eastern nose of Mt. Grande to cut Highway 9. The British 78th Division, British 13 Corps, would be on the right.

The 362d headquarters was established in a ruined farm house, Ca' Clanco, just in the rear of Mt. Grande. The regiment closed at 2247, 25 October, and the remainder of the combat team was to close by daylight.

During the motor move to the new area, the detrucking area of the 1st Battalion near Sasseleone received 35 rounds of enemy 170mm artillery fire. Twenty-eight casualties resulted.

Technician Fifth Grade Franklin E. Carlson, a medical aid man, was hurled from his vehicle by one of the blasts and received severe wounds. Disregarding his own injuries and the intense shell fire, he crawled across the road to a seriously wounded comrade. After rendering first aid to the man, he tried to crawl to the assistance of another casualty, but his own wounds had sapped his strength. Realizing that he could not reach the soldier alone, Technician Fifth Grade Carlson called for someone to carry him. He reached the stricken man and summoned his last ounce of strength to render first-aid treatment. Heedless of his own wounds, which later resulted in his death, he continued to give first aid until he collapsed. These cold words inadequately express the supreme courage and selfless devotion to duty displayed by Technician Fifth Grade Carlson and the other medical men who gained the undying gratitude and respect of front-line fighting men the world over.

The area assigned to the 362d was the only space available and was mostly in the bottom of the Sillaro River valley, at that time but a small stream. But trouble was on the way, for the greatest rains in the area in a hundred years began that night. At first the river was only slightly affected, but between midnight and 0300 of 26 October the waters caught in the large drainage basins at the head of the Sillaro burst out in a tremendous flood, overturning vehicles (even 2½-ton trucks were rolled over and over) and washing away equipment.

Most of the 3d Battalion's vehicles were, fortunately, just above the high-water mark, and the 2d Battalion was able to move its vehicles part way up the slope. The 1st Battalion, however, had no place to

"Carlson continued to give first aid until he collapsed."

which to move its vehicles quickly. The final count showed about 45 vehicles carried away or wrecked in the flood. Eight awards of the Soldiers' Medal were made for heroism during this disaster. Farther upstream the 88th Division also lost many vehicles and much equipment.

The river continued to rise, and at 1030 all wire lines went out. Most of the 1st Battalion was on the wrong side of the torrent, and 1st and 2d Battalions had lost most of their equipment. Strange to say, for a few hours the road to the rear was improved by the rains washing the mud off the roadway, but by evening of 26 October it had become impassable for any but the most modest flow of traffic.

At 1155 the bridge which afforded communication between the regimental CP and the regiment, which was in rear of the CP, went out, and shortly thereafter the only bridge between the three battalions and the Service Company was also swept away.

While it was evident by this time that the attack to Highway 9 would have to be postponed, Mt. Grande still had to be made secure; so at 1220 the 3d Battalion was alerted for movement as soon as a footbridge could be thrown over the Sillaro and the streams entering the Sillaro from the west. The battalion was to draw ammunition, weapons, and rations from the 2d Battalion, which would be able, it was hoped, to get more by mule train during the night. The 2d was alerted for movement as soon as it could be resupplied.

On 27 October the 3d Battalion relieved the 1st Battalion, 350th Infantry. All but two vehicles were hauled out of the river, and most of them were evacuated to the rear over the bridge, which was replaced at 1445. The 1st Battalion and all salvaged equipment was back on the east side of the river by 0130, 29 October. The bridge in the rear of the CP was declared "in," and traffic, with ammunition and rations, once more moved forward to supply the 362d CP and the 88th Division on Mt. Grande, which had been without supply by vehicles for two and a half days.

By 0445 on 31 October, the 1st Battalion had relieved the 3d Battalion, 351st Infantry, on the south bank of the Sillaro and was in contact with the enemy and the left brigade of the British 78th Division. At 1000 the 362d Infantry had relieved the 351st, and by 0116 of 1 November, the 362d's 2d Battalion had moved to reserve position on the mountains at Ca' di Sasso.

At 0500, 9 November, command of the sector passed, and the 362d came under command of the British 1st Division, British 13 Corps, with which it remained until 1400, 12 November. At 2300, 9 November, the 362d passed command of the front line to elements of the British 1st Division and assembled back in the "flood" positions.

The period in line had been passed in the usual manner: enemy raids, patrolling, shell fire, and shell reports, enemy (man and horse) wandering into our minefields, and establishment and maintenance of a smoke screen to hide traffic in the Sillaro valley.

On the evening of 9 November, British battalions started to relieve elements of the 362d and for several days were under its command.

The 346th Field Artillery; Company B, 316th Engineers; and Company B, 316th Medical Battalion, had been detached from the combat team on 26 October and, after serving with the 88th and British 1st Divisions, came back to the 362d during the move from the Mt. Grande area.

Never were men more glad than the men of the 362d when they finally passed out of the dark valley of the Sillaro and the Mt. Grande sector. Enemy action had not been too severe, but the eighteen days spent in that storm-wrecked region by men who at the start were tired and worn out were not the subject of pleasant memories.

On 13 November the regiment closed into Italian barracks at Pistoia. They were the first troops to occupy this rest area, and after the previous

Division engineers did trojan work making roads out of seas of mud like this

Mechanics overhaul engines in the Division ordnance shop

eighteen days, the cold, damp rooms and the windows without glass were, for a time at least, an agreeable change.

The month of October had been significant for the 91st Division. The momentum of the drive through the Gothic Line had carried it forward through the Monghidoro Line, the Loiano Line, and finally the Caesar Line. This offensive involved the Division in some of the bitterest and most difficult fighting of the Italian campaign. Constantly attacking through the jagged Apennines and extremely bad weather, it had met opposition that grew stiffer and stiffer as it approached its goal, Bologna, which Kesselring had orders to hold. Below Pianoro the Division had passed for the first time in its combat history from the offensive to the defensive. Although less spectacular than the breakthrough of the highly-publicized Gothic Line, the breaching of the Caesar Line at Livergnano was just as difficult and just as much of a triumph.

Because of bad roads, bad weather, and a shortage of parts, October was the busiest month the 791st Ordnance Company had had so far.

It supplied the Division during the month with 661 carbines, 81 caliber .30 machine guns, 187 pistols, 252 bayonets, 476 trench knives, 331 M1 rifles, 58 engines, 246 automotive assemblies, and 86 vehicles of all types. This is, of course, not to mention the vast amount of maintenance work done. Technical Sergeant Everett B. Wolfe, a member of the company, was commended by the Brazilian Army for his valuable assistance to them in setting up their supply system. He also received commendations from General Clark and General Livesay for this work.

Trucks of the 91st Quartermaster Company travelled 75,783 miles over muddy, bombed-out roads during the month. The company issued 858,420 rations, 492,995 gallons of gasoline, 22,660 gallons of Diesel fuel, and 500 tons of winter clothing—shoepacs, heavy wool socks, field jackets, sweaters, and bed rolls. It also set up a clothing exchange for the purpose of outfitting RTUs.

The 91st Signal Company, besides installing, maintaining, and operating the Division's communications system, issued 50,000 batteries for radios and flashlights and almost 1,000 miles of telephone wire.

The 91st Military Police Platoon operated control posts and controlled traffic on Highway 65, the main axis for the Division and II Corps.

At the end of the month, the Division, as a part of the Fifth Army, had passed from its prime objective of destroying the enemy in its sector of Italy to its secondary objective: containing his forces and thereby assisting the drive of the United Nations from the east and west against the Reich. Although disappointed that it had not captured Bologna, the Powder River Division was as always, carrying out its assigned mission.

CHAPTER 11

THE WINTER LINE

A BRIEF analysis of the terrain and the conditions under which the Division organized and prepared the defenses of its sector of the Winter Line will give an indication of the tremendous difficulties faced and overcome.

In the first place, the area was ill suited for defense. The forward lines ran perpendicular to a series of corridors, extending deep into friendly territory, which made it impossible to prepare long, unbroken fields of machine-gun fire across the front.

By placing your hand on a flat surface with the fingers slightly extended and designating as north the direction in which the fingers point, you will get a rough miniature of the unusual terrain formation in the area. Your fingers represent the long ridgelines extending north from the Livergnano escarpment, and the area between your fingers indicates the low ground between the ridges. These draws became known as the "bowling alleys." They offered the only natural, covered routes forward, and the Germans were determined to deny us their use. Heavy concentrations of enemy artillery and mortar fire covered them from one end to the other. By this time, however, the battle-wise troops of the Powder River Division had learned that the easiest, most logical approaches were not always the best. Instead of using the draws, they selected the tortuous, inaccessible ridge-lines for their advance.

Along the left boundary of the defensive sector lies the Savena River; on the right flows the Zena. Both of these streams had become wild, swirling torrents, swollen to overflowing by the heavy fall rains. They further complicated the supply problems of outpost positions established on the far banks.

Farther to the west, across the huge gorge formed by the Savena River, lies Mt. Adone. Jutting southward a full kilometer and a half behind the forward positions, this granite-faced monster, rising defiantly to a height of over two thousand feet, dominated the approaches of three division sectors. It towered like an evil giant over all who faced it for six long, weary months. Mt. Adone became the most highly respected and publicized enemy stronghold on the Italian front during the winter of 1944-45.

Keeping this picture in mind, glance down at your hand once more and imagine an ant crawling over your fingers, trying desperately to find something to cover his advance. This illustrates the situation in which the infantry found itself when it was directed to dig in. True, there were folds and slight accidents of terrain which offered some

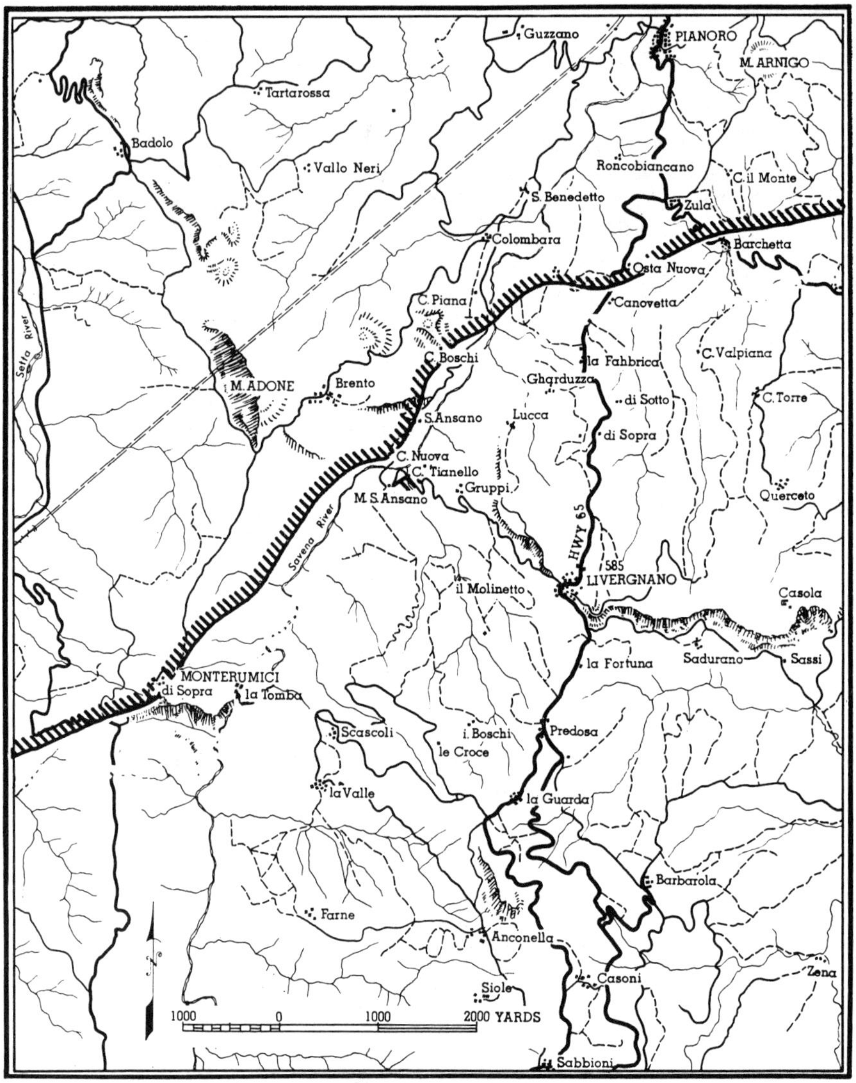

Map 20: Highway 65 sector

protection from flat-trajectory fire, but the gradually sloping ridgelines, extending into no-man's-land, were completely devoid of even the runty bushes that had once sparsely covered their barren, rock-strewn surface. The usual rock-walled Italian houses were conspicuously missing too. The enemy had been very thorough in carrying out a scorched-earth policy as he withdrew. Only smoldering piles of rock and mortar remained where houses had been. Fate lent us a helping hand, however. The thing which had been cursed most vehemently by privates and generals alike—the abominable Italian weather—proved a blessing now. The constant rain and the low-hanging clouds reduced visibility to a minimum while the Division was digging in. When the sun finally managed to break through the clouds, no sign of life was visible above the surface of the ground in the 91st's sector. The Division had holed up for the winter—but it had not gone into hibernation. The troops occupying the forward positions were constantly on the alert against possible enemy action. During the hours of darkness there were supplies to hand-carry forward; communication lines, severed by enemy shellfire, to be spliced; casualties to be evacuated; men returning from hospitals and replacements to be led into position; listening posts to be manned; patrols to be made deep into enemy territory; barbed-wire entanglements to be erected; trip-flares and antipersonnel mines to be set up. Every spare moment was spent improving individual emplacements with an eye for comfort and all-around defense.

Due to the nature of the terrain, the forward positions were lightly manned. The bulk of the Division was kept in a state of highly mobile reserve farther to the rear, where it could be used effectively at any threatened point to block attempted German breakthrough. This gave the men, while not in line, an opportunity to rest and clean up, to catch up on letter-writing, and to get food hot from the kitchens.

Each regiment followed a definite policy for the rotation of its units. The 361st Infantry, which occupied the sector north along Highway 65, rotated its battalions so that each spent six days in the line and three out, with one company held at Livergnano for patrols. On the right, because of the exposed flank, all companies were held in the line constantly.

The 363d Infantry, which had relieved the 135th Infantry, 34th Division, and assumed command of the sector on the night of 1-2 November, rotated its battalions every ten days. In the battalion on the right, companies occupied the line eight days and rested four days. On the left, since the flank was exposed, all companies remained in the line for the full ten days.

Activity on the main line of resistance and to the immediate rear was sharply curtailed. The II Corps ordered that: "(1) No offensive action of more than company strength be made without prior authority from Corps Headquarters; and (2) no troop movement was to be made except during the hours of darkness." No one, regardless of rank or importance of mission was permitted to visit any of the front-line units from dawn to darkness, and even night travel was held to a minimum; thus observed enemy artillery fire was materially reduced.

No attempt was made either to conduct offensive operations or to improve the defensive positions during the daylight hours. The main offensive activity was vigorous night patrolling. Each night six or more patrols were sent out to obtain prisoners and information. Listening posts were established to observe enemy positions, and reconnaissance patrols investigated the enemy-held strongpoints to the front to maintain contact with the enemy. Barchetta, Boschi, Piano dello Spino, Osta Nuova, Canovetta, Mt. S. Ansano, Ca' Nuova, and La Tomba were investigated regularly. Combat patrols captured prisoners and tested enemy strength.

A typical night's activity occurred during the night of 5-6 November, when five patrols were sent out. A patrol from Company A, 361st Infantry, entered Canovetta from the east and found it unoccupied. The patrol drew machine-gun fire from a point just west of the town, however. A patrol from Company B, 361st Infantry, advanced to within fifty yards of Ca' Nuova, where it was fired upon by machine guns and small arms from the buildings located there. A patrol from Company B, 363d Infantry reconnoitered the southeast bank of the Savena River and reported no enemy activity. Men from Company F 363d Infantry, established a listening post north of Gruppi and located enemy machine-gun positions by their firing. The patrol was unable to cross the river because of the swiftness and depth of the water.

At 1100, 11 November, General Livesay issued Field Order No. 27. The 91st Division assumed control of the sector of the 34th Division at 0600, 12 November, with the 135th Infantry remaining in position, attached to the 91st and continuing the defense of its sector on the right. General Livesay ordered that all units conduct aggressive patrolling "to contain the enemy and seek out soft spots in his defenses." He also directed that raids be conducted to obtain identification and maintain the identity of all enemy units on the front. In accordance with these instructions, all three regiments made strenuous efforts to capture prisoners.

On the night of 15-16 November, six of the ten patrols sent out were

A Division OP on Hill 592

ambush patrols. Three of the ten made no contact with the enemy; three were fired on directly; and three reached their objectives, Canovetta, Barchetta, and a house northwest of Ca' Trieste, from which they observed machine-gun and mortar firing. One patrol from Company E, 363d Infantry, as it was returning from Mt. S. Ansano, engaged an enemy patrol in a fire fight. No prisoners were taken by any of the patrols, but contact was maintained with the enemy, and much valuable information concerning his positions and disposition of weapons was obtained.

For daytime observation of the enemy, the 91st Reconnaissance Troop set up two observation posts—one in the vicinity of La Croce and the other on Hill 585 overlooking Livergnano. From 12 to 22 November a third observation post was maintained by the Recon Troop on Hill 521 in the sector of the 135th Infantry. Visibility from these positions was often limited by rain, fog, or cloud patches lying in the valleys. To add to their difficulties, the observation posts were frequently heavily shelled. Despite these handicaps, the Troop sent in valuable daily reports to the Division G-2 on enemy activity. From these and the information brought in night after night from all parts of the Division front by patrols, a comprehensive picture of enemy dispositions emerged.

Facing the Division were the enemy's 4th Paratroop Division, west

Mount Adone as seen from th

›p of Mount delle Formiche

of Highway 65, and 65th Division, east of Highway 65. These units were as active as the 91st Division. Four-man patrols, interrogation of prisoners revealed, were sent out regularly from each enemy company to reconnoiter the territory and houses immediately in front of its position. Frequently enemy and friendly patrols clashed in brief skirmishes, exchanging small-arms fire and grenades, each group trying to take the other prisoner.

The enemy also employed bold and daring ruses, such as ambushes led by English-speaking leaders who tried to convince our men that surrender was the only way to safety. One position in the zone of the 363d Infantry was especially subject to hostile raids. Practically isolated across the Savena River, the companies which occupied the position were subjected to almost nightly raids by small parties.

The enemy also used short-range spies to obtain information behind our lines. The customary technique was to filter Republican Fascists through the front lines, either on specific missions or on general reconnaissance. At other times the enemy used agents dressed as natives of the region to infiltrate through our lines with groups of refugees. These spies were trained by the German intelligence service to obtain information concerning air and ground installations.

In the sector of the 91st Division, the Counterintelligence Corps personnel took extraordinary precautions to combat these enemy efforts to obtain information. Special bulletins were issued to all men of the Division warning them to be alert for suspicious natives loitering near military installations. The CIC investigated the native population of the area and questioned all those suspected for any reason. These and other measures were remarkably successful. During the period 1 to 22 November, six tactical agents, young Italian students especially trained by the enemy intelligence, were apprehended and sent to higher headquarters, where they confessed. Also, two political agents were arrested just north of Loiano. They were returning to Bologna from Florence, where they had delivered a large sum of money to Republican Fascist groups.

In general, since the enemy was primarily seeking information, his activity was limited to patrolling and spying. There was, however, a notable increase in air activity early in November. Planes usually appeared at dusk over the Division sector. Highway 65 was strafed and bombed, and both antipersonnel and demolition bombs were dropped. The damage was small, however, and the effectiveness of the raids was slight.

Counterattacks in strength were rare, the strongest occurring on the

Both tanks and tank destroyers supported the infantry with accurate and deadly fire

night of 18-19 November. Enemy activity had been increasing notably, and on this night, after an artillery preparation, 15 Germans, supported by approximately 55 more, attacked the left platoon of Company F, 361st Infantry, which at the time was occupying the forward positions to the left of Highway 65. Company G, 361st Infantry, occupying positions echeloned to the left rear of Company F, assisted Company F in beating off the enemy thrust, and no ground was lost. Three casualties were sustained during the action. The Division Artillery, less the 348th Field Artillery Battalion, helped throw back the attack by firing 239 rounds according to prearranged plans. Four men of Company F were missing after the skirmish.

Counterattacks in somewhat less strength were launched in the sector of the 363d Infantry on 20 November. The prime objective of these attacks was clearly not to recover lost ground, but to capture as many prisoners as possible and to feel out the strength of our defensive positions.

Artillery fire on both sides was held to a minimum. The enemy ammunition allotment, it was reported, had been sharply curtailed. According to a prisoner-of-war intelligence report, in the German 165th Artillery the allotment was five rounds per gun per day. But in spite of these shortages, large concentrations continued to fall on Livergnano, Predosa, Lucca, and Highway 65 north of Livergnano, as well as on front-line positions.

Activity of the 91st Division Artillery was similarly restricted. From 1 to 15 November approximately 30 rounds per gun per day were allotted. Later on the allocation was halved. These conditions forced the utmost care in the selection of targets and cut down on the harassing missions which could be undertaken. Prudent use was made also of the fire power of the 755th Tank Battalion, the 804th Tank Destroyer Battalion and the 4.2-inch mortars of the 84th and 100th Chemical Mortar Battalions. The regiments were warned to pinpoint targets for the artillery, and the 91st Reconnaissance Troop reported likely targets observed from their OPs. In addition, the 804th Tank Destroyer Battalion set up an observation post of its own at i Boschi, northwest of La Guarda, and selected numerous targets. Visibility from these positions was often limited by bad weather. The artillery observation planes, too, were hampered by the weather, either at the airstrip south of Futa Pass or over the target areas.

Despite these difficulties, the artillery continued to give excellent support to the main line of resistance. One notable mission was fired on 8 November, when an observer caught a company of enemy marching along a road. The 348th and the 916th Field Artillery battalions fired a concentration of 79 rounds on the target and then continued harassing fire on the section. The enemy smoked the area, but three more volleys were fired. Later on, German ambulances were seen in the vicinity.

Activity of enemy personnel formed the bulk of the targets during the period. Houses known to be occupied by the enemy constituted profitable targets, and 69 of them were completely demolished. As a result of the 94 counterbattery missions fired, 17 direct hits were observed, and all areas were covered effectively. Tribute to the effectiveness of the curtailed program of firing carried on by the Division Artillery was paid by a German prisoner, who reported that our artillery had not caused many casualties lately, but had the effect of keeping the men pinned in their dugouts all day.

Throughout the period 22 October to 22 November, in which the Division manned the Defensive Sector, the situation was complicated by the weather. At first the rain was very heavy, and later the damp, Italian cold made life at the front miserable. Roads which were poor to start with had to carry the burden of heavy traffic during the whole time. Especially troublesome was the small one-way route which ran from Sabbioni through Anconella to La Guarda. This was the main supply route for the 363d Infantry, and the fall rains, together with the heavy traffic, caused it to become deeply rutted and very difficult

Whenever possible, hot chow was served to the men

to negotiate. Every day maintenance parties of the 316th Engineer Battalion worked on it, and by 22 November they had installed four culverts and put 231 loads of crushed rock on the roadbed.

Supplying the Division's outpost across the Savena River involved even more difficult problems. Normally the Savena is a peaceful and well behaved stream, capable of being easily waded. But after the rains of October and November, it became a raging torrent, twice its normal width, and four to five feet deep. To relieve the 135th Infantry, men of the 363d waded through the chest-deep cold water, holding to an overhead rope as they crossed. To ferry supplies across the river, rubber boats, manned by Company B, 316th Engineer Battalion, were used. A cable was stretched across and fastened to trees on either bank. A pulley rode the cable and was attached by a rope to the front end of a rubber boat; other ropes went from either end of the boat to the shore. Thus the boat, loaded with men, food, ammunition, barbed wire, or wounded, was towed from one side of the stream to the other, guided by the cable and pulley.

Living conditions for both the front-line troops and the rear installations likewise required constant attention. Every effort was made to house the troops inside buildings or in improvised shelters and to keep the foxholes on the main line of resistance as dry and comfortable as possible. All buildings occupied by the troops were scrubbed and made sanitary, and the men showed considerable ingenuity in making living quarters out of the materials at hand. Magazine pin-up pictures adorned the walls of dug-outs and buildings. They made excellent wallpaper and offered colorful subject matter for foxhole conversations.

While the 91st Division maintained its defensive position with two regiments, the 362d Infantry, which had been attached to the 88th Division on 26 October, was occupying positions in the 88th's sector.

Activity in the sector of the 362d Infantry was much the same as that in the 91st's. Patrols probed the enemy lines nightly, and German artillery and long-range machine-gun fire harassed the regiment's positions. Artillery fire was heavy, especially at the ford over the Sillaro River south of San Clemente. Occasionally stray rounds fell in the regimental installations, and one of them hit the doorway of the building occupied by the regimental headquarters at Fornace. Eighteen casualties resulted. The 362d maintained the main line of resistance for a week, 3-10 November, and then, on the night of 10 November, its relief was begun by the British 1st Division. Command of the sector was relinquished by the 362d at 2200, 12 November. At that time the 1st and 3d battalions were at Villanova, and the 2d was entrucking for Pistoia.

At 0800, 17 November, Field Order No. 28 was issued. It outlined the process whereby the 91st Division was to relinquish control of its sector between 20 and 24 November to elements of the 34th Infantry Division and the 1st Armored Division.

The 361st Infantry was relieved by the 133d Infantry, one battalion on the night of 20-21 November and the other two on the night of 21-22 November. Command of the zone passed to the 34th Division at 0600, 22 November. The 363d Infantry was relieved by Combat Command A of the 1st Armored Division, one battalion each successive night between 21 and 24 November. Command of the regimental sector passed to Combat Command A at 0600, 24 November. Upon relief, both regiments entrucked for the Fifth Army Rest Center at Montecatini for refitting and a well earned rest.

Powder River Doughboys Take A Break

For the first time since its entrance into combat on 12 July, the infantry of the Powder River Division was to enjoy a rest period, while the remainder of the Division stayed in the line. Manning the defensive sector of the Winter Line had not been so difficult or so costly as the great drives of September and October, but it had been tough and tedious. Constantly on the alert against enemy traps, raids, and counterattacks, the front-line troops had not relaxed their vigilance for a moment. The weather had been cold, wet, and disagreeable, and the enemy artillery and mortar shelling had been an ever-present threat. For the Powder River doughboys the opportunities for rest and relaxation at Montecatini and Pistoia were most welcome.

Famous Ponte Vecchio, a familiar sight to men on pass to Florence rest center

Upon relief of the infantry units, the Division Artillery was attached to the 34th Division, but plans for rotating its personnel to Montecatini were worked out. A similar plan was arranged for members of the Special Troops. Thus all men of the Division were given the opportunity to visit the famous Italian resort town.

Montecatini was an ideal spot for the Fifth Army Rest Center. Over two hundred hotels and *pensions,* beautiful parks and boulevards, numerous theaters and cafés, and the famed mineral baths, none of which had been damaged, were at the disposal of the men of the Division. Instead of working for the wealthy tourists from the United States, Great Britain, France, Germany, and even Japan who had once frequented the resort, the civilian population now served American soldiers. In addition, the area's close proximity to many large cities permitted day passes to points of great historical and cultural interest. Tuscany, which includes Florence (Firenze), Siena, Pisa, Leghorn (Livorno), Pistoia, and Montecatini, besides being renowned for its rich contributions to the world in the fields of art and literature, is climatically one of the most favored regions in Italy. The weather is mildly temperate, and the rainfall not excessive.

The 361st and 363d Regiments moved to the Montecatini rest area as they came off the line, 20-24 November. The 362d had moved to Pistoia on 12 November and was bivouacked in old Italian barracks. This area was much less desirable than Montecatini, and considerable work had to be done before the men could be comfortable. For all three regiments the program was the same. The first four days were

Monghidoro, with its shattered church steeple, was an important supply point for the Division

spent primarily in rest and relaxation. Showers and clean clothes were available, as well as the full resources of the Red Cross, Special Services, and the United Service Organizations.

During the combat period, replacements had come to the Division in such numbers that company and platoon organization was weakened by the unfamiliarity of the members of the units with each other. Accordingly, special efforts were made by the officers and noncommissioned officers to get to know their men and to perfect the organization. Thanksgiving Day, spent by the majority of the Division at Montecatini, was particularly memorable for its food and good fellowship.

After the first four days, a moderate program of six hours training a day was conducted. The nomenclature, assembly, and disassembly of weapons were reviewed, and selected personnel were trained in the use of rocket launchers, rifle grenades, and flame throwers.

For the infantry units there was training in the tactics of small units, enemy tactics, and field hygiene; for the headquarters and other specialized troops, emphasis was on the correction of errors noted in combat and the perfection of standard procedures. All men engaged in physical training, close order drill, and classes in military discipline.

When the 362d Infantry was relieved from duty with the 88th Infantry Division and again reverted to control of the 91st on 12 November, it moved directly from the vicinity of Sasselone to Pistoia. In honor of Captain Hugh S. French, the first officer of the regiment to meet his death in combat, the camp at Pistoia was officially named Camp French. The program was comparable to that at Montecatini. The first four days were spent primarily in rest and rehabilitation, and the remainder of the period was devoted to specialized training, especially in the tactics of small units. On 22 November the 362d was ordered into Corps reserve and moved to Villanova, where the training was continued.

Throughout the period, the 91st Quartermaster Company and the 791st Ordnance Company made concerted efforts to prepare the men and equipment of the Division for winter in the mountains. Winter clothing, sleeping bags, and mountain tents were issued to the front-line troops in anticipation of the cold Italian winter months. At Montecatini, the Quartermaster Company set up a motor pool and carried out second- and third-echelon maintenance. The 791st Ordnance Company at Monghidoro stepped up its care of the Division's armament and vehicles. Bad roads and adverse weather added to the work of the automotive section, which made major repairs on 302 vehicles, replaced 44 engines, relined 1,042 brake shoes, and made hundreds of other repairs of all sorts. During this period, the armament section, still at Monghidoro, assisted by a crew of the Fifth Army's ordnance company, inspected and repaired all of the Division's 57mm antitank guns and 105mm howitzers, as well as running a spot check on all other artillery pieces in the Division.

On 26 November, the 1st Battalion, 362d Infantry, was moved, on II Corps order, to Fornelli, ready for immediate use. The remainder of the 362d Infantry remained in the vicinity of Villanova in II Corps Reserve.

Return to the Front

On 30 November the Division began its return to the front line. The 362d Infantry was alerted by the Division Commander to relieve the 133d Infantry, 34th Division. The 1st Battalion relieved the

Observers keep close watch for all signs of enemy activity

133d's 3d Battalion on the night of 1-2 December, and the 3d Battalion completed the relief the following night. Command of the sector passed to the 362d at 0600, 3 December, in accordance with Field Order No. 29. To screen the relief and the return of the 91st Division to the front lines, Companies I and K of the 133d Infantry were attached to the 362d. Until the relief had been completed, all patrols were organized from the personnel of these two attached companies. The screening force was relieved at 2030, 9 December, and the 362d maintained a defense in depth throughout the sector.

One of the major objectives of the force was to maintain contact with the enemy and to secure as much information as possible about him, his weapons, and his defensive positions. The 91st Reconnaissance

Troop returned to its old observation post on Hill 585 overlooking Livergnano, and from it made hourly reports on enemy installations and movements to Division G-2.

At night, listening posts were established, and reconnaissance patrols were sent out to obtain information about minefields, wire obstacles, and the approaches to the Savena River. Prisoners were especially sought, and combat patrols engaged the enemy and set up ambushes each night. The enemy, however, was both wily and elusive, and it was not until the last night of December that prisoners were captured.

On a typical night, 17 December, three reconnaissance patrols and one combat patrol were sent out, and five listening posts were set up. One reconnaissance patrol found a dead German soldier 400 yards northwest of Gruppi and identified him as a member of the 11th Company, 11th Regiment, 4th Parachute Division. Another reconnaissance patrol encountered heavy machine-gun and mortar fire 500 yards farther to the south, on the road halfway between Gruppi and Mt. S. Ansano, and returned. The other two patrols were organized to participate in a coordinated probing of the enemy lines. The combat patrol of one officer and thirteen men, equipped with a flame thrower, attacked an outpost in a draw immediately east of Ca' Nuova. In the subsequent fire fight, they received machine-pistol and machine-gun fire and hand grenades from the nearby houses. One hundred and forty rounds of enemy mortar fire also covered the area. According to plan, the reconnaissance patrol moved out under cover of this fire fight to reconnoiter the river crossing just to the west of Ca' Nuova. At a point fifty yards short of the river, the patrol received heavy machine-gun fire from Ca' Piana, two sixteen-round concentrations of mortar fire, and eight rounds of direct self-propelled-gun fire. Unable to proceed farther, the patrol returned.

At 2300, 14 December, a combat patrol consisting of one officer and eighteen men of Company B, 362d Infantry, entered a house set close against the slope of a hill at C. Tianello. When they discovered that the second story was occupied by about ten Germans, a fierce fire fight ensued, with the Company B patrol on the ground floor and the enemy on the second floor. The patrol began firing up through the ceiling of the first floor, and soon blood began dripping from above. The enemy countered by dropping concussion grenades down through a hole, slightly wounding five members of the patrol. A second patrol was sent to reinforce the first and secure the house. The enemy, now greatly outnumbered, crawled through an upstairs window to the slope of the hill and managed to escape under the cover-

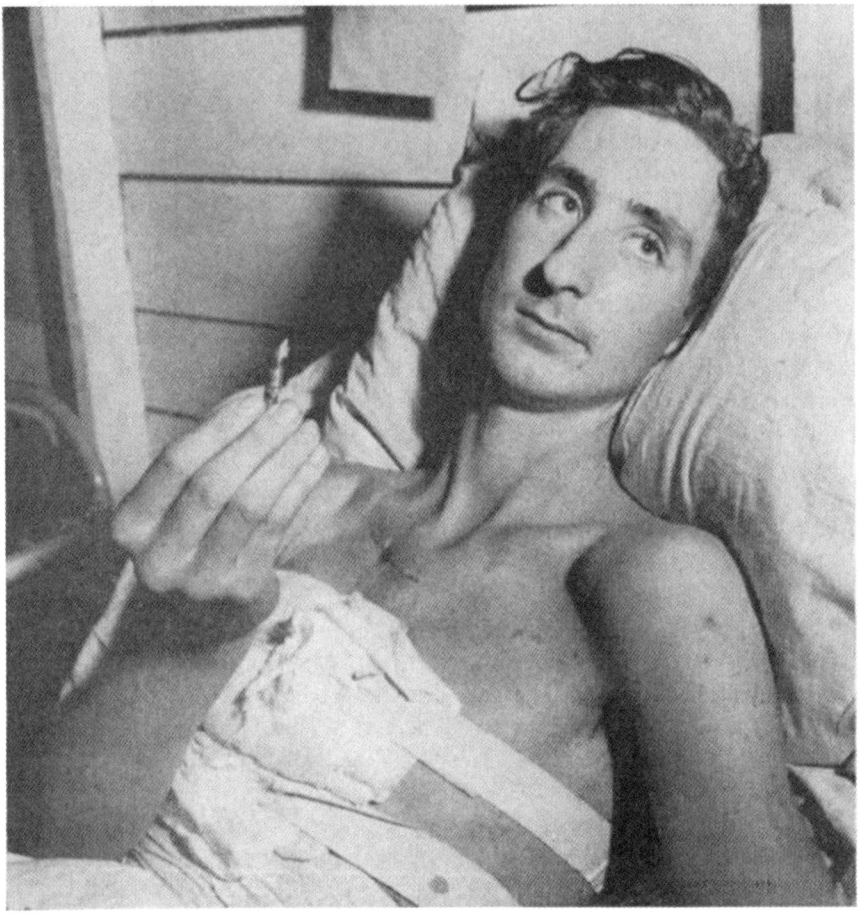

Pfc. Ray F. Shaffer, a machine gunner of Company I, 363d Infantry, holds the machine-gun slug which, after passing through his lungs and heart, was removed from his abdomen

ing fire from other forward enemy positions. Our positions were consolidated and held against repeated attempts by the Germans to dislodge our troops.

While the 362d Infantry occupied the line, the 361st and 363d Regiments were at the large estate of the Marchesa Gerini near Villanova. The 363d left Montecatini on 4 December and closed at Villanova by 1700. The 361st followed the next day and bivouacked in the same area. Both regiments immediately began developing as comfortable quarters as possible. The rain made the ground muddy and wet, and special care was taken to protect the health of the personnel. Walks and roads were gravelled, and close supervision of sanitary installations was maintained.

As soon as living quarters were satisfactorily established, the training, begun at Montecatini, was continued. Ranges for machine guns, mortars, and small arms were set up. One set up by the 363d Infantry was, appropriately enough, on Monticelli, the scene of extremely bitter fighting by the regiment three months before. Both regiments carried on extensive training in the firing of weapons and conditioning by marches and other physical exercise. Classes were held in censorship, orientation, map reading, booby traps, and the detection and neutralization of mines. Appropriate classes were also organized for specialized personnel of the regiments.

While the 361st and 363d Regiments were thus occupied, the Division Artillery was supporting the 362d; the 346th Field Artillery Batbatlion in direct support and the other three battalions in general support. The 316th Engineer Battalion worked on the roadnet of the sector, winterized the tents of the 316th Medical Battalion, and made preparations for the expected deep snow on the mountain roads.

The Pianoro Plan

At the same time the Division was maintaining its defensive sector in the Winter Line. At the command post just south of Loiano, the general and special staffs were planning the broad tactics and multitudinous details of an attack. Each section of the staff was affected when, on 28 November, the Division received from II Corps the Pianoro Plan, an outline of a contemplated attack north across the Savena River, on Mt. Adone, Mt. dei Frati, and the high ground overlooking the Po Valley. According to the plan, one regiment of the 91st was to relieve the left regiment of the 34th Division, assuming command of the zone by 0600 on D-day minus five. Two companies of the 34th Division were to be attached to the Division to screen the entrance of the regiment into the line and to conduct patrol activity. On D-day the 91st was to attack north with two regiments in line. The main effort was to be on the right. The first objective was Mt. Posigliano; then, by an enveloping action from the east, the division was to capture Hill 544 and Mt. dei Frati. From this position, the attack was to continue until the Division, with the assistance of the 1st Armored Division and possibly the 85th Division, captured Mt. Adone, one of the key terrain features in the Corps sector. This accomplished, positions on the First Phase Line were to be consolidated.

Two other phases of attack were outlined in the Pianoro Plan, by which, upon the completion of the Third Phase, the 91st Division would hold Hill 469 and Mt. Mario. In this position II Corps, of

Anconella was a familiar CP location for units occupying the left sector of the Winter Line

which the 91st Division would be the left flank unit, would dominate all the terrain to the north and be in an advantageous position to attack the Po Valley.

For the Division Commander and his staff, this projected plan meant swift, careful planning and preparation. Originally, speedy execution of the plan was contemplated, and II Corps ordered its lower echelons to submit their plans to the Corps commander on 2 December.

The coordination of all the varied preparations for the attack, from Halazone tablets to terrain studies, from the attachment of new units to the Division to anticipation of the problems of supplying replacements, was accomplished by Colonel Joseph P. Donnovin, Division Chief of Staff. By his close attention to the special work of each of the staff sections and his continuous supervision of the normal administrative details of the Division, he knitted the general and special staffs into a smooth-functioning team and freed General Livesay for thorough and patient study of the tactical situation. With the assistance of the G-3 Section, General Livesay planned the actual battle participation of the Division. He decided what troops to use, how to maneuver them in the narrow sector assigned the Division on the line of departure, and how to protect the left flank until Mt. Adone could be enveloped. The results of the planning by Colonel Donnovin and the General Staff and the decisions made by General Livesay were embodied in Field Order No. 30, issued at 0800, 5 December.

On 6 December General Keyes met with General Livesay, his staff, his unit commanders, and selected members of their staffs to discuss all phases of the Pianoro Plan. Immediately thereafter the separate units began reconnaissance and planning for the coming attack. Throughout the remainder of the period, fire plans and the details of the attack—such as disposition of ammunition, use of smoke, and preparation of foot-bridges and road clearance for the units moving north from Villanova—were planned by the regimental commanders and coordinated at the Division command post.

The 91st Division Artillery, commanded by Brigadier General Ralph Hospital, prepared defensive fires for all objectives prior to H-hour as well as a complete plan for supporting fires. Much thought and time was spent planning the role of the artillery in the Pianoro Plan. Although the attack was to jump off without artillery preparation, subsequent firing was very carefully worked out, and the final plan involved 189 separate target areas.

In the midst of this planning, the Powder River Division was honored by a visit from members of the Military Affairs Committee of the House of Representatives on Sunday, 18 December. Arriving in Monghidoro at 1130, they attended church services conducted by Chaplain Dean T. Stevenson, Regimental Chaplain, 361st Infantry. After eating a dinner of K rations with the Combat Platoon of Headquarters Company, they visited an observation post in the vicinity of Loiano, a scant few miles from the front lines. Everywhere they talked with the officers and enlisted men, gathering impressions to report to Congress and the citizens of the United States.

Just as the Division was set to attack on 26 December with the other divisions of II Corps, the enemy attacked in the west down the Serchio Valley. Movement of enemy troops indicated the further possibility of an attack in force to recapture Leghorn, the main supply port in northern Italy. To meet this threat, some units of II Corps were withdrawn and moved to the west, forcing the postponement of the execution of the Pianoro Plan indefinitely. An immediate reorganization of the Division zone was begun to meet any potential enemy offensive.

HOLIDAY SEASON IN THE APENNINES

An interesting development during December was brought about partly by the plans for the attack and partly by the coming cold weather. The Fifth Army, determined not to have a repetition of the serious epidemic of trench foot which had incapacitated so many soldiers in

Christmas mail and presents poured in by the truckload. There was a full mailbag for each man in the hospital.

the Winter Line below Rome, ordered that every precaution be taken against it. The Powder River doughboys had been directed, as a preventive measure, to change their socks every day. But the losses of socks on the front lines had been very great, and there was not an inexhaustible supply in the theater. The problem of salvaging and washing socks, therefore, became of paramount importance if the Division was to be maintained at peak strength. Fifth Army Quartermaster laundry units washed socks for the divisions, but the service was slow, and the haul to the rear areas was long. If the roads should be blocked by snow, the service might even have to be suspended, and the whole preventative program would be seriously hampered. Accordingly, the Division G-4, Lieutenant Colonel Campbell W. Newman, and the 91st Quartermaster, Lieutenant Colonel Beuford A. Pittman, made plans to construct a Division sock laundry. The 791st Ordnance Company improvised wringers and hand-operated washing machines out of oil drums and scrap lumber, and local Italian labor was recruited. The construction of the unit took more time than had been anticipated, but at the end of the month it was in most successful operation.

Men of all faiths attended religious services

On the night of 20 December the 1st and 3d Battalions of the 361st Infantry relieved the 1st and 3d Battalions of the 362d Infantry in place and assumed command of the sector at 0600, 21 December. The 362d moved south to Villanova for rest and rehabilitation. This relief, carrying out the provisions of Field Order No. 31, invalidated Field Order No. 30, and accordingly, at 1800 on 20 December, Field Order No. 32 was issued. It made no change in the Division's strategy as outlined in Field Order No. 30, but made the adjustments necessitated by the relief of the 362d Infantry.

The 91st Division celebrated its first Christmas abroad joyously but simply, for D-day had originally been set under the Pianoro Plan

Staff Sergeant Vincent C. Post, 363d Infantry, plays Santa Claus for the children of Loiano

T/5 Frank Luther presents Pfc. George Griffin with a birthday cake

for 26 December. All through the month, the Division APO had been delivering packages to the men in a swelling flood. On Christmas Eve alone 1,800 sacks were delivered, totalling 28,800 packages. Between 25 November and Christmas Day, the APO handled 212,376 packages for the Division and its attached units, a heart-warming indication that the home front had not forgotten its men.

Religious services were a most important part of the holiday. Fifty-eight church services were held by the Division on the Sunday before Christmas and on Christmas Day. At most services, quartets or choirs sang carols and responses. Midnight Masses on Christmas Eve were packed. The next morning the crowd of 200 men at the Division Artillery Mass was too large for any building; so an altar was set up in the snow outside, and Chaplain Peter Duignan said Mass for the men on the side of a hill.

Sometime during Christmas Day almost every man in the Division enjoyed a superb turkey dinner, and the Service Companies made every effort to bring the food to the very front lines. Those men who could not enjoy the feast on Christmas received it the following day when reliefs were held within battalions. Due to the projected movement, the 363d Infantry had its Christmas on 24 December. At the 91st

Preparing the intermediate and switch lines was a full-time job

Quartermaster Company, a brief ceremony was held in which Lieutenant Colonel Beuford A. Pittman presented awards to three members of the company.

During the rest of the day, parties and celebrations of every sort were held. The members of Headquarters Battery of the 91st Division Artillery produced a variety show called *Buon Natale.* At Loiano, Staff Sergeant Vincent Post of the 363d Infantry played *Papa Natale* to the children of the oft-shelled town and distributed food and candy contributed by the men of the Division. At Villanova a similar party was held on the grounds of the Marchesa's castle. Parties were held in the vicinity of other installations of the Division for the local *paesanos.*

On the front line the day brought business as usual, with one hundred per cent guard during the night and fifty per cent guard during the day. Even here, however, the men celebrated Christmas as best they could. Near Livergnano, three men of the 361st Infantry took their rifles and an ax and brought back a small fir tree, which they set up in their cave and decorated most ingeniously. Another group of men, who were out on a mission Christmas Eve and all Christmas Day, held a simple worship service of their own on their return and exchanged gifts with each other.

With the execution of the Pianoro Plan indefinitely postponed and the withdrawal of some units from II Corps for use elsewhere on the Fifth Army front, the 91st Division's defensive sector was reorganized. On 27 December, acting on orders of the Corps commander, General Livesay issued Field Order No. 33. The 363d Infantry was to relieve the 6th Armored Infantry Battalion and the 81st Reconnaissance Squadron in position on the night of 27-28 December. The relief of the 6th Armored Infantry was completed by 0010, 28 December, by the 3d Battalion, and the 363d assumed command of the sector at 0700. The 363d's 2d Battalion moved during the daylight hours of 27 December to Anconella, where it went into regimental reserve. The 1st Battalion moved to Roncastaldo and then to Monzumo. On the night of 28 December it relieved the 81st Reconnaissance Squadron. Immediately, counterattack plans were prepared for submission to General Livesay by 1800, 30 December, to meet potential enemy thrusts between Monterumici and Vado, Mt. Adone and Monterumici, and directly south along Highway 65.

In accordance with Field Order No. 34, issued at 1200, 29 December, the 2d Battalion, 361st Infantry, relieved the 361st's 3d Battalion (less elements of Companies L and M) and also Companies E and K of the 133d Infantry, 34th Division. Prompt measures were begun for the organization of these positions in depth, and both the 361st and 363d Infantry Regiments began preliminary work on an Intermediate Line and a Switch Position, to be used in the event of hostile penetrations of the front lines. Attached to the Division was an unusual amount of artillery and armored fire power. In addition to the Division Artillery, there were: the 151st Field Artillery Battalion and the 175th Field Artillery Battalion of the 34th Division; the 804th Tank Destroyer Battalion; the 757th Tank Battalion; and elements of the 755th Tank Battalion. The 757th's tanks were held in mobile reserve, while the fire of the other units was coordinated by the Division Artillery to reinforce the front-line positions. The 362d Infantry, which remained in II Corps reserve until 1200, 31 December, moved from Villanova to Loiano on 30 and 31 December.

While these changes in disposition and deployment were taking place, the front-line troops continued to maintain contact with the enemy and to gather information by patrols and observation posts. On 27 December the 91st Reconnaissance Troop reestablished Observation Post No. 1 near Le Croce and, on the following day, Observation Post No. 3 near Ca' di Pagliano, in the sector of the 363d Infantry. Each night reconnaissance and combat patrols were sent out. Snow and bright moonlight complicated the always difficult and dangerous assignments. On the night of 27 December, seven listening posts were established and five patrols were sent out. One of these patrols, moving west of La Fahbrica, engaged an enemy patrol of approximately ten to twelve men in a fire fight. In an attempt to outflank the enemy force, the patrol leader and two of his men got within twenty-five yards of the enemy and exchanged hand grenades with the Germans. The patrol leader was slightly wounded, but he managed to reunite his patrol and return without further casualties.

The next night, 28 December, in the sector of the 363d Infantry, personnel of Company C, 316th Engineer Battalion, took part in an unusual patrol. The 3d Battalion, 363d Infantry, was having great difficulty in supplying its outposts at La Tomba, La Piana, and Varosole, across the Savena River. The men who hand-carried food, water, and ammunition to the outpost positions had to wade the icy river and get soaking wet. The banks of the river were steep, and it took a long time to hand the supplies down one bank and up the other. Besides, the crossing was an enemy mortar target, and time spent there was precious. To alleviate the situation, six engineers, guided and protected by 3d Battalion infantrymen, made their way to the crossing

Pfc. William O. Allison gassing up one of the Cub planes of our air OP

in makeshift white camouflage suits and despite the danger resulting from the inevitable noise, successfully leveled down the banks.

Flank patrols maintained contact with the 88th Division, on the right, and the 6th South African Division, on the left, each night. Once again the Savena presented an obstacle, but the men of the 363d solved the problem neatly. They stretched a telephone wire across the river. Patrols then carried sound-powered telephones to the spot, attached them to the wire, and made contact with patrols of the 6th South African Division on the other side of the river.

The night of 30 December brought a variation in the routine of patrols and listening posts. A hundred per cent alert occurred in the sector of the 363d Infantry when the Germans brought up a broadcasting unit and welcomed the regiment back to the line. After the salutations, three brief musical numbers were played, and a few mediocre jokes were told. The show then ended with a promise of a repeat performance the following night. On New Year's Eve, promptly at 2100, the broadcast began again, but it was cut short abruptly when an ambush patrol of the 363d threw grenades. Although a mortar preparation was fired to box the enemy in, and an attempt was made by another patrol to outflank the position, the enemy managed to escape. Elsewhere on the regimental front that night, an ambush patrol from Company C completely trapped a German patrol, killing or capturing all its members. The three prisoners brought in provided a New Year's-morning present for General Livesay.

The men had to dig through a foot of snow to get san

ɔr revetting the intermediate and switch positions.

During December the role played by the Division Artillery was passive. Firing was at a minimum, and there were two days when no missions were fired at all. Further, visibility was so poor that of 815 missions fired during the month only 281 were fired on observed targets. In an attempt to overcome the difficulties involved in the variation of the weather between the airstrip south of Futa Pass and the forward positions, a landing strip was established nearer the front lines, but it was found impracticable to use because of the high wind and fog. Of the 281 observed missions which the artillery did fire, 23 were counterbattery, 49 were directed at enemy-occupied buildings, and 13 were on vehicles on Highway 65. Six enemy gun-emplacements were permanently silenced, 27 buildings were completely demolished, and 7 vehicles were destroyed.

In an effort to make the unobserved fire more accurate and effective, Artillery Headquarters developed an extensive countermortar program. From direct observation and reports gathered from the infantry units and the Division observation posts, a list of all known and suspected locations were compiled and sent to each battalion. Near the end of the month, the program was extended further when flash observation posts were set up so that enemy mortar and gun positions could be located by intersection.

At midnight, 31 December, the stillness of the night was broken when the enemy, signalizing the passing of the old year, fired tracers into the sky across the entire Division front. From the little enemy-held town of Di Sopra on Monterumici, church bells rang out across the snow, and the men of the Powder River Division, alert in their foxholes, greeted the New Year—which each one hoped and prayed would bring victory.

The New Year—1945

Following the brief display of enthusiasm with which the enemy greeted the year 1945, the tedious business of defensive warfare settled over the lines once again. On the left, the 363d Infantry, with the 1st Battalion on its left and the 3d Battalion on its right, occupied a 4,000-yard front running northeast from Ca' di Marchino to Le Fosse. On the right, the 361st Infantry, with the 1st Battalion on its left and the 2d Battalion on its right, held the line across Highway 65 east to the vicinity of Mt. Belmonte. The positions in the sector of the 361st had been occupied by elements of the 91st Division since October 1944 and had been improved by their successive occupants. The positions in the sector of the 363d, however, had been taken over by

A typical winter scene along Highway 65

the 91st Division for the first time late in December. The regiment, therefore, expended much time and energy strengthening the defensive positions with wire, mines, and trip flares and improving the living conditions of the troops. At night emplacements were improved by further excavation and by adding overhead covers, sandbags, and duckboards. The men built dugouts and constructed heating and lighting arrangements of great ingenuity so that they could dry their clothes, read, write letters, and play cards.

During the day the chief occupation was observing the enemy's activities. Many days this was impossible, because of snow, fog, or valley mist. Smoke, especially in the region of Mt. Adone, occasionally hid enemy activities. Yet, whenever it was possible, the Reconnaissance Troop, which manned three Division observation posts, the artillery forward observers, and the infantry observers all combined to report every movement of enemy personnel and any change in enemy emplacements. Much emphasis was given to shell reports of enemy artillery and mortar firing. On the basis of all these reports, coordinated and collected by Division G-2 and the Division Artillery S-2, most of the artillery firing was done.

Reconnaissance and ambush patrols continued to probe the enemy lines. Close contact was maintained, but the number of prisoners captured was considerably smaller than was desirable for intelligence purposes. As a result, the strength and fire power of raider patrols

These mortars helped put a ring of steel around the outposts in addition to firing harassing missions

were increased, and they were supported by carefully planned mortar and artillery preparations.

On the night of 9-10 January one patrol of eleven men of the 363d Infantry and eight men of the 91st Reconnaissance Troop went to Di Sotto, 600 yards northwest of the outpost position at La Tomba. The group was heavily armed, carrying ten submachine guns, two Browning Automatic Rifles, six M1 rifles, and one carbine. In addition each man carried three hand grenades. To conceal their movement through the snow, the men camouflaged their guns with engineer tracer tape or white paint and wore white parkas. A rather elaborate fire plan was set up in support of the patrol. Four check points were located along the route. Mortar concentrations were laid on these, and the fire was controlled by members of the raiding party, who laid wire as they went. Thus, preceded by a rolling barrage of mortar fire, the patrol made its way to Di Sotto. As they neared their objective part of the group formed a connecting file to the rear to protect the rest of the patrol in case of an enemy attack.

The remainder of the patrol split into three groups and closed in on the town. Only one man, a member of the 91st Reconnaissance Troop, made physical contact with the enemy. When his submachine gun, the oil frozen by the intense cold, failed to function, he ploughed into the enemy in a rough-and-tumble fight. He knocked one man unconscious, but the other, after severely biting his assailant's hand, managed to escape. The noise of the fight brought enemy fire down on the patrol, and they were forced to withdraw under the protective covering of prearranged artillery and mortar concentrations.

One of the enemy devices, which often made our patrol activity fruitless, was his intermittent occupation of positions. Buildings or emplacements where the enemy was known to have been located would be raided, only to be found vacant. Our experience in the town of Barchetta is a case in point. On 4 January a patrol found the town unoccupied; yet four nights later, when another patrol attempted to enter the town, it was met by small-arms and mortar fire each time it tried. The next night Barchetta was again found unoccupied. On 10 January a carefully planned raider patrol, consisting of an officer and 25 enlisted men, made its way to the town and again found no trace of the enemy. Two nights later a patrol observed a group of German soldiers busily engaged in hauling material into Barchetta with carts. On 15 January a patrol arrived just in time to observe an enemy group enter the town from the direction of Zula, stay a short while, and then return in the direction it had come from. On 16, 20, and 23 January patrols checked the town and found it unoccupied, although on the last two occasions machine-fire was received from high ground to the north.

Known enemy positions were raided constantly by patrols of varying size to maintain contact and take prisoners. No evidence was ever discovered to substantiate the rumor, circulated both in Italy and the United States at this time, that the German forces were abandoning their front in Italy.

The enemy was not entirely inactive during this period. Although it appeared that he did not patrol so extensively as the 91st Division did, several attacks in force were made, especially in the sector along Highway 65. On the night of 1-2 January a force of 25 to 30 Germans attempted to raid the positions held by Company E, 361st Infantry. The enemy laid down a terrific concentration of 120mm mortar fire on the company for thirty minutes. Company E countered by directing the entire fire power of Company A, 100th Chemical Mortar Battalion, on Osta Nuova, and the Division Artillery fired a prepared countermortar concentration. Within a short time, the raiding party withdrew—with heavy casualties, it was believed. Thereafter, enemy shelling continued to cover the area, hampering efforts of the medical aid men to render assistance to those wounded in the initial attack.

Another mortar position in close support of front-line positions

On the night of 3-4 January the 2d Battalion, 361st Infantry, beat off another attempt to breach its lines. This time, preceded by heavy mortar concentrations, small enemy assault groups struck all along the 2d Battalion front. Companies E, F, and G, with the immediate support of friendly mortar and artillery fire, repulsed the attacks, but two men of Company G, in an outpost position, were surprised and captured. Later on, the number and strength of these raids fell off, and during the remainder of the period the enemy contented himself with moderate shelling of the Division's positions, occasional patrols and raiding parties, and maintaining his defensive positions. In these ways, he attempted to harass our troops and deny us knowledge of his movements and intentions.

Most of the activity of the Division which might be termed offensive was carried on by the artillery. In view of the greatly extended line the 91st was holding, its artillery was augmented by the 175th and the 151st Field Artillery Battalions, 34th Division; the 804th Tank Destroyer Battalion; the 755th Tank Battalion; and the 100th Chemical Mortar Battalion. Mainly because of the weather, most of the firing was unobserved and harassing in character. However, every break in the weather was exploited to the fullest extent. Enemy personnel, gun emplacements, and buildings were the prime targets, and the results of the firing were highly satisfactory. Two tanks and three medium

This unusual shot of an incoming German 170mm artillery shell was taken the split second of detonation. Three men were killed and several more seriously wounded in the almost direct hit on the truck.

guns were destroyed. Numerous German mortars were blown out of action, and the artillery's countermortar firing made it extremely dangerous for enemy mortars to remain in one position for even a short while. Houses used as German command or observation posts were special targets; no day when observed firing was possible went by without direct hits being scored on one or two buildings.

Sniping by artillery was also developed to a greater degree than at any time previously. The 348th Field Artillery Battalion specialized in this type of firing. Two bicycles and three motorcycles were among the targets hit. Later two self-propelled guns and a jeep were caught on the road. Firing on minor targets was in no sense a waste of artillery ammunition, since it convinced the enemy that not even the least

Preparing defensive positions on high ground overlooking Loiano

important traffic could move unmolested past points of observation.

It was during this period that the closely guarded variable time fuze (VTF) artillery shells were first used widely on the Italian front. At first the device was used in firing only the most important concentrations. The earliest observed concentration using this type of fuze, fired on 4 January, was very effective, with a uniform height of burst of four yards. Thereafter employment of the fuze was gradually extended, and it was used on enemy personnel with demoralizing effect. The 175th Field Artillery Battalion, for instance, fired on a German bridge-building crew, and the forward observer reported that the enemy was still picking up his dead the following morning.

While the 361st and 363d Regiments, reinforced with artillery fires,

Ordnance Company mechanics keep 'em rolling

held the front lines, the 362d, in reserve, and the reserve battalions of the regiments on line, devoted themselves to constructing alternate positions. Organizing a defense in depth, they dug, revetted, and sandbagged positions on the Switch Line, immediately behind the front lines, and on the Intermediate Line, running roughly east and west of Loiano. They laid minefields and strung connecting wire, so that prepared positions would be completely ready if the enemy mustered enough strength to break through the front-line positions. On 19 January, when the 362d moved to the sector of the 34th Division, it was reported that ninety-five per cent of the defensive work was completed.

At 0600, 17 January, General Livesay issued Field Order No. 35, which ordered the 91st Division to relinquish command of its sector to the 88th Division and revert to II Corps reserve. In accordance with the orders of II Corps, the 362d Infantry moved to the sector of the 34th Division as reserve for that division. The move was accomplished without incident, and the 362d closed into its new area at 0930, 19 January. Two battalions of the 351st Infantry relieved the reserve battalion and one battalion of the 361st on the line on the night of 21-22 January. The next night the 361st's remaining battalion was relieved, and command of the sector passed to the 351st Infantry at 2328, 23 January. The 361st then moved to Montecatini for a six-day

This "Rube Goldberg" looking gadget is actually a charcoal-making machine. Charcoal is a vital fuel for the Italians.

period of rest and rehabilitation. On the following night, 23-24 January, the 350th Infantry relieved the 363d, which moved to Madna dei Fornelli to work on Defense Line No. 2. Command of the Division sector passed to the 88th Division at 0700, 24 January.

In Reserve

For the next three weeks, 24 January to February 14, the bulk of the Powder River Division, in II Corps reserve, was scattered about the Italian countryside. The 362d Infantry, initially in 34th Division reserve, was at San Benedetto; the 361st, at Montecatini; and the 363d, at Madna dei Fornelli. The Division command post was at Villanova, with a small advance detachment in Pietramala. The Ordnance Company also remained in Pietramala. The artillery battalions remained in their firing positions in support of the 88th Division, with the exception of Headquarters and Headquarters Battery, which moved to Montecatini. Careful and detailed counterattack plans were formulated to go into effect in the event of a breakthrough in the II Corps sector. The elements at Montecatini were on a thirty-six hour alert, while those at Villanova and north of Villanova were on a twenty-four hour alert.

After approximately six days, in accordance with a plan for the rotation of units within the Division, the 361st Infantry moved to the

Men of Ordnance Company were required to perform a wide variety of tasks

Gagliano training area on 29 January, and the 363d moved to Montecatini on 30 January for rest and rehabilitation.

Behind the front lines, the supporting units of the Division carried on their normal duties. January had proven especially difficult for the 316th Engineer Battalion. The snow was heavy, but the alertness and labor of the Engineers, who cleared and sanded the roads constantly, kept it from blocking the roads. In addition, they prepared demolitions and cratering charges at important points in the Division roadnet. Finally, they checked the mines and protective installations placed during the month of December and assisted with much of the work on the Switch Line and the Intermediate Line. In all, the battalion cared for and maintained an equivalent of 270 miles of roads, laid approximately 6,000 antitank mines and 700 antipersonnel mines, and erected 5,100 yards of barbed wire.

Although the Division was in a static defensive position for more than three weeks and spent the remainder of January in Corps reserve, the flow of supplies and the number of repairs required to maintain it in the field is impressive. A few figures will serve to indicate the tremendous amount of work carried on by the service units during the period. The 91st Quartermaster Company, for example, issued 427,990 rations of all types during the month; 325,625 gallons of gasoline; 145 tons of coal; thousands of gallons of Diesel fuel and oil; and hundreds of units of winter equipment. Throughout the month, the laundry, operating under the supervision of the Quartermaster Company, averaged 3,000 pairs of socks a day. In the 91st Signal Company, the Construction Platoon maintained 450 miles of wire; the Message

Center section handled 12,450 pieces of mail and 28,941 code groups; the Radio and Telephone section handled approximately 22,000 telephone calls and 26,000 code groups; Signal supply issued 21,300 batteries and 350 miles of wire; and the repair sections repaired 214 radio sets, 27 switchboards, and many other items of Signal equipment. The 791st Ordnance Company issued 125 tons of Class II supplies, including 914 tires, 84 axles, 66 engine assemblies, 164 springs, and 68 steering wheels. The Division Ammunition Officer, Captain Ritchie, issued 1,514 tons of ammunition. The repairs section of Ordnance made 298 major assembly repairs during the month.

In all units of the Division the period was highlighted by a large increase in pass quotas. In addition to enjoying the familiar delights of Montecatini, many men visited Florence and Rome. The chance to clean up, to enjoy the sociability of Italian metropolitan life and to visit places of great historical and religious interest was a welcome change from the rigors of the battlefield. For the Powder River men of Catholic faith the visit to St. Peter's Cathedral and an audience with the Pope was the crowning point of their lives. To those of all faiths, the Immortal Edifice and nearby Vatican City were awe-inspiring. The purchase of trinkets, jewelry, and souvenirs to send to friends and families at home was enjoyed by all. The Yanks found the Italians pretty shrewd traders, but money was cheap, and the novelty of buying outside an Army PX made price seem unimportant. The American Red Cross aided by offering free package-wrapping facilities for all who desired them.

On 4 and 5 February the 361st Infantry moved to Piancaldoli, and on 7 February the 362d, minus the 1st Battalion, moved to Montecatini, as the 363d moved to the Gagliano training area. While at Piancaldoli the 361st worked on Defense Line No. 2 and conducted two regimental problems.

HOLDING THE IDICE SECTOR

At 1200, 6 February, Field Order No. 36 was issued, putting into effect II Corps' order that the 91st Division relieve the 34th in the Idice sector. The 361st Infantry was to relieve the 135th between 12 and 14 February, assuming command of the sector at 0700, 14 February. In the new Division sector, the regiments were to be supported by the artillery of the 88th Division, directed by the 91st Division Artillery Headquarters Battery.

The relief was complicated by flash floods. On 9 February heavy rains fell, and, with these and the melting snow, the streams in the

Highlight of a pass to Rome for these men was an audience with the Pope

sector began to rise rapidly, and bridges were washed out. Late in the day the rain changed to snow, and the streams gradually began to subside. Not before a great deal of damage had been done, however. Company A, 316th Engineer Battalion, in direct support of the 361st, immediately began work on bridges over the streams and on the roads of the regimental sector. Although there was some question as to whether the relief could be carried out as scheduled, it was decided that it could be. On the night of 10-11 February, the Division commenced its movement into the 34th's sector.

The approach to the 361st Infantry sector was through the Zena River gorge, and it was necessary to cross this swirling stream eleven times in a distance of three miles from the front. The relief of the 135th by the 361st was accomplished despite enemy activity, weather, and terrain through sheer determination of the troops.

On 9 February, when General Livesay assumed temporary command of II Corps in the absence of General Keyes, General Hospital took

Four shivering members of the 361st view the swollen Zena River

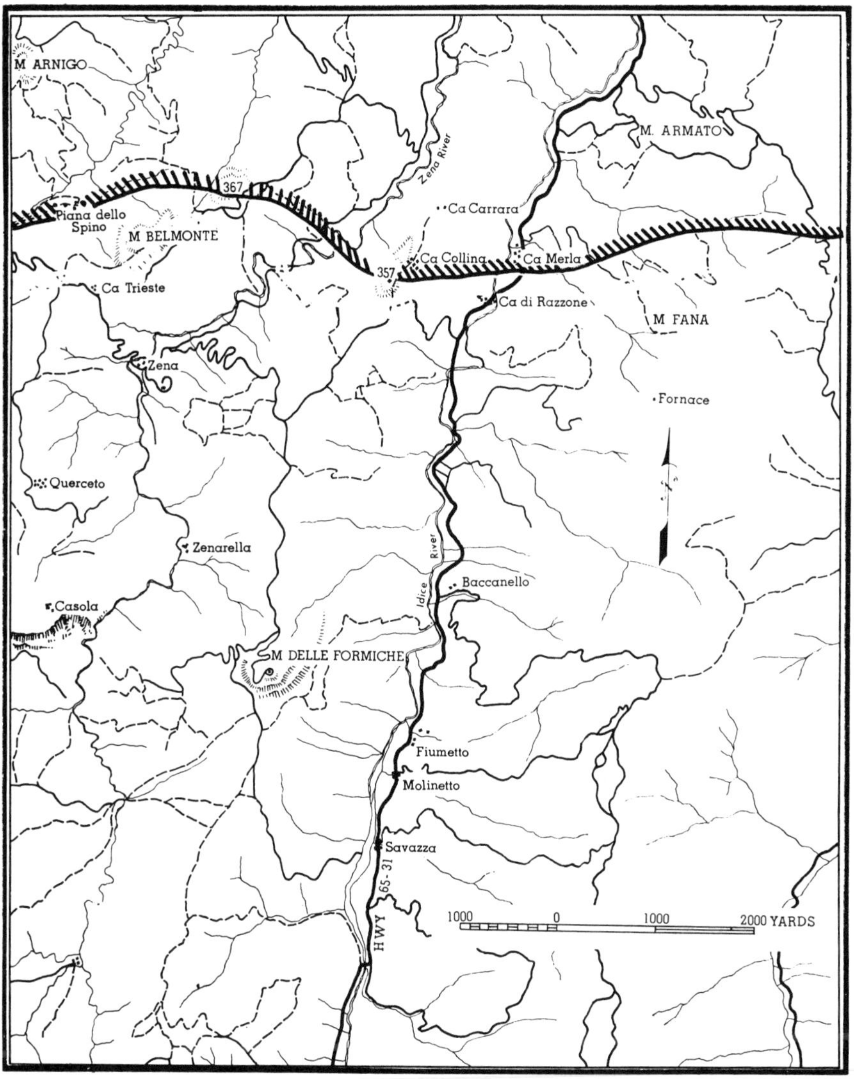

Map 21: The Idice sector

over command of the 91st Division. General Livesay resumed command of the Division on his return, 17 February.

Upon completion of the relief of the 34th Division in the Idice Sector, the 91st had three regiments in line: the 361st Infantry on the left; the 363d in the center; and the 362d on the right. Opposing the three infantry regiments, it was reported, was the German 305th Infantry Division, whose sector roughly matched ours. Opposite the

Easy Street wasn't easy, and could hardly be classed as a street, but the unceasing efforts of the engineers made possible the job of supplying the Division in Idice Valley positions

361st Infantry were elements of the 146th Regiment of the German 65th Division. The sector was relatively quiet, with light artillery fire and a negligible amount of offensive patrolling. The enemy had strong positions, generally along the 33d Northing, in terrain which favored the defense, and he enjoyed excellent observation. It was believed that only ammunition restrictions prevented him from exploiting his advantage. His defensive patrolling close to his own lines was extremely good, and he maintained a high standard of alertness, especially at night.

After assuming command of the sector, the Division plunged immediately into its most pressing problem: roads and transportation. Traffic was ordered cut to a minimum, and a Traffic Control Point was set up and manned by the 91st Division MP Platoon twenty-four hours a day on Highway 65-31, the main supply route in the sector. In the 361st Infantry's sector, where the problem was most acute, a system of road guards and telephones was set up to direct traffic and to see that as little two-way traffic as possible occurred. Collateral benefits of this traffic control system were an increase in the time available for proper maintenance of vehicles, a reduction in the number of accidents, and a forty per cent reduction in the expenditure of gasoline.

The chief benefit, however, was that, with the traffic held to a minimum, the roads were not further rutted and torn up before they could be serviced, and the Engineers were able to work uninterruptedly to get them back into shape. All three line companies of the 316th

Lt.Col. William C. Holley, Division Engineer, flanked by two of his combat engineers

Engineer Battalion set to work immediately, but so many different jobs demanded their attention simultaneously that elements of the reserve battalions of rifle regiments were detailed to assist in the work. Details of riflemen from all regiments worked with 316th Engineers throughout the remainder of February. Roads were drained and surfaced with rock and gravel, bridges were repaired, and in some places treadway or corduroy roads were built. During the last two weeks of February four bridges were constructed and seven were repaired, while 1,389 loads of rock and gravel were dumped on the worst stretches of road. Aided greatly by a period of warm weather and sunshine, by the end of February, Lieutenant Colonel William C. Holley, Commanding Officer, 316th Engineer Battalion, was able to report that the road network in the Division sector had been greatly improved.

The early spring thaw also necessitated work on the front-line positions. A few mortar and machine-gun emplacements were moved, and all positions were surveyed and improved. Counterattack plans were studied carefully, and the alternate positions for the sector were checked.

As always in a defensive position, however, the prime concern of the Division was patrolling in order to maintain close contact with the enemy and capture prisoners and maintaining alert defenses against enemy action. Most of the enemy attacks occurred in the sector of the 362d Infantry, where, on the nights of 14, 17, 18, and 28 February, patrols were driven off by small-arms, mortar and machine-gun fire. On 15 February an enemy penetration was attempted in the sector of

A typical view of the front-line positions occupied by the infantry

the 363d Infantry, but was driven off in a fire fight, with one enemy killed. None of these engagements was in any appreciable force, and all were repulsed by the alert action of the infantrymen in the front lines and the supporting machine gunners and mortarmen.

The patrol activities of the 91st, on the other hand, became more numerous and aggressive. Listening posts and ambush patrols were standing procedure, but the enemy was elusive, and frequently no contact was reported. Larger patrols of up to thirty men were organized to make raids on known enemy emplacements, to capture prisoners, and destroy positions. Accounts of these raider patrols make exciting reading. In addition to the expected opposition of an alert enemy and mines, the patrols were hampered by mud, which made travel both difficult and noisy, and by bright moonlight, in which every movement was apparent to the enemy unless great caution and skill were exercised.

On 17 February a patrol of one officer and five enlisted men from Company E, 363d Infantry, made a raid on suspected enemy machine-gun positions on Hill 357 near Poggio. The group approached without detection and located two bunkers, one occupied by four men and the other by two. The patrol split into two groups and crept to within a few feet of the bunkers before they were discovered. They exchanged hand grenades with the Germans, and a brief fire fight ensued. After ten minutes of sustained action, the enemy sent up a flare. As a result, the patrol received rifle and machine-gun fire from foxholes supporting the bunkers and had to dive over the escarpment to its rear to avoid being captured or wiped out. Although no prisoners were taken, three known casualties were inflicted on the enemy, while the patrol suffered none.

Two nights later, 19 February, a raiding party of an officer, 25 enlisted men, and four-man litter team was organized by the 363d Infantry to take prisoners at a house in the vicinity of Ca' Merla. Careful preparations had been made for the raid. On the three preceding nights four patrols were sent out to reconnoiter the route, to observe the house, and to search for an alternate route of withdrawal. From an enemy deserter, the Regimental S-2 and the patrol leader learned the floor plan of the house, the probable disposition of enemy positions in the area, and the dispositions of the occupants of the house and their armament. On the day of the raid, the patrol rehearsed the plan and the ruse by which the guard at the only entrance to the house was to be captured.

At 0245, 20 February, the patrol, heavily armed—it had 9 Browning automatic rifles, 5 Thompson submachine guns, 8 grenade launchers, 3 hand grenades per man, and knives or bayonets—moved out from Ca' di Razzone. Equipped with three radios and three telephones, it was in constant contact with its command post. Everything went according to plan, and the members of the patrol took their positions surrounding the house. Approaching from each side, four men inched their way along the side of the house to the door, where the guard stood. The patrol leader tossed a rock to attract the guard's attention in the opposite direction, but the ruse did not work. Investigating cautiously, the German sentry spotted two of the patrol members creeping up on him and fired. The two men took cover in a nearby shed while the remainder of the patrol remained hidden in their positions around the house. The sentry's shot, however, alerted the rest of the enemy in the house, and they began throwing grenades from the second story. When the smoke and confusion caused by the exploding grenades died down, the patrol leader searched for the two men who had been fired upon, but without success, for they did not dare give away their positions in the shed. The patrol then withdrew, and later the two men returned safely.

Another large operation occurred on 25 February, when a full-strength platoon from Company E, 363d Infantry, assaulted a strong group of German positions on the ridge 600 yards northeast of Ca' Merla, just south of the 33d Northing. Without mortar or artillery

A patrol armed with carbines and a submachine gun an

mpanied by war dogs starts out on its mission.

Division engineers earned commendation from the II Corps commander for work on roads in the Division sector

support, the platoon made its way to a point not more than fifteen yards from the enemy positions. There they waited for more than a half hour, until 0625, 25 February, the time for the attack. Just as they were about to move forward, they were discovered, and the fight was on. The infantrymen charged over the ridge to the enemy bunkers, where the Germans, having just removed their machine guns from their night firing positions, were preparing to get some sleep. Two bunkers were cleared and ten prisoners captured before the startled enemy began covering the area with small-arms, mortar, and machine-gun fire. All but nine of our men managed to withdraw to the company lines.

These nine men split into two groups and occupied the enemy bunkers. Low on ammunition, but fortunately equipped with a radio, they fired German guns and ammunition they had captured and directed artillery and mortar fire on enemy positions. During the day plans were formulated to rescue them. Just at dusk a curtain of artillery and mortar fire was laid down between the bunkers and the enemy positions, while the nine men escaped back to their own lines. The results of the raid had been highly satisfactory: three machine guns and three bunkers cleared and ten prisoners, a German radio, and the German radio code captured. The platoon sustained five casualties: one man killed and four wounded.

General of the Army George C. Marshall talks to men of the 363d Infantry on his inspection of the Italian battle front

In addition to raiding the enemy's positions for prisoners, the Division arranged broadcasts in German urging enemy soldiers to desert. The number of deserters which had straggled into the Division lines indicated that the enemy morale was low and that such broadcasts should prove effective. Accordingly, on the night of 20-21 February, a representative of the Army Psychological Warfare Branch broadcast the current news and an appeal to desert. Two German soldiers were convinced and made good their escape to the 91st's lines. Both reported that reception of the broadcast had been excellent and that its effect had been considerable. Only fear of reprisals against their families prevented more men from giving up the fight, the deserters asserted. Another broadcast was undertaken the night of 21-22 February, but enemy mortar fire cut the wires and forced abandonment of the project.

In accordance with Field Order No. 36, the Headquarters and Headquarters Battery of the 91st Division Artillery assumed control of the 88th Division Artillery as of 1800, 12 February. The 913th Field Artillery Battalion was in direct support of the 361st Infantry; the 337th Field Artillery Battalion, of the 363d Infantry; and the 338th Field Artillery Battalion, of the 362d Infantry. The 339th Field Artillery Battalion was in general support. Since enemy artillery fire was light during the period and the enemy offensive action was restricted,

General Livesay points out strategic points on the map to General of the Army George C. Marshall, Gen. Joseph T. McNarney, Gen. Mark W. Clark, and LtGen. Lucian K. Truscott.

Visiting authors and artists don combat clothing for a visit through the Division area

the artillery supporting the 91st fired mainly harassing and countermortar programs. Of a total of 910 missions fired, 603 were harassing missions. Only 157 missions were fired on targets of opportunity. The bulk of this firing was directed at enemy-occupied houses, machine-gun emplacements, self-propelled guns, and personnel.

In view of the enemy's extensive use of mortar fire, the chief activity of the artillery was its countermortar program. An organization for this had been perfected during the Division's occupation of the defensive sector along Highway 65 during December 1944 and January 1945. On the Division's assumption of command of the Idice sector, this countermortar fire plan, taken over from the 34th Division, was immediately expanded and verified by photo-interpretation reports. A conference of the Division countermortar officer and regimental countermortar officers was held, and arrangements were completed to make the fire plan more flexible and to bring the selection of targets closer to the sources of information.

Two events of especial interest highlighted February for the Powder River Division. On 14 February General of the Army George C. Marshall, Chief of Staff, visited the Division, accompanied by General Clark; Lieutenant General Joseph T. McNarney, Deputy Supreme Commander, Mediterranean Theater of Operations; Lieutenant General Lucian K. Truscott, Commander, Fifth Army; and General Livesay, then Acting Commander, II Corps. After a brief stay at the Division command post, where he met Major General Charles L. Bolte, Commander, 34th Division, and General Hospital, then Acting Commander,

91st Division, General Marshall visited the 3d Battalion, 363d Infantry. He talked with the soldiers, singling out for special attention those who were wearing Purple Heart ribbons. He made a brief address to the men he inspected, thanking them for appearing for the inspection and for the splendid work they and all the men of the 91st Division had done in the Italian Campaign.

After his tour of the combat area in Italy, General Marshall said in an official statement:

> The difficulties of the mountainous country, with few roads and winter conditions, are very real. The strength of the enemy's defensive positions in such a country is equally apparent. Under these conditions, our United States troops and those of our Allies have done a splendid job and made a great contribution to the war. A large German force has been held in Italy and prevented from bolstering the enemy's hard-pressed troops on the eastern and western fronts.

A week later, on 21 February, thirteen authors and artists visited the Division sector in search of authentic atmosphere and detail. The group, representative of the whole American publishing field, was a distinguished one. Included were Carl Carmer, author of *Stars Fell on Alabama;* Marquis Childs, well known newspaperman and student of international politics; Rex Stout, creator of *Nero Wolfe* and Chairman of the Writers War Board; Stanley Young and his wife, Nancy Wilson Ross; Carol Hill; Joseph Marshall; Mary Cookman and Beatrice Blackmer Gould of the *Ladies Home Journal*; Roger Burlingame; Dorothy Cameron Disney; Toni Frissell, one of America's outstanding photographers; and Harold von Schmidt, illustrator and former president of the Society of Illustrators. The group surveyed front-line installations, visited the Division prisoner-of-war cage, and ate K rations at the command post of the 3d Battalion, 363d Infantry. In interviews published later in *Stars and Stripes* they marvelled at the achievements of American forces in the Italian theater and paid tribute to the courage and endurance of American fighting men.

On the night of 27 February the 2d Battalion, 363d Infantry, relieved the 3d, the relief being completed without incident by 2130. The 1st Battalion remained in position.

During the week 4-10 March, 1,694 enlisted men and 151 officers went on patrols. Although on many nights patrols reported no contact, during the first week and a half of March there were few nights when fire fights did not break out along the Division front. On the night of 2 March a four-man listening post on Hill 358, three hundred yards east of C. Collina, observed no activity; so three of the men crept forward a hundred yards to a slit trench which led to an enemy bunker.

Making their way carefully to the bunker, they surprised a sentry standing by the doorway and immediately opened fire on him. At the sound of the shots, another German came around from the side of the bunker to investigate, and the three men fired on him. After throwing grenades into the bunker, they withdrew without casualties.

On the following night a twelve-man patrol from Company F, 362d Infantry, made its way to the vicinity of Ca' Columbara to set up an ambush. As the patrol was being placed in position, one of the two men already in place stretched out his arms to rest and, to his surprise, touched the barrel of an enemy machine gun. Speaking in German, he demanded that the gunner surrender, and when the enemy refused, he rolled a hand grenade into the emplacement. A fire fight ensued in which the opposition was discovered to be a patrol of approximately the same size as our own. The Company F patrol suffered one killed and three wounded and killed at least one enemy, while wounding an unknown number. Later that night attempts were made to recover the body of the patrol's wireman, who had been killed in the action, but it was found to be booby-trapped, and the patrol leader was ordered to return, because the risk of losing more lives was too great.

In addition to patrolling the Division conducted company-strength raids on enemy positions. These raids were in reality small-scale, limited-objective attacks of great precision and effectiveness. They were planned in detail and even rehearsed, so that no contingency that could be foreseen was left unconsidered. The raid of Company E, 363d Infantry, on 25 February had demonstrated the success possible in this type of action, and the Division conducted more such raids during March.

On the night of 3 March Company K, 361st Infantry, executed a very successful raid. The objectives were the church ruins and cemetery on Hill 367, one of the most dangerous threats to the regiment and one of the most difficult positions to attack because of the excellent enemy observation.

During the forty-eight hours before the raid extensive planning was carried out, and both map and terrain studies were scrupulously undertaken. A reconnaissance patrol went to the area the night before the raid, and on the previous day Company K rehearsed the details by conducting a full-scale problem on a hill in friendly territory.

During the afternoon of 3 March the artillery pounded all known and suspected enemy positions which might support the troops at the church ruins on Hill 367. Neither the ruins nor the cemetery, a second enemy threat in the near vicinity, were fired on, for it was not desired to drive the enemy from the locality.

Leaving the reserve area at 1830, the company went to Vigna, the Company I command post, where the Commanding Officer of Company K also established his command post. From it he was to control and coordinate the raid by sound-powered phone and radio. The church ruins on Hill 367 were slightly less than a thousand meters northeast of Mt. Belmonte. The two platoons which were to be actively engaged in the raid diverged. The assault platoon, composed of the 1st Platoon reinforced with one squad of the 3d Platoon, crept cautiously up the draw between Hill 367 and Mt. Belmonte, south of Gorgognano, while the 2d Platoon, the cover platoon, crept to positions 600 meters to the east, a short distance beyond Company I's outpost positions near Casetta, just south of the church ruins. The night was dark, and it was essential to finish the mission before the moon rose at 2330, for from their positions on the hill the Germans could look directly down on the platoon's route of advance.

The assault platoon reached the bottom of the hill on the northwest side and notified the company commander that it was ready to jump off. At 2157 the cover platoon was in position prepared to support the attack of the assault platoon if necessary, and the assault platoon was orderd to move out.

Forming a skirmish line, the platoon inched its way forward slowly and cautiously, making certain that the enemy was not warned. Edging forward, it covered ninety of the remaining hundred yards to the objective without disclosing its presence to the enemy. Then, just ten yards from the top of the hill, its advance was checked by a barrier of concertina wire, heavily booby-trapped. The platoon leader quickly and silently reconnoitered the obstacle and soon discovered gaps in the wire. He ordered his platoon to slip through the openings and re-form the skirmish line on the other side. At 2312 the platoon reached a point less than ten feet from the top of the hill, almost directly in front of some machine-gun positions. It had taken nearly an hour and a half to cover the last hundred yards. All was quiet. The platoon waited, then crept forward again.

Back at the regimental command post, the regimental commander and his staff waited anxiously for word concerning the progress of the assault platoon. The time was 2315. The night was cloudless, and in fifteen minutes a full moon was due to break suddenly over the crest of a nearby mountain and illuminate the entire countryside with its cold light. Not a word was spoken. Eyes were held almost hypnotically on the operations-map board, while the Coleman lamp which lighted the interior of the small room seemed to roar like a blast furnace. Near the top of Hill 367, the assault-platoon leader dared not even whisper into his walkie-talkie radio for fear of alerting the enemy a few feet in front of him.

Then one man scraped an old German shovel with his foot.

At the sound two German machine guns, one to the right and one almost directly overhead, opened up, and a third gun began firing from the left at a range of 800 yards. The Company K men hugged the ground, and almost immediately, on a call from the platoon leader, supporting tank destroyers silenced the machine gun on the left. As the men hit the ground, they simultaneously, as previously planned, threw hand grenades over the crest of the hill into the German positions. For five seconds the Germans continued to spray the area with machine-gun, machine-pistol, and small-arms fire. Then forty-two hand grenades exploded in unison. There was a deafening roar at the top of the hill; then a dead silence.

"Heave grenades and move in!" The silence was broken by the shout of the leader as he led the platoon over the crest. Jumping almost directly in front of the center machine gun as he cleared the top of the hill, the leader threw a hand grenade into the positions and went on. He fired his submachine gun into the faces of three Germans stunned by the grenade and killed all three.

Meanwhile another member of the platoon had rushed forward, firing an antitank grenade from his M1 rifle at the machine gun on the right. He, too, killed three Germans. As it overran the German position, the entire platoon let out the Division's famous war-whoop, "Powder River—Let 'er Buck!" The enemy became panicky and began running in confusion, some trying only to escape, others stopping to fire and throw grenades from dug-in positions. One German concussion grenade exploded near the platoon wire man, wounding him slightly and blowing three reels of wire from his lap, thus putting the sound-powered phone out of commission. The Company K platoon systematically went about clearing out emplacement after emplacement, shouting for the enemy to come out and then throwing in hand grenades. Some of the Germans tried to escape, but they were quickly cut down.

As the men continued to clear the holes, a machine gun commenced firing on them from the cemetery to the rear of the church several hundred yards away. The entire platoon turned its firepower in that direction, and the enemy fire ceased. At least a dozen enemy were seen retreating, and the platoon did not stop to check on how many were left dead or wounded.

In an attempt to establish contact with the cover platoon at the bottom of the hill, the assault-platoon leader, accompanied by one of his

A reserve unit moves up to take its place in the line and relieve the men in the foxholes

squad leaders, started in that direction. Hearing a gun firing, they called down to it to cease firing, as the hill had already been taken. They were answered by the explosion of a hand grenade nearby. Taking cover in a shell hole, they sprayed the side of the hill. The fire had come from a German gunner who had been instrumental in holding the cover platoon at the bottom of the hill, but after the fire of the platoon and squad leaders, he was heard from no more.

One of the men in the assault platoon had passed a German helmet while the platoon was clearing the holes and went back to get it. As he reached for it, the helmet moved, and he saw a hand pulling it into a hole. Calling for the occupants of the hole to surrender, he covered the position with others of the platoon until three dazed Germans climbed out.

From the top of the church ruins, the radio operator, who had been reporting the progress of the raid since the sound-powered phones went out, reported the capture to his battalion commander in these words: "Here comes one—here comes two—here comes three. We've got three of them, Colonel. We're getting out of here. G'night, Colonel."

The prisoners seemed nervous and indicated that all should get out of the area as their own artillery would begin shelling it at any moment. So, having completed its mission of destroying the enemy positions and capturing prisoners, the Company K platoon made sure there were no live Germans left and, carrying one enemy machine gun with it, began a rapid withdrawal along the route it had come up on.

On the way down, the platoon encountered an enemy outpost position located in a cleverly concealed bunker in front of the church ruins. Without pausing to count the occupants, the platoon quickly liquidated them with grenades and small-arms fire.

By 2330, just as the moon was rising, elements of the Company K platoon had reached covered positions within our lines. The Germans laid a heavy artillery and mortar concentration on the route of withdrawal, but it was too late. The platoon had returned with four prisoners and had accounted for more than twenty dead Germans and probably more, as well as knocking out three machine-gun nests and destroying a formidable enemy position. Of the five casualties sustained by the platoon, two were litter cases and two others were only slightly wounded. The fifth man, wounded in both arms, both legs, and the chest, unconscious from shock, and pronounced dead by an aid man, dragged himself to a Company G position early the next morning and was evacuated.

Tanks and tank destroyers fired on the cemetery to cover the withdrawal. The entire countermortar fire, which began at 1930 and did not stop until 0030 the next morning, was very effective in reducing the amount of enemy fire.

On the night of 8-9 March another raid in force was made in the sector of the 363d Infantry. Three large patrols of the 3d Battalion attacked positions on either side of the road running from Ca' Collina to Ca' Carrara. One patrol from Company L, consisting of an officer and 28 enlisted men, advanced from Ca' Collina to locate and destroy an enemy bunker reported by German prisoners to be in the draw immediately north of Ca' Merla. Just short of its objective, the patrol ran into extensive *Schü* mine fields. A path was cleared through the fields, and the patrol advanced, searching for the bunkers. The bunker was not found, and the patrol was recalled by the Battalion Commander to avoid confusion with other raiding parties starting at dawn.

The second patrol of an officer and 32 enlisted men from Company I and the third patrol of an officer and 14 enlisted men from Company K took part in a coordinated action. Both were attacking enemy-held positions along the high ground to the east of the road between Ca' Collina and Ca' Carrara, the larger patrol advancing from the east and the smaller from the west. The attack jumped off promptly at 0545, 9 March, as planned. As they closed in on their objectives and started up the steep slope of the ridge, the enemy showered them with hand grenades from his positions above. A heavy concentration of mortar fire which followed pulverized the whole area. In the fire fight which developed, the attackers became disorganized, and the large patrol of Company I was driven back.

The patrol from Company K, however, continued its advance up the road toward Ca' Carrara. As day broke, the patrol found itself under the direct observation of a sizable enemy force on Hill 358, just to the east. Both the patrol leader and the assistant patrol leader were wounded by machine-gun fire, the officer mortally. Throughout the day the patrol lay in a little stream, pinned down behind the enemy

Reserve units occupied areas such as this a thousand yards behind the front lines in Speedy Sun Valley

Two lucky doughfeet shared this room with southern exposure

lines. On at least two occasions the group was observed by German medics and was certain it would be subjected to a heavy mortaring. One member of the patrol reported that it was the "longest eleven hours and forty-five minutes I ever spent. Many times I thought my watch had stopped running."

While the patrol lay in the stream bed, aerial reconnaissance was made in an attempt to discover its position. At approximately 1550, 9 March, smoke was laid in the saddle of Hill 358, and one man, who had become separated from the Company I patrol, worked his way back through friendly lines. At 1645 the main body of the Company K patrol began very slowly to inch its way back down the draw. Just as the men had cleared the positions they had occupied during the day, they observed a group of Germans approaching to investigate. After a brief exchange of rifle fire, seven of the patrol made good their escape and reached our lines. Total casualties for the entire action were: one killed, three missing, and seven wounded.

In *Stars and Stripes* of 9 March a three-inch item appeared at the bottom of the front page, reading in part: "Activity on the Fifth Army Front restricted to aggressive patrolling of enemy lines. . ."

During the next ten days, activity in the Division sector quieted

Many and ingenious were the contraptions rigged up for the luxury of a shower. Some even had hot water and privacy—to a degree.

considerably. Enemy shelling was light to moderate in intensity and usually scattered and harassing in nature. Despite continued patrolling, little physical contact with the enemy was reported. Machine-gun and mortar fire drawn by patrols demonstrated, however, that the enemy had not retired. In the early morning of 13 March an enemy patrol tried to infiltrate through the positions of the 361st Infantry near Ca' Trieste, but was driven off in a brief, but fierce, fire fight. Two mornings later, at about the same time, a German patrol probed the positions of Company L, 363d Infantry, with an equal lack of success.

There was no relaxation in the observation of enemy movements and the continued artillery coverage of known or suspected targets. Houses, machine-gun emplacements, enemy personnel, and mortar positions formed the bulk of the targets fired. Better weather permitted more extensive use of the artillery observation planes, and more enemy troops were fired on than had been possible since the Gothic Line breakthrough.

On 16 March, Field Order No. 38 was issued, giving the details of the relief of the 363d and 362d Infantry Regiments by Italian troops known as the Legnano Group. Veteran troops who had been fighting with the United Nations since the fall of Mussolini, the Legnano Group, commanded by General Umberto Utili, was composed of the Special Regiment, the 68th Regiment, the 11th Artillery Regiment, and the 51st Mixed Engineer Battalion. As part of the Italian Corps of Liberation, the Group had fought during July with the Polish Corps in the vicinity of Ancona, and as part of the Fifth Army, had captured Pergola, Gagli, Urbania, and Peglio during August. After a period of rest and reorganization, the Corps of Liberation was divided into two combat groups, the Folgore and the Legnano.

The Legnano Group was attached to the 91st Division on 18 March. During the period 18-21 March, the 68th Regiment was to relieve the 362d Infantry, and between 21 and 23 March, the Special Regiment was to relieve the 363d Infantry. The reliefs were accomplished according to schedule, with the 68th Regiment assuming command of its sector at 0030, 21 March, and the Special Regiment assuming command in its sector at 2335, 22 March. Following their relief, the 362d and 363d moved to rear areas for rehabilitation and intensive training.

At approximately the same time the 91st Division Artillery was relieved for the first time in over six months. At the beginning of March the battalions were in support of the 88th Division. Then they occupied the Defensive Sector on either side of Highway 65, while Headquarters and Headquarters Battery was directing the fires of the artillery battalions of the 88th Division in support of the 91st in the Idice Sector. On 6 March, with the relief of the 88th Division by the 34th in the Highway 65 sector, the Headquarters and Headquarters Batteries of the 88th Division Artillery and 91st Division Artillery exchanged missions and assumed command of their firing battalions. Thus the 91st Division Artillery was in support of the 34th Division, and the 88th Division Artillery was in support of the 91st Division.

From 20 to 28 March the 91st Division Artillery was relieved and returned to the Gagliano–Scarperia area for rest and rehabilitation. Beginning on the night of 20 March, when the 346th Field Artillery Battalion, having completed its 197th consecutive day in the line, moved to Gagliano, the relief was completed on 28 March, when the 916th Field Artillery Battalion closed in Scarperia. The armament platoon of the 791st Ordnance Company performed the semi-annual inspection of the Division howitzers, and each howitzer was calibrated. Small arms and vehicles and all other equipment were inspected, and necessary repairs and painting were accomplished. In the training areas chief emphasis was placed on physical conditioning by marches and competitive athletics. Volley ball courts and baseball diamonds were cleared and marked off, and after the winter months on the line, the men began enjoying the warm Italian sun.

On 21 March, after its relief by the 68th Regiment, Legnano Group, the 362d Infantry moved to Pietramala, where it plunged into strenuous training. Prime emphasis was placed on training for offensive action, weapons training, physical training, and mine warfare. Squad, platoon, and company problems in the attack were executed. Highlighting the training period were studies of recoilless weapons and flame throwers. The 363d Infantry moved to Barberino, where essentially the same program was put into effect. Companies B and C of the 316th Engineer Battalion bivouacked near the Sieve River and trained in the erection and dismantling of Bailey bridges. All units, with the assistance of the Division Special Services Office, included as much organized athletics and recreation in their programs as possible. Movies were shown nightly, and the Red Cross set up special dayrooms in Pietramala, Gagliano, and Villanova, where the troops could read, write letters, and enjoy snacks.

It should be noted that the unremitting efforts of the 316th Engineer Battalion in working on the roads during the winter months were recognized in a commendation by General Keyes. On 19 March he wrote to General Livesay, in part:

> I wish to express my admiration for the fine work the Engineers of your Division have done during the past winter. Their development of trails into main supply routes where no roads existed was an outstanding accomplishment. Their perseverance in this and other difficult tasks during the winter months is recognized as an important factor in the attainment of our present readiness for future operations.

On 21 March the Powder River Division was host for the second time to Mrs. Clare Boothe Luce, Congresswoman from Connecticut. During her brief visit, she inspected the 2d Battalion, 363d Infantry, and elements of the 752d Tank Battalion. After talking personally with the men from Connecticut, she spoke to the battalion as a whole. Refusing to speak from a soap box, she stood on sandbags and in an impromptu speech praised the men for the splendid job they were doing.

The following day General Kuo of the Chinese Army, escorted by officers of II Corps, visited the 91st Division command post and the headquarters of the 361st Infantry. The general visited the regimental observation post, from which he viewed Mt. Belmonte and the current

Congresswoman Clare Boothe Luce talks to one of her Powder River constituents

front lines. The same afternoon Mrs. Mary Bradley, a correspondent of *Collier's*, visited the 361st Infantry area.

During the last ten days of March the Division maintained its defensive positions in the Idice sector with three regiments in line: the 361st Infantry, the 68th Regiment, and the Special Regiment. Although 27 patrols were sent out between 21 and 31 March, no prisoners were captured, and no casualties were sustained. Contact with the enemy was maintained. Enemy artillery, mortar, and long-range machine-gun fire harassed the front lines, but did little damage. The artillery of the 34th Division, now in direct support of the 91st, began a campaign to soften up the enemy positions, and on 31 March excellent results were reported by the air observation post on a special program on enemy-occupied houses selected from prisoner intelligence reports and fired by all battalions. The 125th Field Artillery Battalion expended 147 rounds on eight houses. The 151st Field Artillery Battalion fired 215 rounds and scored 12 direct hits. The 175th Field Artillery Battalion expended 269 rounds on seven houses, with 23 direct hits. In addition, the customary countermortar and harassing missions were fired.

As of 0100, 1 April, the Legnano Group was detached from the 91st in anticipation of the relief of the Division on 5 April, and at the close of March all energies were being bent toward success in the coming offensive.

CHAPTER 12

THE PO RIVER VALLEY CAMPAIGN

APRIL, the final month of war on the Italian Peninsula, was the most spectacular month of action in the history of the 91st Division. At the beginning of the month elements of the Division were in line in the Idice Sector; other elements were in training at Barbarino, Pietramala, and Gagliano; and the Division Rear command post was at Villanova, a few miles north of Florence. At the end of the month the Division was assembled at Treviso, hundreds of miles to the north and east, having participated in the complete and final destruction of the German armies in Italy. It was a month of tense, eager preparation, brutal, difficult fighting to crack the enemy's Winter Line and capture Mt. Adone, and swift, slashing pursuit of the enemy fleeing across the Po Plain. On 2 May the final goal was achieved: the German forces in Italy surrendered to the Allies. The Fifth Army, first American army in the field, had been first to bring the enemy to his knees. The 91st is proud of its contributions to that great triumph.

Planning the Knockout

The opening two weeks of April were devoted to final preparations for the coming offensive and to training and rehabilitating the troops so that they would be at their highest combat efficiency on D-day. On 1 April the 361st Infantry was occupying a 4,000-yard front in the Idice sector, while the 362d at Pietramala and the 363d at Barbarino were carrying on strenuous training programs. Prime emphasis in the training was on physical hardening, familiarization firing of weapons, and a review of basic subjects, especially those developing alertness and self-reliance. At Gagliano the Division Artillery concentrated on a thorough preparation of its equipment. Howitzers were given the semi-annual check, and every worn part was replaced. All howitzers were calibrated, with excellent results; test concentrations grouped perfectly. Each howitzer was comparatively calibrated, and a VE (velocity error) correction made. The gun position, lateral observation posts, and impact area were carefully surveyed. At the same time motor maintenance crews brought all vehicles up to first-class condition. Winter equipment was turned in, and all impedimenta was reduced to the minimum.

Plans were going forward for the shift of the Division to its attack zone, and on the nights of 2-3 and 3-4 April the 361st was relieved by the 133d Infantry, 34th Division. During the relief, the 2d Bat-

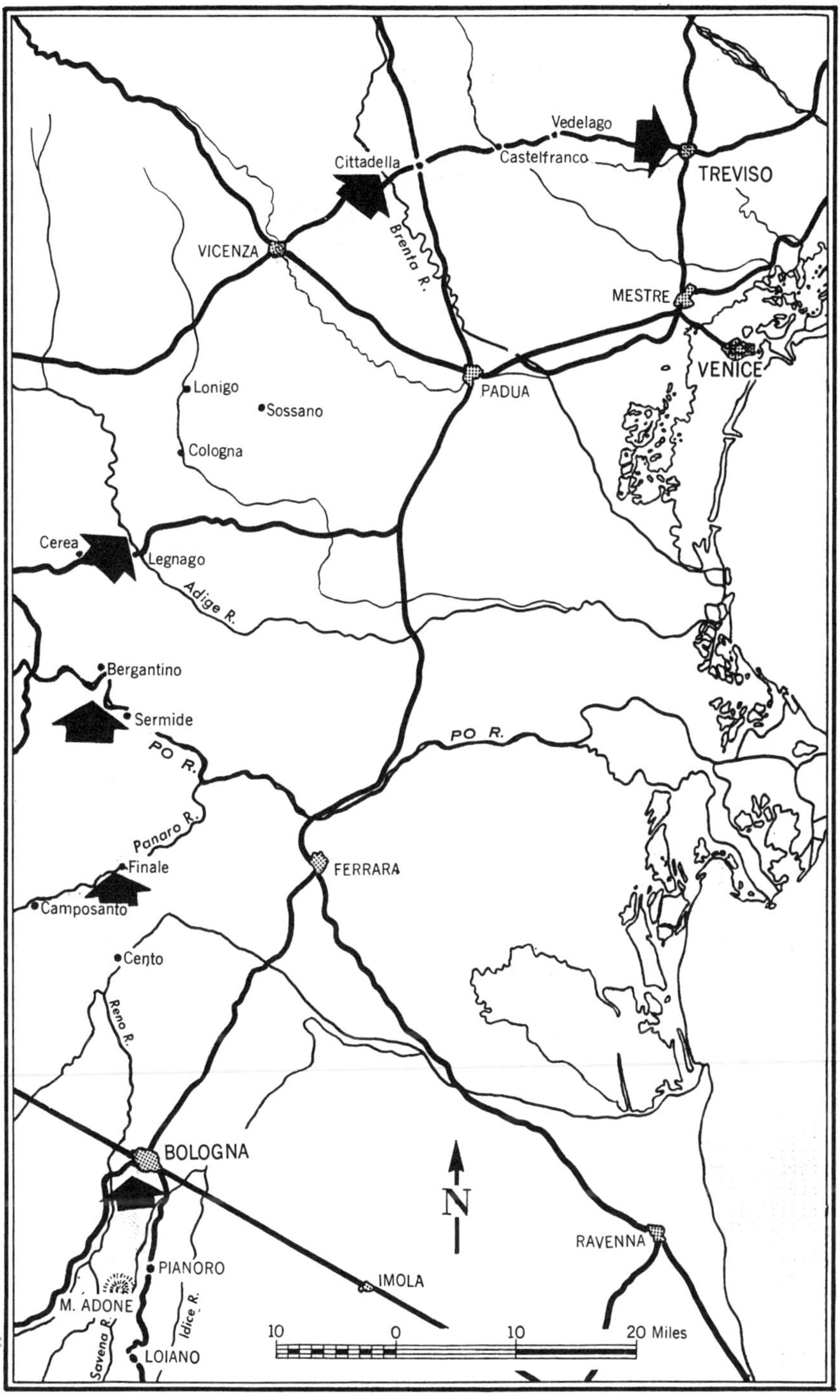

Map 22: From Mount Adone to Treviso

Two members of the Division's air OP maintenance crew keep busy at a landing strip north of Florence

talion, 349th Infantry, 88th Division, was assembled near Macerataio to be used in case the enemy attempted an attack. The relief was completed as planned, however, and command of the regimental sector passed to the 133d Infantry at 2255, 3 April. It was not until 1400, 5 April, however, that the 34th Division assumed full command of the 91st Division sector.

The 361st Infantry assembled near Le Croce and immediately undertook rehabilitation of troops and intensive training for the offensive, emphasizing the same subjects as the 362d and 363d Regiments.

Meanwhile, other elements of the Division began to move to their battle positions. To facilitate the deployment, the 1st Battalion, 362d Infantry, relieved the 1st Battalion, 168th Infantry, 34th Division, on the night of 2-3 April, and on the night of 3-4 April the 2d Battalion, 362d Infantry, relieved the 2d Battalion, 168th Infantry. Command of the sector was assumed at 2330, 3 April, under the operational control of the 34th Division. Relief of the 3d Battalion, 168th Infantry, by the 3d Battalion, 362d Infantry, was completed by 0130, 5 April, so that the 362d was occupying the sector astride Highway 65 with three battalions in the line.

On the nights of 8-9 and 9-10 April the 363d Infantry moved to the vicinity of Hill 484, north of the Livergnano escarpment, and occupied with two companies the sector assigned to it, previously held

Division flyboys

by the 1st Battalion, 362d Infantry. On the night of 9-10 April advance elements of the 361st Infantry moved to La Guarda and, relieving the 3d Battalion, 362d Infantry, assumed command of the sector on the left. The 91st Reconnaissance Troop established observation posts on Hill 520 and Hill 603, and the Division command post moved to the vicinity of Loiano on 11 April. The Division assumed command of the sector at 1100, 11 April, at which time the organic elements of the Division which had been temporarily attached to other units reverted to Division control. The 362d Infantry, minus the equivalent of one company which was being used as a screening force, went into Division reserve at Loiano. The Powder River Division was back in its old, familiar zone.

Active defense of the sector was undertaken, with vigorous patrolling to maintain contact with the enemy and reconnoiter routes of approach into enemy territory. At the same time preparations for the offensive went forward rapidly. Engineers of the 316th Engineer Battalion began to remove our demolitions and minefields, make necessary repairs on the roads within the Division sector, and sweep forward areas for mines. The Division Artillery revised its hostile mortar list as rapidly as interpretation of aerial photographs could be made. Throughout the period before D-day the list was under constant revision as new information was received.

On the night of 11-12 April, Division Artillery, reinforced by artillery from II Corps, fired a heavy concentration on all known enemy gun and personnel positions, expending 2,484 rounds of ammunition. After the preparation, flares were sent up to simulate a signal for attack.[1] During daylight periods throughout the preparatory phase, the Division Artillery concentrated fires on enemy hillside caves and other positions and made every effort to blow up known mine fields. The latter technique was proved inadequate, however, when the actual attack was made against the German Winter Line.

The main energies of the Division were concentrated on implementing and perfecting plans and preparations for the offensive announced in Field Order No. 40. The 91st had drawn a gargantuan assignment. The main effort of II Corps was to be directed west of Highway 65 to break through two strong enemy lines; the first (Brown Line) anchored on the Mt. Sole ridge, Mt. dei Frati, Pianoro, and Possio dei Mori; the second (Black Line) anchored on Tre le Mandre, Praduro,[2] Mt. della Capanna, and Pianoro. The key to the maneuver was Praduro, from which Bologna could be approached up Highway 64 in the valley of the Reno River and enveloped from the west.

The most formidable obstacle in the path of this advance was Mt. Adone, which formed the apex of a double V system of ridges bounded by the Setta and Savena rivers. From their positions on Adone the enemy had unrestricted observation of an arc running from Farne and i Balzi through La Guarda to Livergnano. Further, Highway 65 was under hostile observation from Mt. Arnigo, to the east. It can be seen that the advantage of terrain in all its aspects was held by the enemy. The 91st was to attack an area of successive cross corridors up steep slopes of which the enemy controlled all the dominating features.

In this terrain, naturally suited to the defense, the enemy had the entire winter to build up strong defenses, emplacements, wire obstacles and minefields. He had indicated during the winter months that he was conserving fuel, ammunition, and stores, and there was every reason to expect that he would put up a stubborn defense before he dropped back to his secondary lines. Although one of the divisions manning his line, the 65th, was not rated a superior fighting force, the 157th Mountain Division, defending the Mt. Adone sector, was

[1]This was a part of a week-long series of deceptive fires planned by II Corps to gain deception as to the time and place of attack and to soften up enemy defensive positions. The volume of these fires may be gauged by the fact that one-third of the 45,533 rounds fired from 5 to 15 April were deceptive preparations. The first prisoners captured after D-day reported that these fires were terrifying, the worst they had encountered during the war.

[2]This community, at the junction of the Reno and Setta rivers, is sometimes referred to as Sasso Bolognese and is so marked on the 1/25,000 map.

An engineer removes a demolition charge placed by the Germans in the road south of Pianoro

composed mainly of mountain-wise Austrians and Bavarians whose morale was notably good and whose capabilities could not be taken lightly.

Mt. Adone was the main II Corps objective, and responsibility for its capture was assigned to the 91st Division. After careful appraisal of the problem involved, General Livesay issued Field Order No. 40 on 5 April. The 363d Infantry, on the right, was to attack and secure Mt. Arnigo and the town of Pianoro. Once it had taken the mountain, it was to turn it over to the 34th Division, clear Hill 357 to the northeast, and then, swinging to the northwest, advance toward Mt. della Capanna.

The left sector, dominated by Mt. Adone, was assigned to the 361st Infantry. Running eastward from Adone for nearly 1,200 yards is the Mt. Castellazzo ridge and escarpment. A thousand yards north of Adone lies Mt. dei Frati, and over 2,000 yards to the northeast of Mt. dei Frati toward Pianoro lies Mt. Posigliano. These four mountains were the key positions in the sector; not only did they afford excellent enemy observation, but they dominated the Division's supply lines and avenues of approach. To capture Adone, the strategy adopted was to use the mountain as a hinge and, while maintaining

Medium artillery gun crew sends one on the way

very strong presssure there, to secure Mt. Castellazzo, Mt. dei Frati, and Mt. Posigliano. From these positions, it was believed, Adone could be successfully stormed. After these mountains had been captured, further objectives—Trei di Monclungo and Mt. della Capanna—were designated.

The 362d Infantry, in Division reserve at Loiano, was to be prepared to move quickly and have plans formulated to assist either the 361st or the 363d in gaining their objectives. Throughout the rest of the field order, General Livesay emphasized aggressive action, constant pressure on the enemy, and speed in exploiting any breakthrough.

All the fires of the Division were massed to support the attacking infantry. Division Artillery was augmented by fire of the 757th Tank Battalion, one company of the 804th Tank Destroyer Battalion, one company of the 100th Chemical Mortar Battalion, and one battery of the 432d AAA Battalion. Initially the Cannon Company and all mortars and heavy machine guns of the 362d Infantry were emplaced for maximum assistance to the attacking regiments. In addition to all these weapons, the Division shared with the 88th Division on its left the support of the 423d Field Artillery Group (II Corps Artillery), made up of 155mm Long Toms and 240mm and 8-inch howitzers. The fire of these weapons was carefully coordinated, and every effort was made to obtain maximum advantage of the volume of fire. Every gun was to have a target, and every known target was to be covered.

The crew of an 81mm mortar warms up for the all-out attack

In support of the 361st Infantry in its attack on Adone, Colonel Broedlow could call on the following weapons:

236 105mm howitzers
112 155mm howitzers
1 Battery of 90mm guns
60 Pieces of heavy artillery: 155mm Long Toms, 240mm and 8-inch howitzers
6 105mm cannons of the 361st Infantry
6 105mm cannons of the 362d Infantry
11 Tank destroyers
12 Tanks
1 Platoon of 37mm and .50-caliber guns of the AAA battery
9 57mm antitank guns
8 .50-caliber machine guns of the special platoon
2 Platoons (eight guns) of 4.2 mortars
1 Platoon of 4.5-inch rockets
6 37mm guns

The same impressive array of weapons was also in support of the 363d Infantry.

Finally, the Division's assault was to be supported by powerful air attacks. On 15 April 756 heavy bombers were to bomb targets along Highways 64 and 65. Three hundred medium bombers were to attack bridges across the Reno River, while fighter bombers attacked enemy gun positions in the sectors of the 88th and 91st divisions. The following day, 16 April, the attack by the heavy bombers was to be repeated, while concentrated fighter-bomber attacks were to be made on Mt. Posigliano, Mt. Arnigo, and Mt. Adone. Also on 16 April first priority for "Rover Joe" [3] air support had been accorded the Division, and an air-liaison jeep was assigned to each assaulting regiment.

To improve the jump-off positions, on the night of 13 April patrols from both the 361st and 363d Infantry Regiments occupied advance positions which gave certain tactical advantages to the Division. When a patrol of the Intelligence and Reconnaissance Platoon, 361st Infantry, sent out the previous night, reported the enemy strongpoint at Ca' Nuova, across the Savena River, to be lightly held, a combat patrol was organized for the night of the 13th to seize the position and capture prisoners. The patrol searched the town and, finding it unoccupied, left one squad of Company F to hold the position. The same night two squads of Company K, 361st Infantry, occupied S. Ansano, 400 yards to the northeast of Ca' Nuova. On the following evening, 14 April, the holding forces in these two positions were built up to platoon strength. The advantage of occupying the positions is at once apparent, for it gave the 361st Infantry footholds across the Savena River and enabled the engineers to clear roads and trails in more forward areas.

In the sector of the 363d Infantry the problem was more specific. In order to support the attack on Mt. Arnigo by tanks, it was necessary to move them to forward positions, either up Highway 65 or from Ca' Trieste westward along the Zula–Zena road. The latter was a better route, but since the bridge at Barchetta had been blown, the bridge had to be repaired or bypassed before D-day.

Accordingly, on the night of 13-14 April Company K, 363d Infantry, accompanied by two squads of Company C, 316th Engineer Battalion, moved to Barchetta. No difficulty was experienced in entering the booby-trapped town, but after the occupation, the enemy shelled and mortared the position heavily for two hours. The engineers, however, managed to study the blown bridge and found that, since the crater in the streambed was approximately 75 feet wide and the banks almost sheer and over 25 feet high in places, it would be impossible to repair the damage by hand labor. It was decided to bring a bulldozer up the following night under the cover of darkness and artillery fire and construct a crossing.

---

[3]Rover Joe was the air-ground liaison system developed by the British Army. A radio detachment went forward in the Division area and maintained contact with the light bombers. When we wanted a target bombed, we told the air representative. He, in turn, transmitted to Rover Joe, and in many cases we could get a point bombed within a half hour.

Mortar crews, such as these, fired in close support of the attacking riflemen

The following night an infantry screen was moved north of the road from Ca' Trieste to Barchetta, and the engineers swept the road. After removing sixty of our antitank mines, they moved the bulldozer up, and work began. First a culvert had to be built. The work progressed swiftly until halted by the lack of a second length of pipe. By the time this was brought up, the enemy had apparently discovered the activity and laid in an artillery and mortar concentration which was described as "terrific." The dozer and the culvert material were left in a depression near the site, and work was suspended until the next night.

On the night of 15 April the dozer cut down one bank, and the other was blasted, but once again mortar and artillery fire became so heavy that the mission could not be completed. The bypass had to be finished by 1200, 16 April, and so, under the observation of the enemy a scant hundred yards beyond the front lines, the engineers worked on in the morning of 16 April and finished the job by 1100. When it came time to use the bypass, there was a moment of great tension. The first tank stalled in the middle of it. If it had been improperly handled, the tank might have torn up the bypass so that it would be unusable. But the tank started slowly and proceeded carefully out of the fill, and the remaining four tanks of the platoon followed without difficulty. Colonel Magill's decision to get this job completed before the jump-off proved to be a wise one.

The Jump-Off

On 15 April General Livesay announced 0300, 16 April, as the time for the Powder River Division's attack. Before the Division lay Bologna—gateway to the Po Valley—and the strongest array of enemy defenses on the entire Italian Front. Virtually every square yard of the area fairly bristled with machine guns, mortars, minefields, artillery emplacements, and every other cunning device and obstacle that could make the assault costly. But the long, weary months of bitter fighting and waiting in the mountains had served to strengthen the will and determination of the attackers. Initially the Division would attack on a relatively narrow front of less than four kilometers. As the attack progressed, however, the zone would be gradually widened until the second phase (Black Line) was reached. There the Division would be attacking on a front of over eight kilometers.

The final preparations for the big offensive were completed on 15 April, and at 0200, 16 April, when the first huge concentrations of artillery were loosed on the enemy, the long winter stalemate drew to a close.

The artillery preparation fired in support of the jump-off was extraordinary in two respects: its tremendous volume and the unusual attention given to enemy mortar positions. For the first fifteen minutes a regular preparation was fired. Then a half hour was devoted to close-in countermortar fires. The last fifteen minutes were devoted to all targets. "Normal" preparatory fires consisted of 162 targets, none of which received less than thirty rounds. All known or suspected enemy installations were covered, with special attention to those guarding immediate objectives. The countermortar program consisted of 61 separate targets, which were blasted by over 1900 high-explosive shells.

On these preparatory fires the Division Artillery expended 3,911 rounds. Corps Artillery thickened the fires and conducted a counterbattery preparation of its own. Heavy artillery, which had remained silent heretofore, including one battalion of twelve 8-inch howitzers, cut loose with all the ammunition it could spare. For the Division Artillery all ammunition restrictions were lifted, and the amount of fire was limited only by the physical endurance of the cannoneers and howitzers and the number of targets available. From 1800, 15 April, to 1800, 16 April, the Division Artillery and its attached units fired 1,823,400 pounds of ammunition. This exceeded the previous record expenditure for a single day, made while the Division was cracking the Gothic Line in September 1944, by over 300 tons.

The infantry, poised and confident, was in position behind the line of departure ready for the big attack to commence. The men listened with grim satisfaction as the friendly artillery opened up. It sounded like a thousand runaway express trains as it crashed through the night air, dealing death and destruction to every living thing in its angry path. Never before had the Powder River doughboys seen so impressive a demonstration of firepower. The seconds ticked slowly by—ten seconds, five, four, three, two, one—then H-hour. Across the entire Division front the infantry moved out—not spectacularly, but dramatically.

On the left, the 361st Infantry jumped off with two battalions abreast, the 2d on the left and the 3d on the right. The 2d had as its objective the capture of Mt. Adone and Mt. dei Frati; the 3d was ordered to drive to Mt. Posigliano and following its capture, to turn the Mt. Adone position. Fighting for Mt. Adone revolved about Brento and the Mt. Castellazzo ridge; for Mt. Posigliano, around Ca' Boschi. Both were strongly fortified positions.

In the attack on Brento, Company E moved out from S. Ansano and almost immediately ran into a minefield. The enemy, in his position almost directly overhead on the Mt. Castellazzo escarpment, was alerted by the sound of the exploding mines and the heart-rending cries for medical aid. He began to pour fire of every sort down on the company. When the company attempted to move forward and spread out in order to reduce the effectiveness of the enemy fire, it was forced to pass through even deadlier minefields and suffered heavy casualties.

Company G, advancing from the southeast through a gap between Mt. Adone and Mt. Castellazzo, made better progress, but was finally checked by hostile fire so intense that the company executive officer reported that he was unable to get a telephone to the company commander, a scant 300 yards up the trail ahead of him. Any movement immediately brought down machine-gun, mortar, and small-arms fire from the vicinity of Brento. Enemy snipers, in excellently camouflaged positions, fired with deadly accuracy at the head of the attacking column. It did not seem possible that anything living could have withstood the terrific air and artillery bombardment that had been directed upon the area, and yet the German defenders were there and were fighting back desperately, like wounded tigers at bay, disputing every inch of ground sought by the attackers.

At 0700 a bombing mission was flown on Mt. Adone, and direct tank and tank-destroyer fire steadily pounded Brento. Company supporting fire bounced ineffectively off the pillboxes in Brento, and

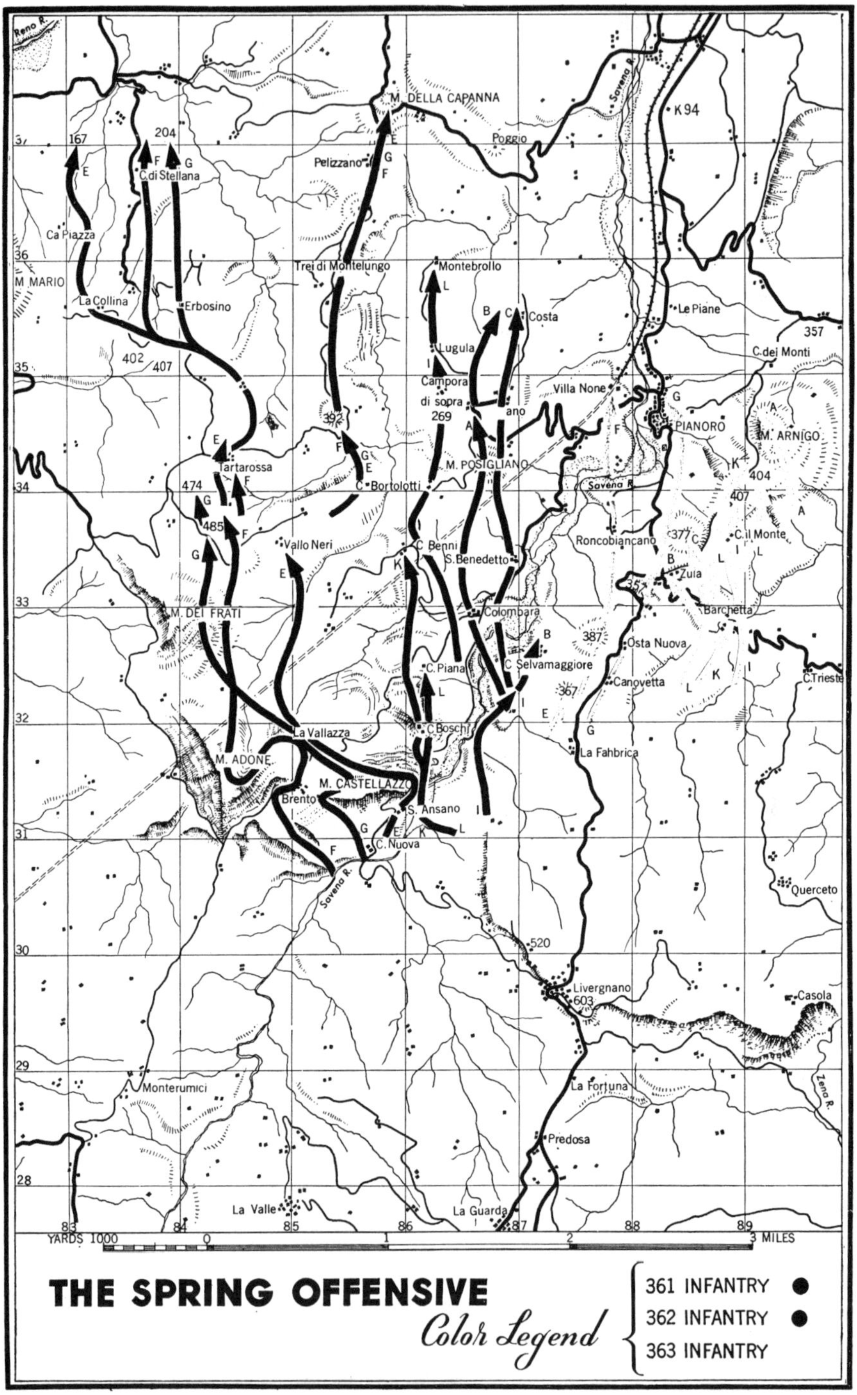
THE SPRING OFFENSIVE
Color Legend
361 INFANTRY
362 INFANTRY
363 INFANTRY
YARDS 1000
3 MILES
Reno R.
M. DELLA CAPANNA
Poggio
K94
Savena R.
Pelizzano
C.di Stellana
Ca Piazza
M MARIO
La Collina
Erbosino
Trei di Montelungo
Montebrollo
Costa
Le Piane
C.dei Monti
Lugula
Campora di sopra
Villa None
PIANORO
M. ARNIGO
Tartarossa
M. POSIGLIANO
Bortolotti
Vallo Neri
C. Benni
S. Benedetto
Roncobiancano
C.il Monte
Zula
Barchetta
M. DEI FRATI
Colombara
Osta Nuova
C. Selvamaggiore
Canovetta
C. Piana
C.Trieste
La Vallazza
C. Boschi
La Fabbrica
M. ADONE
M. CASTELLAZZO
Brento
S. Ansano
C. Nuova
Querceto
Livergnano
Casola
Monterumici
La Fortuna
Zena R.
Predosa
La Valle
La Guarda

Map 23

twenty direct hits by tank-destroyer fire succeeded only in knocking off a corner and blowing open a door of the most important emplacement there. Later the fire of the 8-inch howitzers blew it apart.

On the right, the 3d Battalion, 361st Infantry, made good progress until it reached the low ground in the vicinity of Ca' Nuova. Here Company I, the lead company of the 3d Battalion, ran into heavy sniper fire and direct fire by self-propelled guns firing down the Savena River bottom from Colombara and San Benedetto. Although some artillery was received in this area, the main hostile resistance impeding the advance, as in the zone of action of the 2d Battalion, was small-arms, machine-gun and mortar fire. In support of the drive Mt. Posigliano was bombed and strafed at 1415, and fires were started there.

The 363d Infantry, attacking with two battalions abreast—the 2d on the left and the 3d on the right—made slightly better gains. The 2d took its objective, the high ground between Ca' Selvamaggiore and Canovetta to the left of Highway 65, within an hour. A half hour later it had seized Canovetta. The advance was continued under cover of a tremendous artillery and mortar concentration, which set the next objective, Hill 367, on fire. One tank reached the top of the hill at 0400, but the intense heat forced it to withdraw. After the fire had burned itself out somewhat, the hill was bypassed, and the advance continued toward a second hill (387) 259 yards west of Osta Nuova.

But by this time daylight had come, and the enemy poured fire into the 2d Battalion's exposed left flank. In the vicinity of Osta Nuova minefields also held up the advance. Approaching Osta Nuova, the 1st Platoon of Company G crossed one minefield safely before it hit a pillbox whose approaches were mined. Two men were killed instantly at this point. One was blown in half, and the other lost both legs clear to the hips. Several others were wounded, among them the company commander. Nine more were wounded as the Germans lobbed grenades from an aperture on the lower left side of the highway. Trenches zigzagged down under the road and connected to a 15-feet-square room dug inside the hill 200 yards south of Osta Nuova. Staff Sergeant James L. McDermott and three other men charged the pillbox, threw four grenades and pumped twenty-four shots into the apertures. They killed all four Germans inside. A machine gun in the pillbox never got a chance to fire.

Technician Third Grade Dennis Plummer, a company medical aid man, had run to the first man hit while the German grenades were still falling, and after he had fixed up the wounded, the platoon pushed on toward Osta Nuova. Approaching the shattered house, the advance elements drew rifle grenades and *Panzerfausts* from the enemy. The area around the house was mined and fortified with barbed wire, and the infantrymen had to jump from shell hole to shell hole to get close. Suddenly a German opened up from a hedgerow just above the group with a machine gun. Sergeant McDermott said, "I could hear the SOB work the belt—it jammed."

The sergeant tossed a grenade over, and when the German fled, he was shot. Two more Germans tried to run down from the house to get the machine gun into action again. Sergeant McDermott shot one in the stomach, the other in the arm, but both escaped back to the house. When it was almost daylight, the platoon was able to see that a trench with overhead cover in parts surrounded the house. Apertures for firing were in the top. One German stuck his head up and fired several shots from a machine pistol. When the riflemen opened up on him, he ducked down again. Then he fired several flares from the same hole. A bazooka failed to get the German out, and when daylight came, he bobbed up again, fired, and threw mashers. Sergeant McDermott waited for him to try it once more and, as the man raised up to throw another grenade, shot him between the eyes with a captured Luger.

Meanwhile, the 3d Platoon of Company G came up and was committed around the right of the house in a flanking maneuver. The remaining Germans in the house gave up. Fourteen prisoners were taken, including the two who had attempted to get to the machine gun. Five more who had gone back and reoccupied the pillbox came in with a white flag after the house was taken. Another two came down the highway from Bologna driving a ration cart drawn by two horses. Private First Class Clyde M. Hendrickson shot one of the horses, upsetting the wagon and spilling the rations over the road. The two men surrendered. At 1000 three more prisoners were dug out of an adjoining pillbox. The prisoners were all from the 145th Regiment, 65th Division.

The cellar of the destroyed house at Osta Nuova came to be known as the "Wine Cellar." While the remainder of the house had been reduced to a pile of rubble, the underground portion was intact and had been used as a bunker under which the Germans had dug another series of rooms. In the Wine Cellar Company G found enemy maps—Osta Nuova had been used as an artillery observation post—of the exact type that Company G and the rest of the regiment were using, perhaps run off on the same press. They had all the 363d Infantry's former positions charted, including the assembly area from which Company G jumped off in the attack.

The 3d Platoon, Company G, after assisting in the capture of Osta

Nuova, pushed on to Ca' Casella, 400 yards farther up the highway, while the 2d Platoon occupied the high ground just south of Casella, Hill 368. Neither objective was contested.

Company F, initially in reserve for the 2d Battalion, moved up between Companies E and G, but by 1300 all forward movement of the battalion was halted by an increase of enemy artillery and mortar fire and because the companies, with the exception of Company F, had expended almost all their ammunition in the morning's fighting.

In the 3d Battalion's sector, the tanks had crossed the fill at Barchetta and were in position to support the attack with fire and movement. Two platoons of Company L, which had suffered many casualties soon after the jump-off, had reorganized and were pressing the attack to capture Hill 377. With the infantry, one of the tanks started up the road leading to Zula to knock out a machine gun firing from there. It hit a mine in the shoulder of the road, threw a track, and turned over. A second tank cut off the right side of the road to avoid a similar fate and climbed the ridge to Hill 377. As the two platoons of Company L assaulted the hill, the Germans intensified their machine-gun and artillery fire from the ridge. A German bazookaman in a foxhole in the forward slope of the hill spotted one of the tanks as it approached and hit it from fifty yards, the tank bursting into flame. Two of the tankers scrambled out and tried to help the third, a sergeant, who was trapped inside. They were joined by another sergeant, and the tank company commander left his tank and ran over to help, but was cut down by the German machine gun at Zula before he reached the burning tank. The crew members and their helper were taken prisoner before they could get the sergeant out of the tank, and he burned to death in the wreckage. Additional machine guns opened up from Hill 377 and swept the two platoons, while artillery raked the area and forced them and the three remaining tanks to withdraw to their former positions.

Company I, reorganizing near Ca' il Monte, was also joined by tanks and resumed the attack toward Hill 407. Moving swiftly across the open slope, the tank-infantry teams approached almost to the crest of the ridge by 1430. One tank was knocked out by enemy fire, and friendly artillery, not realizing that the attackers were so far advanced, covered the hill with a concentration. Before this fire could be stopped, Company I and the tanks moved back down the slope to Ca' il Monte to avoid additional casualties on the naked hillside. When the fire was finally raised, one platoon advanced most of the way up the slope again to positions just short of Ca' Nova di Sopra. The remainder of the company reached a draw just south of the houses on the slope by 1800.

During the night the 1st Battalion, 363d Infantry, moved up from its reserve position in the vicinity of Barchetta and went into position between the 2d and 3d Battalions in the center of the regimental sector. At 0600, 17 April, the 1st and 3d Battalions launched a coordinated attack. By 0630 Zula and Hill 377 had fallen to the 1st Battalion, and the 3d Battalion had captured Ca' Nova di Sopra. After the 1st Battalion went into Zula, tanks and tank destroyers were immediately moved into position, giving direct fire support on targets on the left flank and to the front, including Posigliano and Pianoro.

At 0730 the 2d Battalion launched its attack to seize the hills west of Highway 65 within the regimental zone of action. With the capture of Zula, these positions had been outflanked, and by constant artillery and infantry pressure Hills 367, 387, and 357 were captured. In the meantime, the 3d Battalion, on the right, continued to meet strong enemy resistance from well dug-in positions on Hill 407, and the advance was constantly harassed by 88mm-gun fire from Pianoro. By noon, however, Company K had taken Hill 407, and Company L had advanced to Hill 404, just to the north.

During the afternoon and night of 17 April three enemy counterattacks in force were repelled. The story of the defeat of the final counterattack, as related by Private First Class Ben C. Kelly, Company I machine gunner, shows vividly the courage and resourcefulness of the men of the Division:

> We had automatic weapons, a limited supply of ammunition, and a total of two hand grenades. The enemy launched a counterattack at our dugout positions at about 2300, and it lasted until 0300. As the enemy rushed us, we fired until our ammunition was almost gone. Private First Class Morell P. Murray, another Company I machine gunner, threw one of the grenades, wounding several of the charging Jerries and momentarily frightening off the German attack. Having only one grenade and very little ammunition left, we were in a bad spot. Murray rolled up mudballs and, when the Germans got close, threw one. It was a clear moonlight night. The enemy could see the mudballs coming and, thinking them to be hand grenades, ran. Murray repeated this several times, throwing mudballs and rocks, and finally the last grenade, while the other men fired the automatic weapons. The Germans gave up the attack finally. Next morning there were nine Germans lying where we could see them very well.

Another outstanding example of heroism in the sector of the 363d Infantry was taking place on the saddle between hills 404 and 407 on 17-18 April. When assigned the mission of securing the right flank of Company K's objective, the saddle between the two enemy-held hills, Lieutenant Leroy A. Bastron led his platoon of eighteen men forward under heavy mortar and artillery fire, through a minefield, and up the slope of the hill to the saddle. Observed for the first time by the enemy as he prepared to defend the hill, he and his men were subjected to an intense artillery and mortar barrage, which pounded their position incessantly for twenty-four hours. The fire was so intense that for the entire 24-hour period the platoon was out of contact with the remainder of the battalion, and it was impossible to supply them with food, water, or ammunition.

Despite the fire, however, Lieutenant Bastron moved fearlessly among his men, selecting positions and arranging his defense so that all approaches were under observation and covered by fire. All day 17 April, while the enemy swept the hill with machine-gun and rifle fire and pounded it steadily with high-explosive mortar and artillery shells, the platoon held steadfastly to its position.

At dusk, when he observed a group of enemy about the size of a platoon forming for an attack, Lieutenant Bastron alerted his platoon. Firing steadily and throwing grenades, the Germans moved toward the saddle from three directions. When they called up for him to surrender, Lieutenant Bastron stood up and, after a rather strongly worded refusal, ordered his men to open fire. For fifteen minutes, while the enemy threw hand grenades over the top of the ridge into his position, he stood up in his hole and fired his carbine and threw hand grenades down on them. He killed three Germans and wounded or possibly killed several more. During the fight, one of his men was killed and two others were wounded by a rocket, which also destroyed the platoon's only machine gun. Again and again the enemy attempted to reach the platoon's position, only to be driven back each time by the accurate fire of Lieutenant Bastron and his men.

The Germans swept the hill with machine-gun fire from three directions all night, and just before daybreak they made their final attempt to drive the platoon from the hill. While their machine guns were trained on the position, their riflemen began to move up the hill, firing and throwing hand grenades.

Realizing that this was the decisive moment in the battle, Lieutenant Bastron left the cover of his slit trench and moved from man to man, encouraging them and urging them to hold on. As the Germans crept closer, he threw his last grenade down on them. Under the cover of their machine guns, they attacked again and again, but were driven back each time by the determined and inspired platoon. When they were thrown back for the last time, Lieutenant Bastron and his men had only a few rounds of ammunition left.

All morning long mortar and artillery fire continued to fall on the position. Machine guns from bunkers located to the front and on

"Lieutenant Bastron killed three Germans and wounded or possibly killed several more."

either flank sprayed the entire area with their red-hot missiles of death. Finally, at 0900, two tanks from the tank company supporting the battalion made their way to within a hundred yards of Lieutenant Bastron's position. Disregarding the enemy fire, he left his slit trench and moved up the hill to his left toward a group of bunkers. For two hours he moved from bunker to bunker directing tank fire on them. On two occasions when the tanks were unable to drive the Germans from their bunkers, he moved into the emplacements and killed or captured the occupants. When the soldiers occupying a large bunker in the center of the position refused to come out, he directed tank fire on them. As the tank opened up, the Germans attempted to escape. The first man out of the bunker was shot through the head by Lieutenant Bastron, and the remaining nineteen immediately surrendered. Moving around to clean out the remaining positions, he captured six more prisoners.

Lieutenant Bastron's display of coolness under fire, his indomitable courage and aggressiveness, and his intrepid leadership were an inspiration to his men in holding a vital terrain feature against overwhelming odds.

In the sector of the 361st Infantry, the fighting on 17 April was equally bitter. All three companies of the 2d Battalion maintained pressure on Brento; Company G from the front, Company F from the

With the advances, prisoners who had been at a premium all winter became more plentiful

southwest, and Company E from the southeast. Company F, moving during darkness in an attempt to outflank the enemy strongpoint at Brento, made good progress until daylight and at 0530, 17 April, had reached a house just 150 yards southwest of the town. At almost the same time, 0505, Company E had fought its way up the eastern end of the Mt. Castellazzo escarpment and was battling westward toward Brento. All during the rest of the day, Companies F and G were engaged by the enemy in fierce fire fights and were unable to move. They were subjected to a steady mortar pounding in which the enemy, in a vain attempt to burn them out, used many white phosphorus, as well as high-explosive, shells. Company E continued to infiltrate very slowly westward along the ridge line on Mt. Castellazzo.

In the 3d Battalion sector of the 361st Infantry, Company I attacked toward Ca' Selvamaggiore at 0245, 17 April, but Companies K and L, checked by heavy enemy artillery, mortar and machine-gun fires, were forced to maneuver for better positions before they could launch their attack. By daybreak Company I, despite heavy casualties, had reached a point just north of Ca' Nuova, and Companies K and L had begun to move toward Ca' Boschi. Throughout the day the advances were small and painfully won. At 1200 the 916th Field Artillery Battalion and the 361st Infantry's Cannon Company fired a preparation on La Vallazza ridge and the draws behind it. This temporarily neutralized the enemy opposition from that direction, but, after an advance of sixty yards, the companies were again held up in a heavy fire fight.

Once again artillery, tank-destroyer, and 37mm-gun fire was turned on the enemy, and Rover Joe air bombardment was pin-pointed on Ca' Piana, from which considerable enemy self-propelled-gun fire was being received. Under this overwhelming artillery and air support, Companies K and L slowly inched forward. By 2100 they had fought their way to positions just short of Ca' Boschi and stopped to reorganize and prepare for an attack on the town the next day. Throughout the day of 17 April enemy mortar fire had drenched the infantrymen of both attacking regiments. The mortar fire received by the Division broke all previous records. The countermortar fire of the Division Artillery, which fired approximately 2,500 rounds, finally succeeded to some extent in neutralizing it, however.

During the night 17-18 April the 1st Battalion, 361st Infantry, moved from its assembly area at La Fortuna and took up positions on the right flank of the regimental sector just south of Ca' Selvamaggiore. Its mission was to seize Mt. Posigliano and turn the enemy line southwest to Mt. Adone.

On 18 April the 361st Infantry, with all three battalions in line, attacked at 0400 in a powerfully coordinated drive which resulted in gratifying successes. The 1st Battalion seized Ca' Selvamaggiore without opposition and continued its advance across the Savena River toward its objective, Mt. Posigliano. By 0600 it had occupied both Colombara and San Benedetto after only brief fire fights, and by 1030 Company C had pushed up Mt. Posigliano to within 300 yards of the crest. Here it received a small enemy counterattack and considerable fire from the tank at Pianoro which had been harassing the advance of the 363d Infantry on the right. The counterattack was successfully repulsed by small-arms and mortar fire. A heavy artillery concentration on Pianoro silenced the tank, and the 1st Battalion's mortars fired countermortar fire which substantially reduced the heavy enemy mortar fire falling on their positions. Advancing slowly and employing tank fire to knock out hostile machine guns, Company C occupied the crest of Mt. Posigliano.

At the other end of the line, the 2d Battalion, 361st Infantry, broke through the enemy defenses and swiftly followed up its successes. Throughout the night of 17-18 April all three companies maintained heavy pressure on Brento and the Mt. Castellazzo ridge. Reorganizing after a night of continuous mortar shelling, Company F attacked and by 0445, 18 April, had entered Brento, where it was immediately engaged in a sharp fire fight. An hour was spent in mopping up the town. Some of the defenders attempted to escape by taking cover in a draw

German prisoners being questioned by a Division interrogator

on the north slope, but machine-gun and rifle fire cut them off. One German was killed, and twenty-four surrendered. Most of the prisoners, members of the élite 157th Mountain Division, were young, between eighteen and twenty-five, well kept, and cleanly shaven.

While Company G, advancing from the south, and Company E, advancing from the east, methodically cleaned out the Mt. Castellazzo ridge, Company F reorganized and then pushed on to seize La Vallazza. Having taken its objective, it pushed slowly up the remainder of the slope to Mt. Adone, searching caves and dugouts as it advanced. By 1000, 18 April, it had taken the main objective of II Corps, the 91st Division, and the 361st Infantry. The hundreds of elaborate pillboxes, dugouts, and machine-gun positions and the huge caves where the enemy had had light artillery pieces and observation posts were mostly vacated. Abandoned German equipment was everywhere. Thousands of shell craters, gouged by artillery and aerial bombardment over a six-month period, pocked the entire southeastern side of the mountain. Brento itself had been reduced to a heap of rubble.

At 1142, 18 April, the American flag was raised on the 2,161-foot peak by three members of the regimental Intelligence and Reconnaissance Platoon and a civilian scout. Two Germans, asleep in their nearby dugout and undiscovered by the probing Company F men, awoke

Six members of the 361st Infantry raise the U.S. colors on top of Mount Adone

shortly afterwards to see the Stars and Stripes waving above them and surrendered without resistance. On the hills and in the valleys on either side of the 361st Infantry, Allied troops could observe Old Glory flying high above them. Some of their leaders later informed Colonel Broedlow of the beneficial psychological effect the sight had on their men. Above all, the raising of the flag symbolized the cracking of the German Winter Line by the Powder River Division.

Not content with this outstanding success, the 2d Battalion pushed on after a brief rest and at 1445 took Mt. dei Frati. By 1830 Company G had secured Hill 485, and Company F, which bypassed 485, had captured Hill 474. Company E, on the right flank of the 2d Battalion, fought its way to Vallo Neri. Positions were stabilized for the night with Company F in position southwest of Tartarossa, Company E at Vallo Neri, and Company G on Mt. dei Frati and Hill 485.

In the center of the 361st Infantry's sector, the 3d Battalion advanced against lessened enemy resistanre. At 0635 the key position of Ca' Boschi fell, and three hours later the 3d had reached Ca' Piana. Mopping up operations proceeded, and the battalion continued its drive forward to Ca' Benni. During the day evidences of the terrific pounding the enemy had taken from the air, artillery, and infantry were seen everywhere by the advancing troops. Throughout the entire regimental sector unburied enemy dead were found; a very unusual neglect for the ordinarily highly efficient Germans. During the day over 250 prisoners had been captured.

In the sector of the 363d Infantry, equally satisfying gains were scored on 18 April. On the left, the 2d Battalion, 363d Infantry, in an attack coordinated with its left adjacent unit, the 1st Battalion, 361st Infantry, jumped off at 0400 toward Pianoro, on the west side of Highway 65. While the infantry advanced, tanks moved into position behind hills 404 and 407 and began shelling the town. At 0600 the 3d Battalion, also supported by tank fire and further strengthened by the attachment of Company A, began cleaning out enemy positions on Hill 407. This was a painfully slow operation and consumed most of the day.

For over two and a half days the enemy, firmly entrenched in bunkers and narrow zig-zag trenches on the top and reverse slope of Hill 407, had been using this position as an observation post to direct the fire of heavy artillery, mortars, and machine guns. The accuracy and intensity of this fire was largely responsible for holding up the advance of the 3d Battalion, 363d Infantry.

When two frontal attacks on the strongpoint by Companies I and K had been unsuccessful and the strength of the two attacking companies had been reduced materially by extremely heavy casualties, Company L, commanded by Lieutenant Harry L. Brown, was assigned the mission of capturing the hill. Almost three days of continuous, bitter fighting had also reduced the strength of Company L; 54 men were all that remained. But after personally directing the fire of two tanks that were supporting the attack, Lieutenant Brown formed his company in an assault formation and led them around an adjacent hill (404) and through a draw in an attempt to attack from the rear. As the company advanced, intense mortar and artillery fire fell in the draw, and four men were wounded.

Lieutenant Brown and his company advanced, and enemy machine guns and small arms opened up on them. Calling for his men to follow, Lieutenant Brown, by crawling forward under fire for a few yards and then rushing forward for a few yards more, firing steadily as he advanced, reached a point fifty yards from the enemy positions. The fire then became so accurate and intense that it was impossible to advance any farther. Because his supply of ammunition was exhausted and his company had suffered heavy casualties, he moved the men 200 yards

"Lieutenant Brown boldly led the final assault by . . . throwing hand grenades and firing his carbine into the enemy positions."

around the hill to a sheltered position where he could effect a reorganization and replenish his ammunition.

After each man had secured a new load of ammunition, Lieutenant Brown quickly formed the company, now reduced to 26 men, into an assault formation and continued the attack up the hill by fire and movement. As the remnants of Company L reached the top, Lieutenant Brown boldly led the final assault by jumping forward, throwing hand grenades and firing his carbine into the enemy positions. In the fierce hand-grenade and hand-to-hand fighting which followed, he and his men succeeded in killing 9 Germans, capturing 15 more and securing the bitterly contested objective.

In an attempt to dislodge Lieutenant Brown and his company, the enemy concentrated heavy mortar fire on the top of Hill 407. Realizing that his men were almost completely exhausted by the intense physical and mental strain to which they had been subjected during the past three days, Lieutenant Brown went from man to man, talking cheerfully to each and encouraging all of them to dig in and protect the hard-won position.

His cheerfulness, self-confidence, and boldness in the face of heavy enemy fire inspired his tiny band to seize and hold Hill 407, and with this strongpoint neutralized, the 3d Battalion's attack plan was successfully completed for the final assault on Mt. Arnigo.

Meanwhile, the 2d Battalion cleared Roncobianco, crossed Campastrino Creek, and occupied a group of houses 500 yards southwest of Pianoro. After weathering a terrific shelling by a hostile tank located at Villa None, Company F pushed forward to Villa None, where it was engaged in a brisk fire fight until darkness fell. Company G, advancing along the western slope of Hill 407, assisted the 3d Battalion in clearing the hill of hostile resistance by capturing 30 Italian Fascist soldiers. It then advanced to a house 200 yards south of Pianoro. Because of prisoner reports that Pianoro was heavily mined and booby-trapped, the battalion commander directed Company G to clear the town by patrols, while the balance of the company covered from the flank. Patrols moved through the town, reported it clear of enemy, and set up a road block to the north. The enemy threw in one artillery concentration after another on Pianoro without finding Company G's positions.

At dusk the 3d Battalion, 363d Infantry, was still engaged in mopping up hills 407 and 404, and Company A was preparing to assault Mt. Arnigo. This key terrain feature had been bombed by Rover Joe on both 17 and 18 April. In preparation for the final assault, eight tanks were brought up to extreme forward positions just behind Company L. Six tank destroyers, in the sector of the 34th Division, reinforced the direct fire of the tanks by firing from Mt. Belmonte to the southeast.

With this tank and tank destroyer support—plus a terrific half-hour artillery and heavy 4.2-inch-mortar concentration and fire on ground targets by 90mm antiaircraft guns and 155mm guns—Company A experienced little difficulty in gaining the objective behind the curtain of a dense smoke screen. That the enemy had been literally blasted off Mt. Arnigo was confirmed by the dazed and trembling prisoners rounded up in the vicinity. Actually Company A did not receive any hostile fire until the troops commenced pulling prisoners out of the houses and dugout positions on the northern slope of Mt. Arnigo.

At the same time that Company A launched its assault on Mt. Arnigo on the evening of 18 April, the 2d Battalion pushed its attack on Villa None. The 2d Platoon of Company F drew the important assignment of reducing this strongly defended enemy position which had been holding up the advance of the 2d Battalion, 363d Infantry. One squad deployed as a holding force, while the other two attempted a double envelopment. The strongpoint at Villa None was composed of two fortified houses and several dugouts and pillboxes joined by connecting tunnels. It was defended by an estimated company of enemy supported by two medium *Sturmgeschütze* tanks armed with 10.5-cm guns. As the flanking squads approached, they came under heavy fire from the tanks, plus mortar, rifle and machine-gun fire. One of the tanks began to advance down the road toward the left squad; the time was about 2300. Staff Sergeant Johnny D. Lake, platoon guide, who had accompanied the left squad in its advance forward, realized that one man would have a better chance than two of approaching the tank without being detected. On his own initiative, he took a bazooka and five rounds of rocket ammunition and moved alone up the road, under the machine-gun, mortar, and 10.5-cm fire.

Without supporting fire and totally disregarding his own personal safety, he got to within 25 yards of the tank, which by this time had halted. Standing up in a waist-deep ditch so that he could get a better view, he took careful aim and fired a round, which hit the tank but did not destroy it. Locating his position in the dark by the rocket flash, the crew immediately swung their machine-gun fire on him. He fearlessly remained where he was, however, coolly reloaded, and fired a second bazooka round. This, too, was a direct hit, but apparently did little damage against the heavy armor. It succeeded, however, in drawing more automatic-weapons fire from nearby pillboxes and dugouts and the other tank. Sergeant Lake remained standing, despite the heavy fire being directed at him, reloaded again and fired a third round. This, another direct hit, set the tank on fire and caused a great deal of confusion among the crew, who attempted to withdraw their blazing vehicle in spite of its crippled condition. Closing in for the kill, Sergeant Lake sent another bazooka round crashing into the fiery hull.

At this time the second tank moved up to support the withdrawal of the first. It directed its fire on Sergeant Lake, but despite this added menace, he fired his last bazooka round, which burst inside the first tank and caused it to explode. So powerful was the explosion that the entire motor assembly was tossed 25 yards, while fragments flew more than 250 yards and knocked out the second tank. Three of the second's crew were injured, and the remainder fled.

After Sergeant Lake had destroyed the two tanks, his platoon was able to close in and take its objective, capturing nineteen prisoners. Company G bypassed Pianoro on the right. Although the town itself had not been entered because of the numberless mines and booby traps, it had been completely neutralized, and Company F was in position to look right down into it from the northwest.

One of the great days in the history of the 91st Division was 18 April 1945. Mt. Adone, Mt. dei Frati, Hills 474 and 485, Mt. Posigliano,

"Sergeant Lake fired his last bazooka round, which burst inside the first tank and caused it to explode."

Mt. Arnigo, and Pianoro had fallen to the Powder River Division because of the unflinching courage of the gallant infantry and the untiring devotion of its supporting units. Through the combined efforts of a great fighting team, the German Winter Line had been smashed and the way opened for the capture of Bologna and the sweeping offensive of the Po Plain. Twice during the day the achievements of the men of the Division were commended by General Keyes. His first message to General Livesay read:

CONGRATULATIONS ON THE CAPTURE OF MONTE ADONE PLEASE GIVE MY COMPLIMENTS TO THE 361.

Later in the day a second message was received:

CONGRATULATIONS ON THE CAPTURE OF MONTE POSIGLIANO AND MONTE ARNIGO KEEP UP THE FINE WORK.

The brilliant successes of the Division had proved costly. The 316th Medical Battalion reported that "the taking of Mt. Adone . . . resulted in such an intensity of casualties as to necessitate using some of the clerks, technicians, and cooks as litter-bearers." Of the 770 battle casualties suffered by the Division during the month of April, 432

Tankers move up Highway 65 through the rubble that was Pianoro

occurred during the three-day period 16-18 April. On 17 April, 230 casualties were suffered, making it the busiest day for the Medical Battalion since the Gothic Line campaign. The majority—62—of the 95 exhaustion cases evacuated during the month occurred during the four-day period 16-19 April. It is worthy of note that whereas the exhaustion cases of the Gothic Line campaign were frequently replacements suffering from panic, the patients treated in the current fighting were mainly old combat soldiers, many of them wearers of the Purple Heart, whose endurance had been worn down as much by the strain of prolonged combat duty as by any single operation.

With the capture of Pianoro and Mt. Arnigo, the main effort of II Corps shifted to the west, with the object of cutting Highway 9 west of the Reno River, isolating Bologna, and opening Highway 64 into the Po Valley. Once Bologna was isolated, it was expected that its capture would be greatly facilitated. Concurrently, strong pressure was to be maintained from the south. As part of this shift, the 363d Infantry was relieved by the 133d Infantry, 34th Division. The 1st Battalion, 133d Infantry, passed through the 3d Battalion, 363d Infantry, on Mt. Arnigo and continued to attack Hill 357, which it captured by daylight 19 April. The 3d Battalion, 363d Infantry, assembled in the vicinity of Barchetta. The 1st Battalion, after remaining on Mt. Arnigo during the day, moved back to the assembly area on the evening of the 19th. The 2d Battalion spent the day patrolling and mopping up in the

Medical Corps men evacuate wounded

Pianoro area. Company G sent patrols into Pianoro and north on Highway 65 as far as Le Piane, while Company F cleaned out the area in the vicinity of Villa None and the mouth of the railway tunnel. By early afternoon Pianoro was completely cleared.

With the II Corps' main effort swinging to the west, the force of the 91st offensive was concentrated on the Division's left flank. At 0500, 19 April, the 350th Regimental Combat Team, 88th Division, was attached to the 91st Division on the left of the new sector with Mt. Mario as its main objective. In the center was the 362d Infantry. This regiment was given the mission of capturing Trei de Monclungo and Mt. della Capanna and, by taking hills 402 and 407, assisting the 350th Regimental Combat Team in taking Mt. Mario. This accomplished, the 362d was to strike out for the high ground between Hills 167 and 204, just to the north of the 37th Northing. During 18 April arrangements to move the 362d Infantry forward were completed, and at 2145 the 1st and 2d Battalions began to move into position for the attack. At 0445, 19 April, the 2d Battalion, 362d Infantry, jumped off without artillery preparation, followed at an interval of four hundred yards by the 1st Battalion.

Throughout the day the advance in the 362d's narrow sector was hampered by congestion on the limited roadnet and by opposition, which materialized in unexpected strength. Thus the original plan, to have elements of the 91st Division cross the Reno River and pass through the 6th South African Division after it had captured Sasso Bolognese, was never put into effect. Initially the 2d and 1st Battalions, 362d Infantry, found elements of the 350th Regimental Combat Team extended into their sector. As they advanced, Company F encountered strong enemy opposition at Tartarossa. Company E moved around the left flank of Company F, and, after a sharp engagement, the German strongpoint was reduced.

At 1345 the 350th Regimental Combat Team was reported on Mt. Mario, and, the sectors once more clearly established, the 362d Infantry began to move forward more rapidly. In succession, Erbosina, La Colina, and Ca' Piazza were taken and by 0230, 20 April, the 2d Battalion had occupied its objective, with Company E on Hill 167 and Companies F and G on Hill 204. The 1st Battalion, following in close support, occupied an assembly area 800 yards to the rear.

The most difficult and disheartening fighting of 19 April occurred in the sector of the 361st Infantry, which was attacking north from its position southeast of Mt. Posigliano, through Ca' Benni and Vallo Neri, to Mt. dei Frati. In this sector the enemy had two strong, prepared defense lines: the Lina Line (Hill 269–Guzzano–Villa None–Pianoro–Ca' dei Monte–Hill 357) and the Hermann Line (Ca' di Stellano–Pelizzano–Tre Poggio–K-94 on Highway 65). Holding fanatically to their positions and backing them with the strongest artillery support encountered since the offensive started, the enemy contested every inch of ground while they covered the withdrawal of their forces north of Bologna. Against this stubborn defense, the 361st Infantry, alternating battalions in the attack, pounded with trip-hammer blows.

The 361st's 1st Battalion moved out from Mt. Posigliano at 0415. In support of the attack a very heavy artillery concentration was fired on Guzzano, with the entire Division Artillery, plus elements of II Corps Artillery, firing an intensive countermortar preparation. The advance of the 1st Battalion continued steadily until it reached the vicinity of Guzzano, one of the strongest points in the Lina Line. Company C, in the lead, had reached a point just north of Guzzano when it became engaged in a fire fight. Company A was southeast of the town with Company B following. At 0845 a counterattack of approximately 45 Germans was launched at the left flank of Company C from the direction of Lugula. With the help of Division Artillery, this enemy thrust was beaten back. The artillery concentrations also reduced the hostile mortar fire to which the company was being subjected. But the enemy held tenaciously to his positions, and by noon little advance had been made.

On the left, the 2d Battalion, although weary from three days and nights of continuous hard fighting and inadequately supplied as a result of the poor roadnet in the sector, jumped off slightly before 0800. Company F, leading the advance, reached the vicinity of Ca' Bartolotti, where the forward movement of the battalion was impeded by a huge gorge. One platoon of Company F managed to work its way down the almost perpendicular canyon wall and up the other side to seize Hill 392 by 1130. In the meantime, Company G, striking off to the right, encountered numerous barbed-wire obstacles in its path. The commanding officer, attempting to locate a gap in the wire through which the company might pass and continue the advance, proceeded several hundred yards in front of his company, where he became exposed to heavy enemy machine-gun fire covering the barbed wire entanglement. One company scout who had accompanied him was killed instantly, and the company commander himself was seen rolling over the crest of a small hill with machine-gun bullets kicking the dirt up all around him. Cut off from his company deep in enemy-held territory, he was thought to have been killed when he failed to return. All attempts to locate him were in vain. Actually, he had rolled over the hill into a German dugout, where he remained the rest of the day, surrounded by enemy troops unaware of his presence. At this point the 2d Battalion was ordered into temporary regimental reserve to rest, resupply, and prepare to continue the attack during the night. All during the afternoon, however, the assembly area of the battalion received a steady artillery pounding.

In view of the widening sector, at 0550 on the morning of 19 April Colonel Broedlow ordered the 3d Battalion to advance between the 1st and 2d Battalions. This push in the center was more successful. By 1100 Company I was approaching Campana di Sopra, and shortly after noon it had taken Lugula, thorn in the side of the 1st Battalion. Several machine-gun nests which had been firing into the flank of the 1st Battalion were destroyed. Immediately, however, the enemy laid in heavy 150mm-artillery fire on Lugula, and the 3d Battalion pushed on. Although hostile artillery fire continued to pour on the troops, they continued to advance, and by 2015 Company L had seized and occupied Montebrollo. Here the 3d Battalion consolidated its gains and held for the night.

Throughout the day artillery fire had backed up the attack notably. Guzzano, Lugula, and Costa had been almost constant targets. Nearly twenty Rover Joe fighter-bomber missions had been called for during the day to isolate enemy positions and soften them up. Yet the progress, in view of an apparent crumbling of enemy resistance to the west, had been dishearteningly limited.

At 2400, 19 April, the 2d Battalion, 361st Infantry, with Company E in the lead, pushed out from Ca' Bartolotti. The violent enemy artillery shellings of the afternoon had evidently covered a withdrawal, for the battalion advanced against no opposition for over three and a half kilometers to capture Mt. della Capanna by 0500, 20 April. A group of German soldiers, evidently stragglers, was surprised and captured in the vicinity of Pelizzano, and at daylight the number of prisoners was greatly increased when an enemy force at Calegare surrendered without opposition. For the third time in two days General Keyes commended the 91st Division and the 361st Infantry. He wired:

CONGRATULATIONS UPON THE CAPTURE OF MONTE DELLA CAPANNA MY COMPLIMENTS AGAIN TO THE THREE SIX ONE KEEP UP THE DRIVE.

In the meantime, at 1800, 19 April, Field Order No. 44 had been issued, giving new boundaries for the 91st and establishing new objectives. Once again the sector shifted westward, with the left boundary of the Division crossing the Reno River at Sasso Bolognese and running northward between the 80th and 81st Eastings. In the left sector the immediate objective of the 362d Infantry was La Costa.

In accordance with these new directives, the 2d Battalion, 361st Infantry, held on Mt. della Capanna. The 361st's 1st Battalion moved out from Costa. Skirting Mt. della Capanna, it passed through the 2d Battalion and mopped up scattered pockets of enemy resistance in the vicinity. As it advanced northward, resistance became increasingly stiff. When a prisoner reported that an enemy force of two companies was holding positions on Hill 379 (Mt. di Mori), the advance of the 1st Battalion was halted while tank and artillery fire were laid into the position. After two hours the battalion resumed its advance and by 1350 had reached positions in the vicinity of Pieve del Pino and Caponara.

At the same time the 3d Battalion, 361st Infantry, had been ordered to move toward the northwest and seize Mt. Belvedere. Advancing rapidly against slight hostile resistance, Company K occupied the mountain a few minutes before noon, 20 April. When word was received that the 362d was to proceed to Vizzano, the 361st's 3d Battalion pushed west cautiously across the Reno River. By 1515 forward elements of the battalion had reached Pal dei Rossi. Having advanced beyond the regimental sector, according to the revised instructions of the afternoon, the battalion's advance was halted by the regimental commander. After dusk the 3d Battalion was shuttled back to Pieve

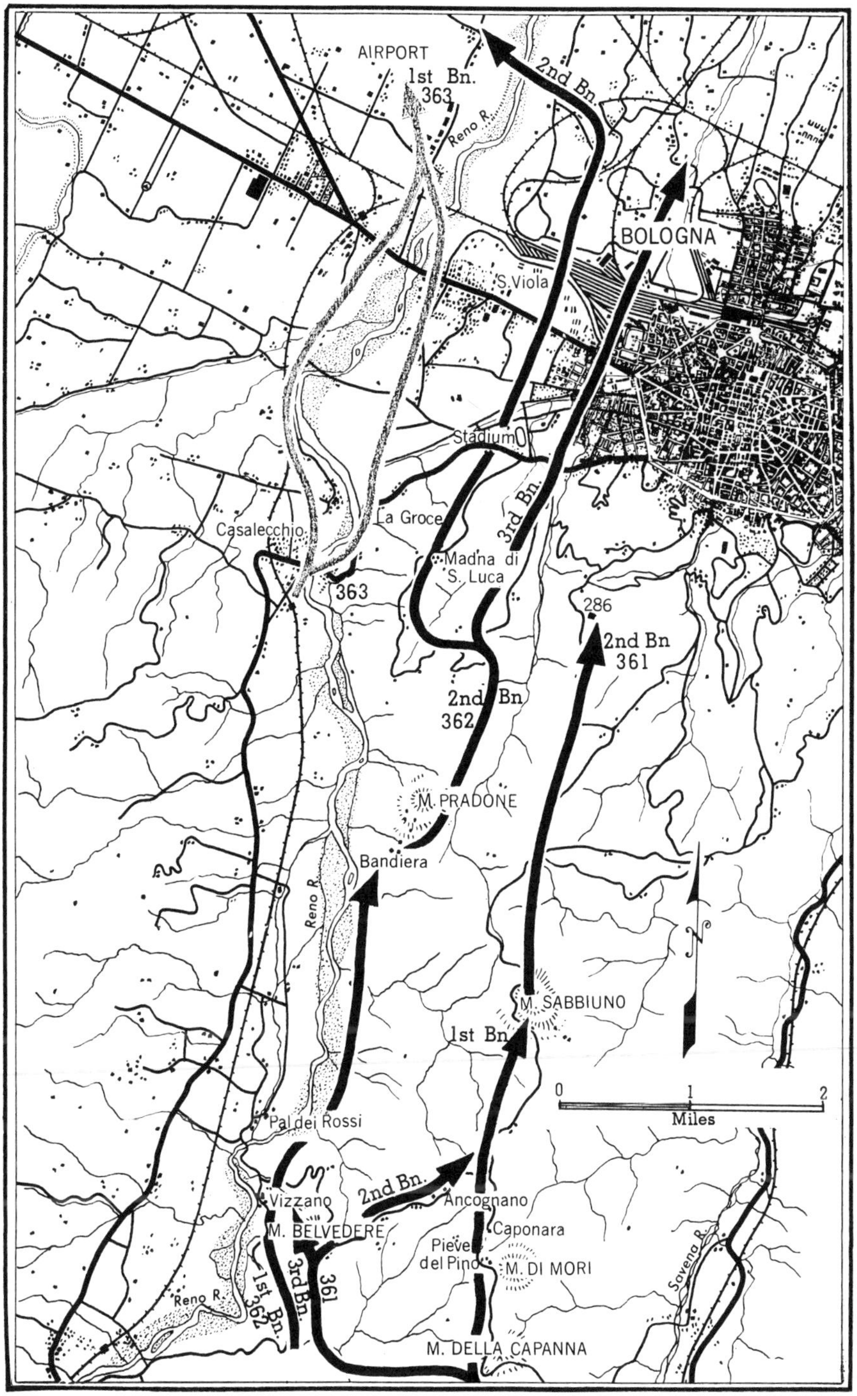

Map 24: The Bologna sector

General Livesay outlines the pursuit plan to Colonel Magill and his staff along a road south of Bologna

del Pino, behind the 1st Battalion, in the new sector which had been assigned to the 361st Infantry. Meanwhile, the 2d Battalion, which had followed the 3d Battalion in its rapid advance to Mt. Belvedere, swung to the northeast and by 1600 had occupied Ancognano.

At 2400, 20 April, the 1st Battalion, 361st Infantry, continued its attack against Mt. Sabbiuno. Moving rapidly through the darkness of early morning against light resistance, Companies B and C were firmly established on the objective two hours later. After advancing 800 yards beyond Mt. Sabbiuno, the 1st Battalion closed into an assembly area. Keeping up the steady offensive by alternating lead battalions, Colonel Broedlow ordered the 2d Battalion to pass through the 1st and continue the rapid pursuit northward. Delayed only by roadblocks, the 2d Battalion reached its objective, Hill 286, shortly after 0800, 21 April. Here, in accordance with Field Order No. 45, the 361st Infantry assembled and went into Division reserve.

From Hill 286, the last high ground south of Bologna, a view of the entire city could be had. After six long months of fighting and waiting just short of the prized goal, the Powder River warriors were a little disappointed that they were stopped short of a triumphant entry into the city. But the hard grind of the past five and a half days of constant fighting and driving over mountainous country had been tough, and the break was a welcome relief. Immediately steps were taken to rest the weary troops, clean equipment, and formulate plans to motorize the regiment, using organic transportation to keep it as mobile as possible.

Troops of the 363d move south on Highway 65 past La Fahbrica after being relieved by the 34th Division

Despite the shift of the Division's left boundary as specified in Field Order No. 44, the 362d Infantry, because of the swiftly moving situation, did not cross the Reno River at Sasso Bolognese, but remained on the west bank and drove north. In accordance with verbal orders of the Division Commander, patrols were sent to Vizzano on the morning of 20 April and by noon the town was completely occupied by the 1st Battalion. New objectives were then set up: Bandiera, on the southern slope of Mt. Pradone, and La Groce, on the east bank of the Reno River one kilometer northeast of Casalecchio along Highway 64. While the 1st Battalion advanced north, followed by the 2d Battalion, the 3d Battalion assembled in the vicinity of Mt. Belvedere, there to join tanks and assemble vehicles to form a motorized task force.

Moving steadily forward, the 1st Battalion reached Bandiera at 0010, 21 April. The 2d Battalion then passed through the 1st and marched steadily forward throughout the night. By 0650 it was reported on its objective, Madna di San Luca. At 0730 the lead company, Company G, marched into the outskirts of Bologna. The company commander asked permission to enter the city. This was refused at first but at 1145 Company G received orders to proceed to the stadium on the western edge of the city. Simultaneously troops of the 34th Division entered Bologna from the south.

While the 361st and the 362d Regiments were fighting northward between Highways 64 and 65, the 363d, after a day in reserve near Barchetta, began, on the morning of 20 April, an end run behind the Division and up Highway 64. The 1st and 2d Battalions moved by truck, while the 3d started out on foot to cut down the length of the shuttle haul. By late afternoon the 363d had assembled in areas along crowded Highway 64. Early the next morning, 21 April, the regiment was shuttled to a point just south of Casalecchio. At 1000 the 1st Battalion moved out, followed by the 2d two hours later, to enter the line again on the left of the 362d Infantry.

Companies A and C moved north on the west bank of the Reno River, Company B on the east bank. Company B cleared La Groce at 1130, 21 April, just as the 362d Infantry was preparing to enter the western edge of Bologna, and then moved on to San Viola. By 1400 the 363d's 1st Battalion had secured its objective—the Bologna airport—against very light resistance. By the end of the day the entire regiment was assembled in the vicinity of the airport. The 362d Infantry, after marching through the western suburbs of Bologna amid crowds of happy citizens, crossed the Reno River and grouped in assembly areas a kilometer and a half northeast of the airport.

The Drive To The Po

With the fall of Bologna, the first phase of the spring offensive of the Po River Valley Campaign was concluded and the second phase, the drive to the Po, began. The character of the fighting changed completely. No longer were the day's gains measured in feet or yards, but in miles. Having fought through the rugged Apennines and reached flat ground for the first time since September 1944, the infantrymen no longer had to fear emplacements or the enemy hidden behind the next ridge. Nor did they have to discover and clean out hundreds of caves and dugouts hewn from solid rock. No longer did the enemy enjoy every advantage of observation. No longer did the road seemingly wind upward all day. And now there were the tanks. For the first time in the entire Italian Campaign, here was real tank country, and the tankers were eager to show the doughboys the true capabilities of armor.

The second phase of the spring offensive began the morning of 22 April, when the 362d and 363d Infantry Regiments, moving out of their positions on the outskirts of Bologna, began the drive to the Po.

A South African soldier shares his candy with Powder River doughboys along the road about a half mile south of Bologna

On every front, elements of the Fifth and Eighth Armies were surging forward in pursuit of the enemy. The German high command in Italy had completely lost control of the situation. American and British airmen bombed and strafed the columns of retreating Germans, while on the ground motorized infantry columns and armor moved swiftly to keep the enemy off balance and prevent him from reorganizing to make a stand.

Channels of communication and supply were cut. Groups of enemy —some large, some small—were isolated, overrun, and taken prisoner. At first enemy trucks, artillery pieces, prime movers, and vehicles of every size and description littered the road, burned or wrecked by air or ground action. Later, as the Allied forces moved even more swiftly and the enemy flight became increasingly disorganized, vehicles in working condition were often abandoned when they ran out of gasoline or developed some minor mechanical failure. Stores of ammunition and equipment of all sorts were captured intact. Horses commandeered by the enemy to pull artillery, wagons, and other vehicles wandered along the highways or roamed the fields. Down every high-

The road to Bologna was now open. Division infantrymen poured through the suburbs, into the city, and spread out to mop up the Po Valley beyond.

way, marching to the rear, came streams of Germans, hands on head, for whom the fighting was all over. Often marching under guard that would have been totally inadequate had the enemy not been demoralized, the Germans walked docilely to the hastily improvised prisoner-of-war inclosures. It was an exciting and memorable week—bringing the final destruction of the enemy's capacity to wage war.

In order to pursue and destroy the enemy in its sector, the 91st Division organized motorized columns, protected by armor, to race ahead of the main body of the Division, bypassing or reducing local resistance and seizing important cities, road junctions, bridges, or fords. All the organic equipment of the Division was loaded to full capacity. Tanks and tank destroyers carried a squad of men each, and thirty trucks were requisitioned from II Corps to augment the Division's equipment. On the average, five infantry battalions were motorized at all times. Thus, in the manner of the German *Blitzkrieg* of the opening days of World War II, the Powder River Division raced toward Treviso.

The chief obstacles in the path of the swiftly moving advance were the four major rivers—the Panaro, the Po, the Adige, and the Brenta—which cut across the Division's zone of action. In every case bridges over these rivers had been pounded for months by the Air Forces, and most of them were either destroyed or seriously damaged. Bridges

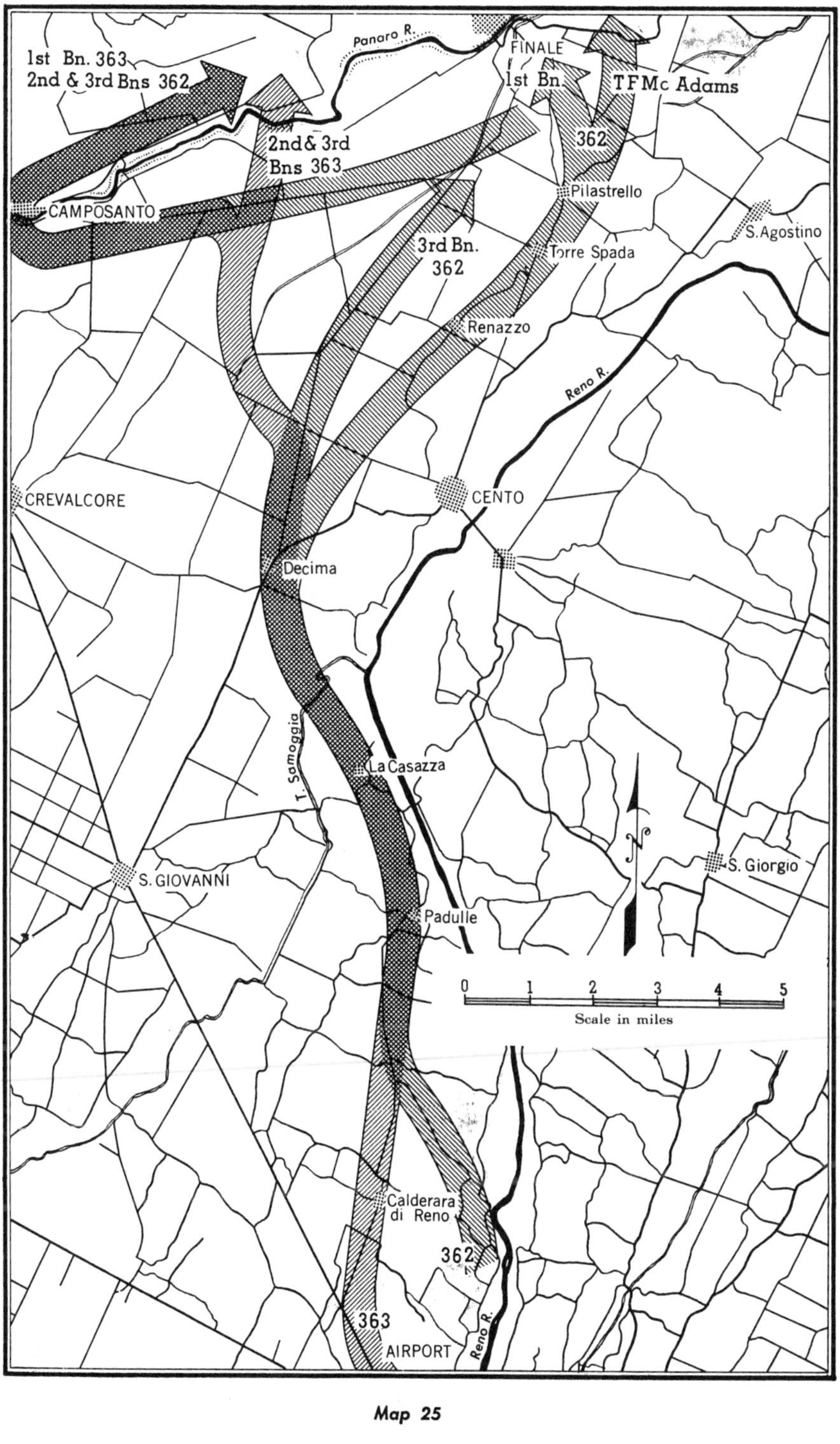

1st Bn. 363
2nd & 3rd Bns 362
Panaro R.
FINALE
1st Bn.
TF Mc Adams
362
2nd & 3rd
Bns 363
CAMPOSANTO
Pilastrello
S. Agostino
3rd Bn.
362
Torre Spada
Renazzo
Reno R.
CREVALCORE
CENTO
Decima
T. Samoggia
La Casazza
S. Giorgio
S. GIOVANNI
Padulle
0
1
2
3
4
5
Scale in miles
Calderara
di Reno
362
363
AIRPORT
Reno R.

Map 25

The citizens of Bologna turned out in mass to welcome their liberators

which the enemy had repaired and ferries which he had used in his retreat were frequently blown out by demolitions which he touched off at the last minute or by the fighter-bombers closely supporting our advance. Hence, at both the Panaro and the Po there was difficulty, especially in getting the vehicles and armor of the advance elements across the water barriers. Here again the engineers were magnificent. The drive was never halted, and the pressure on the enemy never ceased.

During the night of 21-22 April the 1st Battalion, 363d Infantry, moved in the left sector to Calderara di Reno, whence it moved out at 0500, 22 April. Although during the night enemy planes dropped fragmentation bombs in the area of both the 362d and the 363d Infantry Regiments, no casualties were suffered, and the attack was not in any way delayed. From their positions along the southern edge of the broad, flat Po Valley, the Powder River doughboys could see the fires of burning German equipment which dotted the whole countryside as far north as the eye could see—proof that Allied air might was playing havoc with the retreating enemy columns.

The 3d Battalion, 363d Infantry, riding tanks and tank destroyers, moved out at 0600 from the vicinity of the airport, followed at an interval of 2,000 yards by the 2d Battalion in trucks. The 3d Battalion, 362d Infantry, moved out at the same time. The mission of the Division was to drive north to the Panaro River, screened by elements of the 6th South African Division, and there to prepare to advance to the Po. Initially the zone of action assigned to the Division ran due north, bounded on the right by the Reno River, bypassing Cento and continuing straight across the Panaro at Finale. The left boundary, running generally parallel to the right boundary, crossed the Panaro River midway between the cities of Finale and Camposanto.

In the left sector, the 1st Battalion, 363d Infantry, pushed rapidly forward, meeting little or no resistance. Five kilometers southwest of Cento the column, now headed by the 3d Battalion, was halted briefly at the Samoggia Canal. Thirty feet wide and forty feet deep from the top of its embankments, the canal stretched across the entire Division front. While Company C, 316th Engineer Battalion, rapidly constructed a culvert and bypass by cutting down the dykes, the infantry dismounted and proceeded forward on foot. As soon as the bypass was completed, the tanks and tank destroyers crossed the canal and rumbled into Decima, where the infantry climbed aboard again. The 2d Battalion continued to follow the 3d at a close interval. Neither battalion met more than light resistance, and by 2400, 22 April, both had reached their objective—the south bank of the Panaro. The 3d Bat-

This bulldozer of Company C, 316th Engineers, constructs a fill and by-pass over one of the Po Valley's many canals so that transportation can keep up with the rapidly advancing foot troops

talion found a crossing site within an hour, and construction of a footbridge was begun.

By 0630, 23 April, the 3d Battalion, 363d Infantry, had crossed the Panaro River and was moving out on foot toward the Po. Their vehicles and armor, unable to cross with the foot troops, turned west to Camposanto, in the sector of the 88th Division, where they crossed on a bridge which had been captured intact. The vehicles of the 2d Battalion were on their way to Camposanto, and the troops were crossing the footbridge. The 1st Battalion, which had been moving forward by shuttle movement behind the leading elements since its relief by the 3d Battalion, was in an assembly area near the river.

In the right sector, the 362d Infantry advanced with almost equal ease and rapidity, although slightly more enemy resistance was encountered. Initially the 3d Battalion led the advance. At 0900, 22 April, the 2d and 1st Battalions, following the 3d Battalion in that order, received instructions to pass through the 3d Battalion and continue the drive northward. In accordance with orders from II Corps, the 2d Battalion was organized into a task force and given the mission of driving an infantry-armor spearhead to Bondeno, located on the Panaro River fifteen kilometers northeast of Finale. It was to secure the city and the crossing sites over the river at that point. The objective of the 1st Battalion was the south bank of the Panaro River just west of Finale.

By 1410 the 362d's 3d Battalion had reached its objective—a road junction just south of La Casazza—where it halted. Two hours later the 2d Battalion, now a motorized task force under the command of Major Ronald E. McAdams, moved out to Padulle to pick up its armored attachments. As finally constituted, Task Force McAdams comprised the 2d Battalion, 362d Infantry; Company A, 757th Tank Battalion; Company A, 804th Tank Destroyer Battalion; Company B, 316th Engineer Battalion; Battery B, 346th Field Artillery Battalion; and one platoon of the 362d Infantry's Antitank Company. After moving a short distance, the task force was halted by lack of a crossing over the Samoggia Canal. Patrols were sent out to reconnoiter for a crossing site, but none could be found immediately. Finally, at 2030, after considerable delay, the task force moved out, led by a platoon of tanks and tank destroyers, with a platoon of troops riding the armor.

As the column rolled forward near Decima in the bright moonlit night, several German planes bombed and strafed it. One antipersonnel bomb hit its mark in the midle of the column, killing one man and wounding six others. By 2300 the force had reached Renazzo. Here the leading jeep was fired on, and reconnaissance patrols were sent out to determine the strength of the enemy. Infantry searched the town and houses along the road, capturing 25 prisoners and killing at least 3 of the enemy.

At this point word was received that the mission of the task force had been changed and that its objective would now be Finale. Accordingly, the column shifted its direction and began to move again. At the road junction just south of Torre Spada, an enemy tank was destroyed after a brief duel with the American armor. As enemy groups came out of lateral roads seeking to turn north toward the Panaro River, it became apparent that the enemy was attempting to retreat along the road Task Force McAdams was using.

A little later, at the same crossroads, a large force of Germans was engaged in a sharp fire fight in which all elements of the column joined. To the rear of the column, for example, Company B, 316th Engineer Battalion, accounted for 11 killed and 60 captured. When the scrap was over, the task force had succeeded in inflicting severe losses on the enemy while suffering no casualties itself. About 150 Germans were estimated to have been killed, and 293 were captured. Three tanks and eight other vehicles were destroyed.

Following this brisk clash with the enemy in the vicinity of Torre Spada, the head of the column moved forward again, but succeeded only in reaching Pilastrello when another fire fight broke out. At the rear of the column in Torre Spada, the battery of the 346th Field Artillery set up its guns and fired on the enemy positions. Infantry and armor finally reduced the opposition, and the task force assembled near Pilastrello. Strong patrols advanced to the Panaro River, but met no further hostile opposition. In its operation during the night of 22-23 April, Task Force McAdams had killed over 200 Germans and captured 423 prisoners. Five German tanks were destroyed, and five more were seriously damaged. One entire enemy personnel section, with its records, and a paymaster complete with payroll had been captured. In addition the task force took uncounted matériel, including trucks, prime movers, motorcycles, telephones, and a code converter in perfect working order.

While the 2d Battalion, 362d Infantry, protected the right flank of the regiment, the 1st Battalion assaulted the town of Finale from the south. Here the enemy was gathered in force to defend the bridges and keep open the escape route for the large group of enemy armor assembled in the area. The 1st Battalion encountered strong opposition and was engaged in a fierce fire fight most of the day. At 1400, 23 April, the 2d and 3d Battalions were motorized, and at 1600 the 3d Battalion led out to Camposanto to cross the river.

In the meantime, the 1st Battalion, 363d Infantry, heretofore in reserve, crossed the Panaro and moved eastward along the north bank toward Finale. At 1630, 23 April, the leading elements of the battalion came under long-range machine-gun fire. The advance slowly continued as the infantry cleaned out the houses along the road. With the help of supporting armor from the 757th Tank Battalion, the assault companies made gradual progress against numerous machine-gun emplacements, pillboxes, and German snipers holed up in Finale's houses. Joining numerous small-arms fire fights, in which it was supported by mortar and artillery concentrations in addition to close air support, the 1st Battalion methodically reduced pocket after pocket of resistance. By 2055 the main enemy force in Finale was reported overcome, and mopping-up operations continued. Company B made contact with the 1st Battalion, 362d Infantry, coming up on the right flank, and secured one usable, but damaged, bridge over the Panaro.

On 23 April a third element of the 91st was operating near the right boundary of the sector, north of the Panaro River. Led by the 3d Battalion, two battalions of the 363d Infantry advanced northward and were checked only when sizable enemy forces were encountered. In the vicinity of the crossroads at La Campa, several enemy columns were caught coming westward along the lateral roads in an effort to escape toward the north. After this area had been cleaned out in the early afternoon, the force resumed its advance up the road, which was bordered by a large drainage ditch, Fossa Reggiana. At Fina Vecchia the column came under fire of an enemy self-propelled gun located in

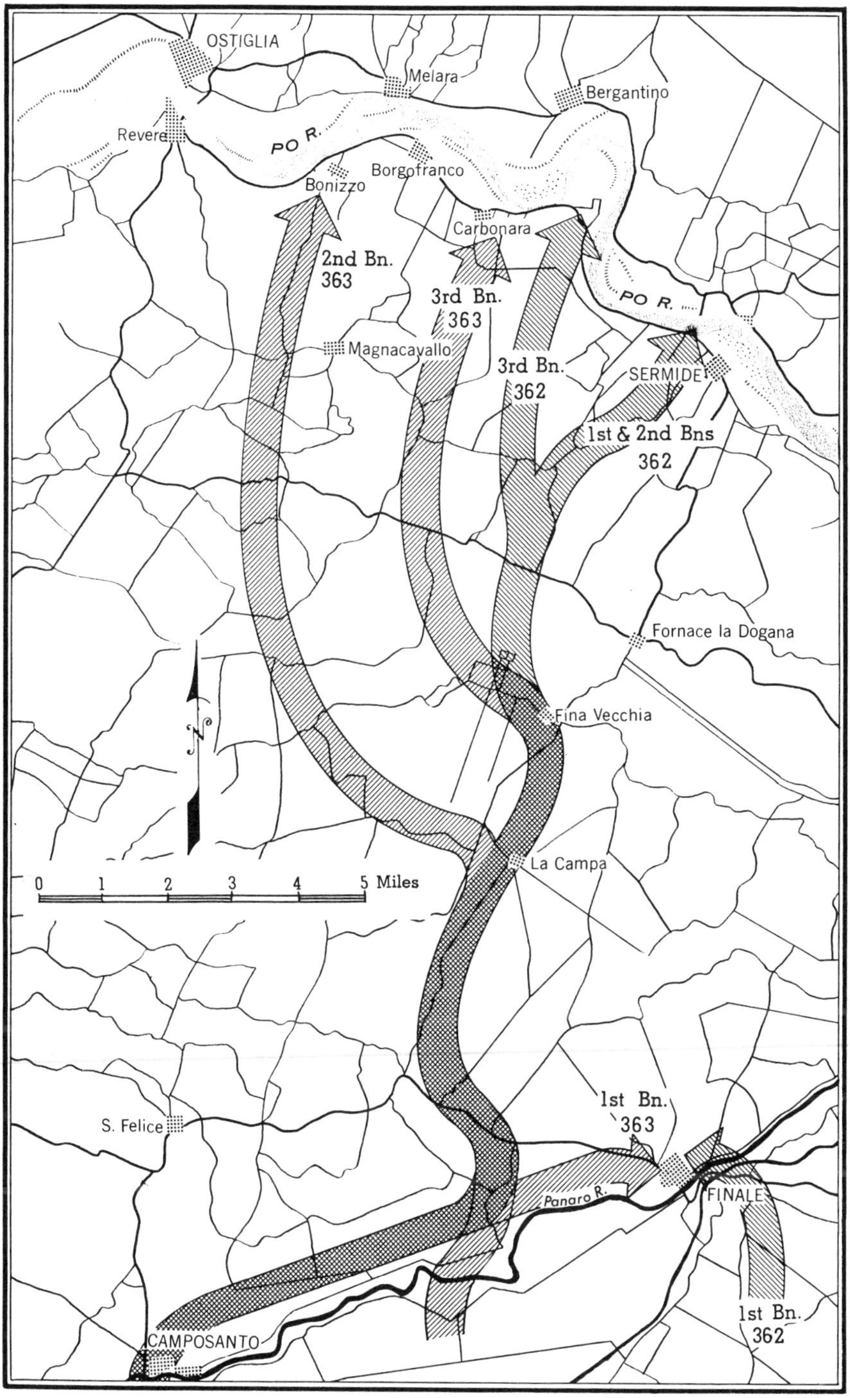
OSTIGLIA
Melara
Bergantino
Revere
PO R.
Bonizzo
Borgofranco
Carbonara
2nd Bn.
363
3rd Bn.
363
PO R.
Magnacavallo
3rd Bn.
362
SERMIDE
1st & 2nd Bns
362
Fornace la Dogana
Fina Vecchia
La Campa
0 1 2 3 4 5 Miles
1st Bn.
363
S. Felice
Panaro R.
FINALE
1st Bn.
362
CAMPOSANTO

Map 26

Staff Sergeant Robert Souse, a Division engineer, levels his transit for the construction of a Bailey bridge

the vicinity of Fornace la Dogana. Resistance was heavy, and the infantry immediately dismounted from the armor and deployed to engage the enemy. Two friendly tanks were destroyed by the fire of the enemy self-propelled gun. Other tanks, however, scored direct hits on two of the enemy's ammunition dumps, setting off ammunition and thousands of multi-colored flares. At least one German self-propelled gun was destroyed, and every building within a radius of two miles was razed.

Late on the night of 23-24 April the enemy disengaged himself and presumably withdrew across the Po, for the next morning the 3d Battalion, 363d Infantry, encountered no resistance. By 0800, 24 April, it reached the Po River at Carbonara. As the armor and infantry approached the town, they met Germans waiting around to surrender. There was no time for the advance elements to pick them up, although they did send 75 back. Hundreds of others were taken by the rear of the column. As the advance elements entered the town itself, more enemy troops streamed out of houses with white flags. When 150 to 200 had been accumulated, the 3d Battalion set up an inclosure. One of the local *paesanos* told riflemen who were rounding up prisoners that there were 75 to 80 more a short distance away who wanted to surrender. Then, seeing how busy the American soldiers were, the Italian brought them in himself. Over 800 Germans were captured by the 3d Battalion alone on the morning of 24 April.

Meanwhile, the 2d Battalion had had its share of fire fights as it

The closer the Division got to the Po River the greater the destruction of German armies and equipment became

advanced behind the 3d Battalion. While the 3d was engaged in its fire fight at Fina Vecchia, the 2d remained at La Campa, 4,000 yards to the rear, until midnight, when seven tanks joined it. With instructions to reach the Po as rapidly as possible, the 2d Battalion swung west to bypass the 3d's fire fight and then proceeded northward. By 0800, 24 April, the hour that the 3d Battalion reached Carbonara, the 2d reached the river at Bonizzo.

Because of heavy traffic passing over the bridge at Camposanto and the fact that its battalions had to move through the sector of the 363d Infantry and part of the sector of the 88th Division and then back again, the 362d Infantry had not cleared the Panaro until late in the afternoon of 23 April. At 2005, 23 April, the 3d Battalion, leading the advance of the 362d, had reached the road junction northwest of La Campa when a long column of the 6th South African Division brought it to a halt. Dismounting, the 3d continued north on foot. By 0800, 24 April, the battalion had reached a point four and a half kilometers north of Fornace la Dogana and was progressing toward the Po. The 2d and 1st Battalions, motorized, passed through the 3d Battalion and, with no opposition, advanced to the river. At 1155, 24 April, the 2d Battalion, 362d Infantry, detrucked at Sermide, and at 1345 the 1st Battalion reached positions south of the river, also in the vicinity of Sermide.

While the 362d and 363d Infantry Regiments had spearheaded the Division's drive to the Po in three days, the reserve regiment, the 361st, moved north by bounds, when transportation was available, to remain in close support of the leading elements. With brief stops at assembly areas west of Cento and in the vicinity of Colombarina, the regiment closed at Magnacavallo, approximately four kilometers south of the Po.

Sealing the Gap

With the seizure of the south bank of the Po, the second phase of the spring offensive of the Po River Valley Campaign was concluded and the third and final phase—crossing the river and sealing the Brenner route, the main enemy exit from Italy—began.

The Po constituted a formidable barrier to the Division's advance, but it had been an equally formidable obstacle to the retreating Germans. The carnage and destruction which lay everywhere along the south bank of the river and lined every road leading to it were tremendous. Allied aircraft had hammered the retreating columns unmercifully, and they had left behind charred and wrecked vehicles and mountains of supplies, guns, and equipment, including such essentials as maps and medical stores. Many abandoned vehicles, however, had been only slightly damaged or had merely run out of gasoline. These were quickly overhauled, and re-marked with the star in the circle, insignia of the Allied forces, and a unit designation, they were pressed into service. Horses, too, roaming at will in the area south of the river, were caught and put to work by enthusiastic infantrymen.

The Division pushed on across the river without halting. All bridge equipment of II Corps was utilized in the zone on the 91st's left, and the Division was ordered to scramble across as best it could, At 2300, 24 April, General Livesay issued Field Order No. 47, which outlined the next phase of the campaign. The objectives of II Corps were to clear the west bank of the Adige River north of Legnago and to seize Verona, destroying the enemy within the zone. The mission of the 91st Division was to continue its attack northward in the right zone of the Corps.

The 362d Infantry was directed to establish the Division bridgehead across the Po and clear the entire bridgehead zone. After the 362d had secured the route north from the river, the 361st was to pass through it. The 362d, however, was to be prepared to advance to Legnago, make contact with the British Eighth Army, and send patrols north along the Adige River to secure crossing sites. The 361st was to attack rapidly north, prepared to exploit crossings over the Adige River. The 363d was to cross the river on order and attack north within its zone. The Division Reconnaissance Troop was ordered to reconnoiter well forward in the direction of Legnago,

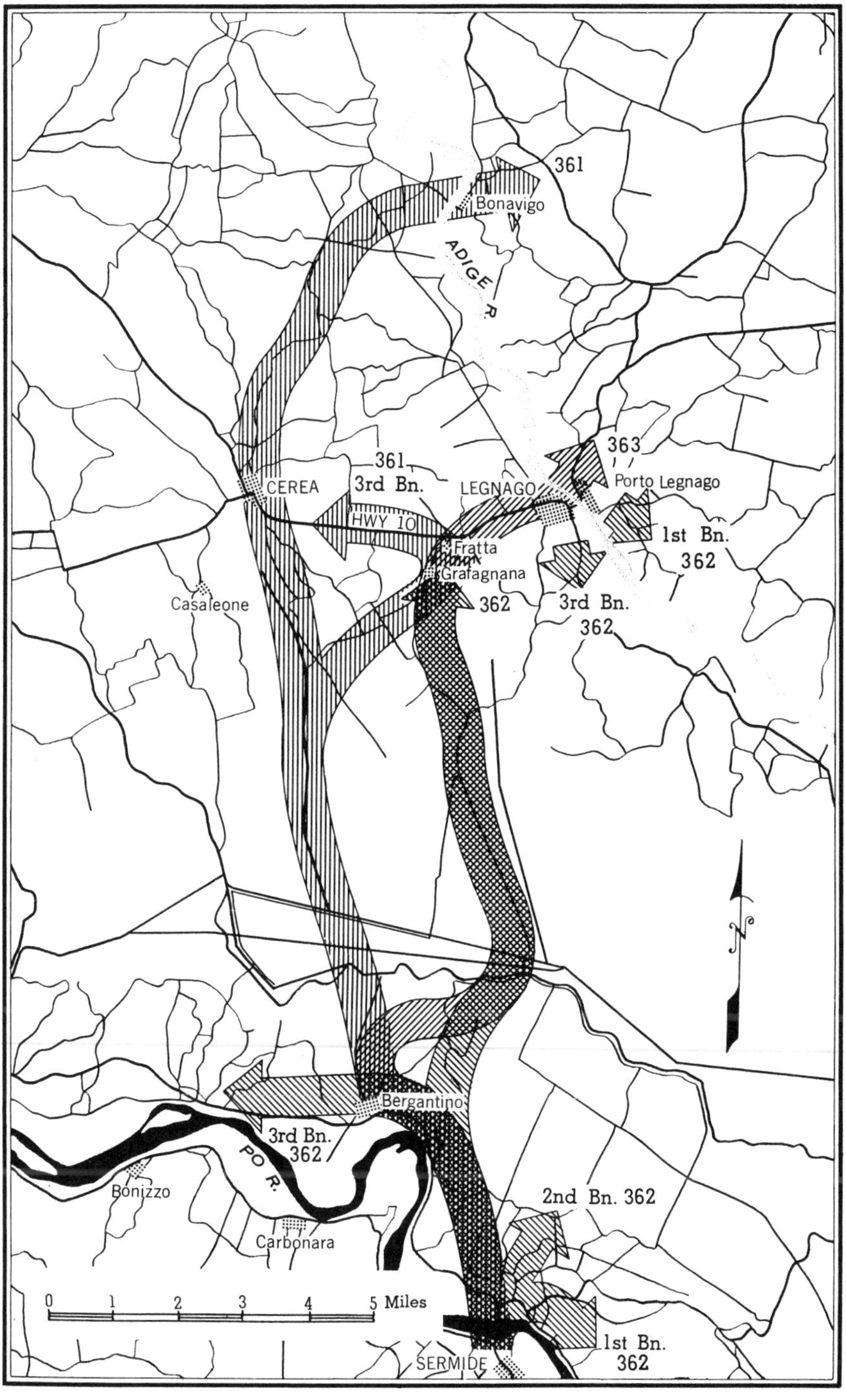
361
Bonavigo
ADIGE R.
363
Porto Legnago
361
3rd Bn.
LEGNAGO
CEREA
HWY 10
Fratta
Grafagnana
1st Bn.
362
362
3rd Bn.
362
Casaleone
Bergantino
3rd Bn.
362
PO R.
Bonizzo
Carbonara
2nd Bn. 362
0 1 2 3 4 5 Miles
1st Bn.
362
SERMIDE

Map 27

General Livesay gives the attack order to Colonel Broedlow while General Williamson, Colonel Donnovin and Colonel Fletcher look on

Cerea, and Bovolone. The importance of rapidly pressing the attack to the Adige River, allowing the enemy no respite or opportunity to reorganize, can readily be understood in view of the fact that the Adige River positions were known to be a main German line of defense.

Reconnaissance of the Po, with the arrival of assault boats, revealed that it would be possible to cross the river much earlier than had been anticipated. Shortly before 2200, 24 April, two companies, B and G of the 362d Infantry, crossed in assault boats near Sermide. Thereafter the infantrymen poured across the river in a steady stream, and the forward elements fanned out rapidly to establish the bridgehead.

A serious threat to the bridgehead was narrowly averted by the heroic action of Private Teddie B. Rexall. With one of the first groups to cross the Po, Private Rexall had pushed rapidly forward to the lateral road running parallel with the north bank of the river. There he saw a German scout car approaching his position and opened fire with his light machine gun, killing the driver and wrecking the vehicle.

A short time later, after his company had infiltrated several hundred yards through the enemy lines, he and one companion, occupying a flank security post at a road junction north of Castelmassa,

"Rexall manned the light machine gun in the road in full view of the approaching Germans."

spotted a force of 50 enemy troops approaching their position. Directing his comrade to take cover in a ditch, he manned the light machine gun in the road in full view of the approaching Germans. Holding his fire until the enemy had reached a point within 75 yards of his position, he opened fire, dispersing the German column. The enemy recovered quickly and opened up with machine-gun, machine-pistol, and rifle fire. Struck twice in the legs, Private Rexall continued to man his weapon until his belt of ammunition was expended and he was forced to return to the cover of the ditch.

Realizing the importance of keeping his weapon in action, Private Rexall, despite his wounds, procured a full box of ammunition and dragged himself back to his gun. As he was crawling forward, an enemy sniper opened fire on him, wounding him twice more in the legs. He was weak from loss of blood, but he crawled the remaining ten yards to his gun and quickly loaded it. His first burst silenced the enemy sniper who had been firing on him, and, with unswerving determination, Private Rexall continued to hold off the enemy force until friendly reinforcements arrived.

The incident ended with all the drama of a Wild West thriller. As his platoon leader reached his side, Private Rexall fired the last

Corps engineers throw a ponton bridge over the Po at Revere

round from his belt of ammunition and sprawled unconscious beside his gun. He had killed four of the enemy and seriously wounded five more. His heroic and determined action, in the face of overwhelming odds, had so demoralized the Germans that they surrendered almost immediately on the arrival of more American troops.

Outside of the enemy action encountered by Private Rexall, only slight opposition was met that day, and by 0330, 25 April, all combat elements of the 362d Infantry were on the north bank of the Po.

With the issuance of Field Order No. 47, the regiment, since there was to be no bridge erected in the sector of the 91st Division, continued to advance steadily on foot. At 1800 the 362d's 1st Battalion was directed to swing its attack westward to seize the bridge across the Adige River at Legnago, and by the morning of 26 April it had reached positions in the vicinity of Fratta, a short distance southwest of Legnago, with the 2d and 3d Battalions moving into assembly areas in close support. So rapid had the 362d's advance been that it had been possible to prevent the demolition of bridges over the numerous canals and waterways which criss-crossed the low country beween the Po and the Adige rivers. Resistance to the advance from the Po to the Adige had been almost non-existent.

The 361st Infantry was the second regiment of the Division to cross the Po. At 0500, 25 April, the 1st Battalion began to cross the river on boats and rafts in the vicinity of Sermide. It used the crossing site which had been used by the 362d Infantry a few hours previously. By 1055 the entire battalion was across and advancing rapidly northward. By mid-afternoon all three battalions of the 361st had crossed the river. Cleaning out occasional pockets of resistance and ferrying the Tartaro River north of Bergantino caused the only pauses in the rush to the regimental objective, Cerea.

At this point, during the night of 25-26 April, began what Colonel Broedlow described as "the night of horrors in the 361st Infantry." Two battalions, the 1st and 2d, were in the vicinity of Cerea, while the 3d, which had swung to the right, was near Grafagnana. After a change in orders from General Livesay, Colonel Broedlow sent the 3d, which was moving on Legnago, northwest on Highway 10, where it would close up to the 1st and 2d at Cerea. This thrust by the 3d caused the enemy to withdraw supply columns and some artillery west on Highway 10 in an attempt to escape north through Cerea. Colonel Broedlow's own version of the subsequent events runs as follows:

> Company A first engaged the enemy supply column, which was proceeding at a very rapid rate of speed without lights. The company fired on two staff cars loaded with officers, destroying the cars, killing several officers, and wounding many others. At the same time the leading trucks were fired upon and destroyed. This engagement at the head of the column caused the rear vehicles to close up and Companies A, E, and G proceeded to chop the column to pieces with bazooka and machine-gun fire. Many of the vehicles were loaded with ammunition and caught fire, causing heavy explosions all through the night, and fires burned fiercely, destroying some hundreds of vehicles loaded with ammunition and other supplies. Tracer ammunition of all colors from several burning dumps on the southwest edge of Cerea kept the sky illuminated until daylight. It was one of the eeriest nights of the campaign.
>
> The enemy troops with the column endeavored to protect it and fought savagely throughout the night. There was no battle line, and as far as I could determine it was practically a free-for-all. The enemy was in front of us, in the rear, and on both flanks. The only difference was that we understood the situation, and the enemy did not. It is believed that the enemy opened fire many times on his own troops. An enemy plane flew over the scene of this fighting several times, but did no bombing or strafing in that area, possibly because of lack of information. At first light of day the battalions quickly reorganized, and I believe I will never again see a regiment in such high spirits after a night of horrors. Our losses were insignificant, and the destruction to the enemy was appalling. Some Germans escaped by using civilians as a screen. We could not fire on them for fear of killing innocent people.

Men of the 362d Infantry cross the Adige River by means of a hastily constructed catwalk on a blown bridge

Corps heavy artillery rolls across the Po River in close support of the Division

After this weird engagement, the advance of the 361st Infantry to the Adige River was taken up once again. During the morning of 26 April the vehicles and attached armor of the regiment, having crossed the Po, caught up with the infantry, and the battalions were again motorized. Prisoners were captured in large numbers as the 361st mopped up the southwest bank of the Adige. Resistance was almost cleared by 1400, and by 1800 the regiment began to cross the river at a ferry site in the vicinity of Bonavigo. Although several times the DUKWs were grounded on the shallow river bed, all the foot troops had been taken across by 0500, 27 April. This was mainly due to the outstanding work of the 316th Engineer Battalion, using small boats equipped with outboard motors. Once again the armor and vehicles were forced to wait for the construction of a bridge, but the foot-slogging doughboys pushed out in a heavy rain toward Zimella. So rapid was their advance that by 0800 they had taken Minerbe, and by noon elements of the 3d Battalion, in the lead, had seized Zimella.

The 363d Infantry was the last regiment to cross the Po and the first to reach the Adige. The 2d Battalion commenced the move at 1300, 25 April, followed by the 1st at 1630 and the 3d at 1800. Once across the river, the troops moved westward to Bergantino to return to their sector. Just north of the town the 1st and 3d Battalions went into assembly areas, while the 2d marched forward toward Legnago. By 0330, 26 April, the 2d contacted elements of the 362d Infantry, made arrangements to pass through them at Fratta, and continued the advance into Legnago. As the battalion approached the city, increasing resistance was encountered, but it was quickly reduced without serious difficulty. The 3d and 1st Battalions meanwhile closed south of Legnago. Along Highway 10, northeast of Fratta, an ambulance convoy of seven vehicles was captured, with wounded, medical personnel, and a nurse.

By noon of 26 April Legnago had been cleared of snipers, and the crossing of the Adige by the 363d was begun. As the first troops of Company G loaded in the DUKWs to start the crossing, enemy machine guns on the far bank raked the levees on the west bank of the river. Despite this, under cover of our small-arms fire, the first two DUKWs took the water. Midway in the stream, the first DUKW became stranded on a sand bar and was immediately taken under fire by an enemy machine gun. Two men were wounded. The occupants jumped into the water and crouched behind the DUKW as enemy bullets riddled the hull. By this time the second DUKW had hit the same sand bar farther down. The timely appearance of friendly tanks on the river bank forced the enemy machine guns from the opposite bank as the third DUKW successfully completed the crossing. Lieutenant General Lucian Truscott, commander of the Fifth Army, General Keyes, and General Livesay were all present to witness the start of this rapid, successful crossing, which was to determine whether the enemy could occupy and defend from the previously prepared positions behind the Adige. As the rest of the regiment rapidly followed across, they fanned out to protect the bridgehead. The 362d Infantry moved up to cross behind the 363d. The Division engineers by mid-afternoon had brought up ferrying equipment from the Po, and the ferrying of small vehicles continued through the night. The Corps engineers brought up heavy bridging equipment and started the vehicle bridge below Legnago. The troops poured across, while the German artillery ranged over the river with air-bursts. Fortunately, the enemy was without observation of the crossing sites and the air-bursts did not come over the troops working at the river edge. In clearing the town of Porto Legnago some pockets of stubborn enemy resistance were encountered by both the 1st and 2d Battalions, 363d Infantry, bringing the first group of serious casualties to the regiment in the Po Valley. The Division forward command post, which had moved just west of Legnago, was heavily shelled toward nightfall, with several casualties resulting. Throughout the night, despite some breakdowns in the overworked ferry motors, supplies and jeeps were pushed across the river. The 362d completed the crossing.

While the 362d remained to cover the bridgehead, the 363d moved

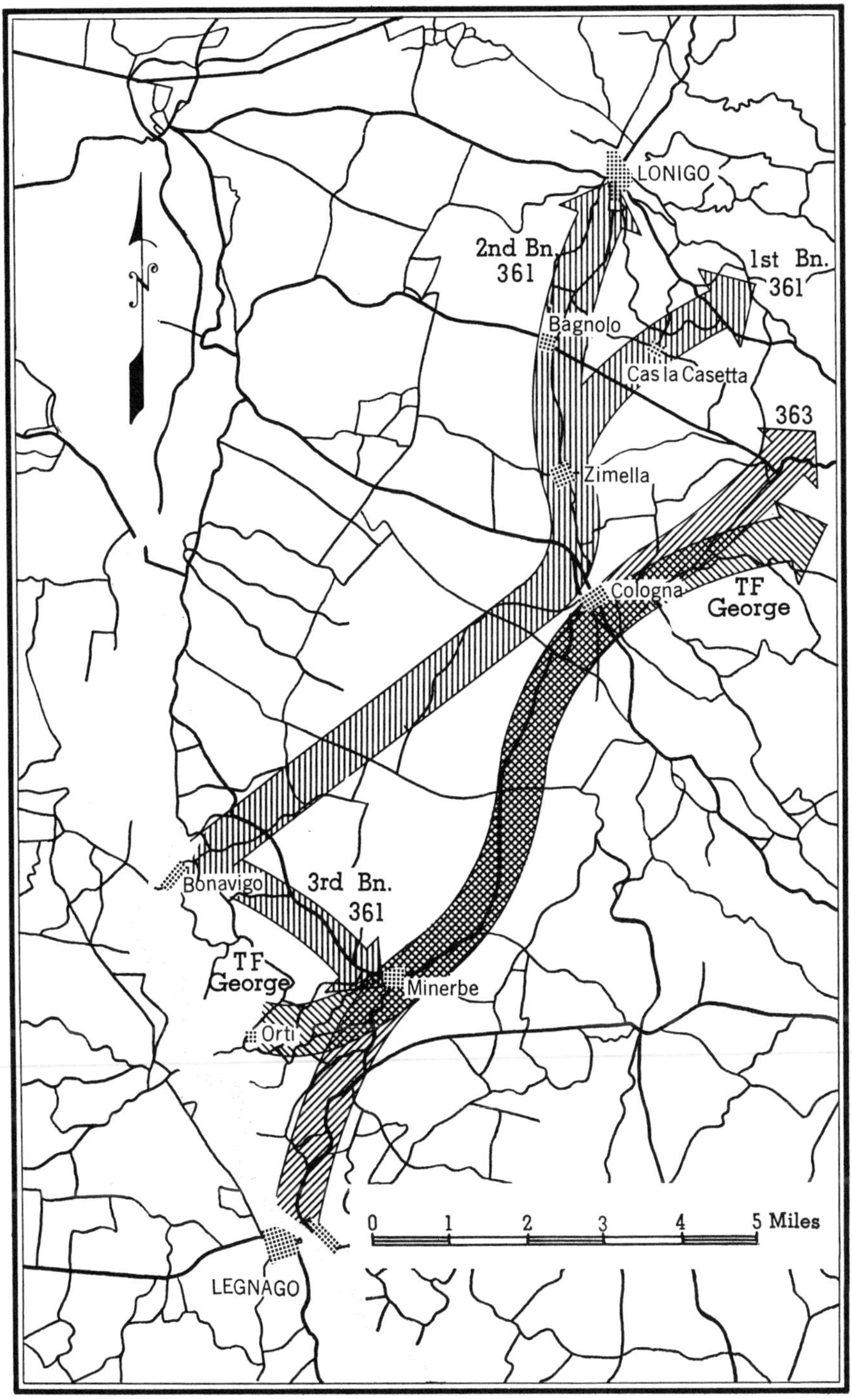

Map 28: Legnago to Lonigo

Mortar men made good use of mules to carry their heavy weapons during the rapid advance

out on foot in a column of battalions at 0845, 27 April, with the 1st Battalion leading. Throughout the day the troops moved through Minerbe and Cologna almost without incident. Dusk found all battalions moving into the town of Sossano, where it was decided to set up an all-around defense. The rapid movement had caught the enemy completely off balance, and he was not aware of the presence of American troops in the town, which proved to be an important supply center for the Germans. The commanding officer of the 3d Battalion had assembled his company commanders on the street to give them the orders for the troop dispositions for the night when three truckloads of Germans drove into their midst. In the free-for-all that followed, with the officers firing from doorways and around corners, the enemy fled, abandoning their vehicles. All of the casualties were on the enemy side.

About the same time in the dusk just west of Sossano, the regimental command party was halted along the road with elements of the 2d Battalion. The radio jeep belonging to Rover Joe was at the tail end. Suddenly a large vehicle ground to a stop just behind it. A 2d Battalion rifleman sitting alongside the road looked up, quickly fired one round, and wounded the driver of the vehicle. In an instant fifteen members of the crew of a tractor-drawn 88 hit the ditch on either side of the road, with their hands in the air. The sergeant in Rover Joe's jeep looked around and remarked, "What a hell of a place for the Air Force to be."

From here on, the night was one of confusion as the enemy proceeded to put in roadblocks along the road in the rear of the battalions. One regimental messenger hit an enemy roadblock and lost his vehicle. An ambulance was halted going to the rear, and the assistant driver was held as a hostage while the ambulance returned to the aid station with the German ultimatum that they would block evacuation unless an agreement was made to evacuate German wounded in our vehicles. In the early morning a mess sergeant was killed while bringing up rations. The trains and troop-carrying vehicles which were now across the Adige were blocked from getting to Sossano.

Came dawn of 28 April, and the answer was supplied through capture of a German officer who had the German field order and overlay issued on the 27th. The 363d had managed to move through the five successive defense lines between the Adige River and Sossano just a few hours before the enemy was ordered to occupy them. The reduction of the roadblocks and enemy resistance west of Sossano was accomplished when the train commander put gunners on all of the vehicle-mounted .50-caliber machine guns and came storming down the road with the guns spraying all likely targets.

The 362d remained in positions protecting the bridge site at Legnago until 1430, 27 April, when Task Force George was organized at Orti. Commanded by Lieutenant Colonel George W. Richardson, the task force was composed of the 3d Battalion, 362d Infantry; eight tanks of the 755th Tank Battalion; twelve tank destroyers of the 804th Tank Destroyer Battalion; the 1st Platoon, 91st Division Reconnaissance Troop; and one detachment, 91st Division Signal Company. The objective of the task force was to seize and hold the city of Vicenza. At 1600 reconnaissance elements were sent out to reconnoiter the road-net east of Cologna. With no more than minor skirmishes, the armored cars of the Reconnaissance Platoon advanced to the outskirts of Coleredo. Here they ran into considerably more opposition, and in the ensuing fire fight the platoon leader was killed, one of the very few casualties suffered by the Division during this final phase of the Po Valley Campaign. One armored car returned to join the task force, which by now had moved out from Orti. No opposition was met by the force as it moved through Cologna, nor any at Coleredo. At Piazza Vecchia an enemy strongpoint was quickly reduced by tank and tank

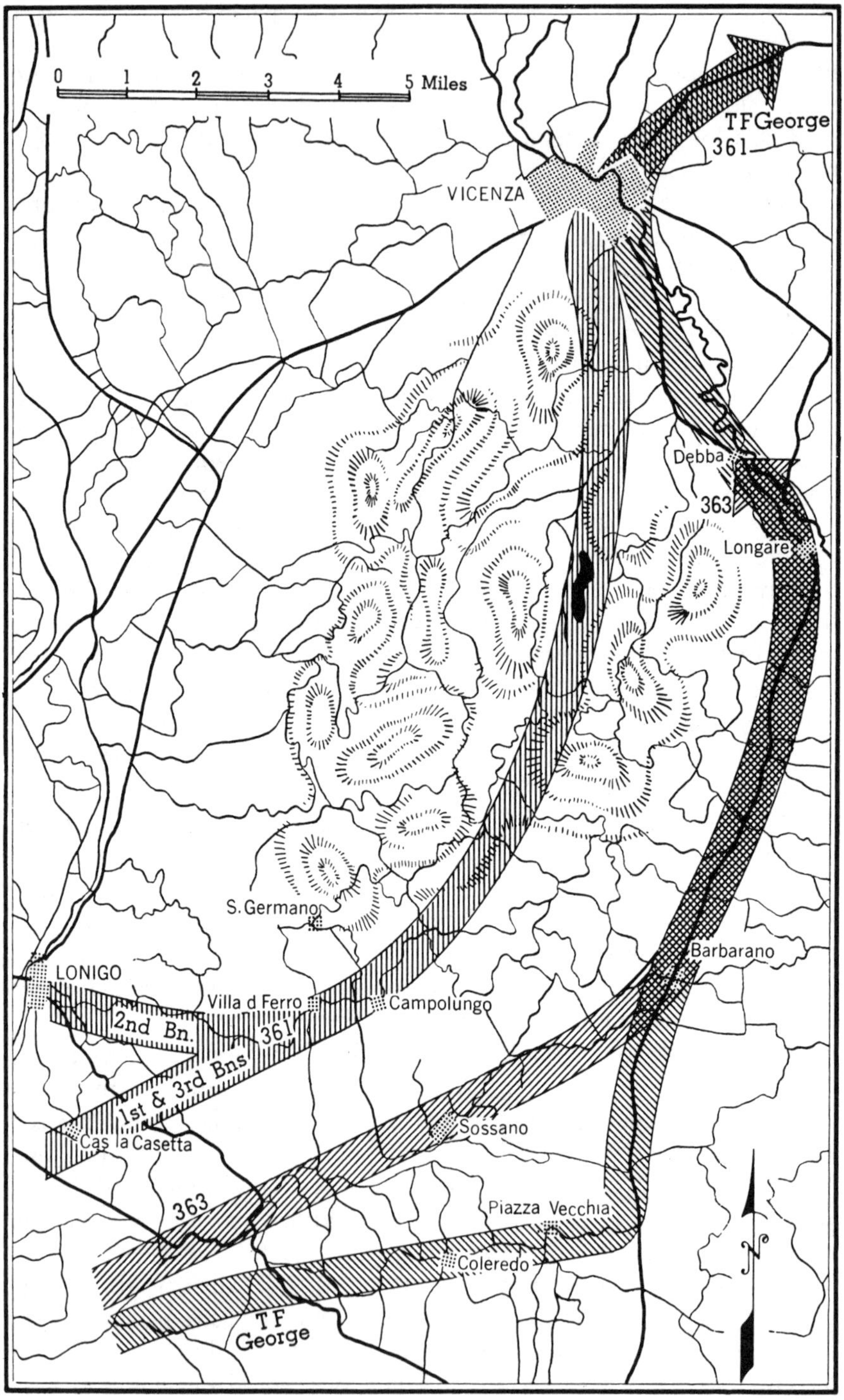

Map 29: Lonigo to Vicenza

destroyer fire, and the column advanced to the road junction two kilometers east of Piazza Vecchia, where it turned north.

Just north of this road junction resistance began to appear more frequently. First a bicyclist was cut down by machine-gun fire; then, a little farther north, seven vehicles and a small force of German infantry were engaged in a more serious skirmish. Several prisoners were taken, and an ammunition dump was destroyed. Five hundred yards south of Barbarano the task force overran an enemy motor pool, which was destroyed by tank and tank-destroyer fire. Each time the column stopped to clean up a pocket of resistance, its rear was fired upon. There were times when there were two separate fights going on within the column at the same time.

It was during one of these fire fights, when the column had been halted by fire from the fields on either side of the highway and from a group of buildings to the front, that General Livesay won the fourth Oak Leaf Cluster to his Silver Star. In the words of the citation:

> General Livesay's presence and calmness at a time when his troops had encountered a surprise attack gave them added confidence. He personally directed tank-destroyer fire on a nearby house containing an unknown number of enemy troops. When the enemy fire diminished he led the column in his own vehicle toward its next objective. His energy and aggressive leadership were responsible for the rapid crossing of the Brenta River and the capture of the Division's objectives.

From Longare to Vicenza no further resistance was encountered, and at 1725, 28 April, the column entered the city. In less than twenty-four hours Task Force George had covered 29 miles, but just before it entered Vicenza it was chided for having been so slow. Men in one of the leading armored cars spotted a lone man coming toward them. At a hundred yards he was identified as British. As he drew abreast of the car, he turned and, in a Cockney accent, said, "Where the 'ell 'ave you been? I've been waiting two bloody years for you blokes." Swift or slow, the task force had captured 200 prisoners and had left 153 wounded Germans along the road because there had been no way to transport them.

In Vicenza the task force contacted the 1st Battalion, 361st Infantry, which had entered the city along a parallel route somewhat north and west of that taken by the task force. The force rested in Vicenza awaiting further orders, which were soon received by liaison messenger from the Division command post at Cologna. They instructed the task force to proceed along Highway 53 to the Brenta River in the vicinity of Fontaniva and, after crossing the river, to fan out and secure the bridgehead.

Accordingly, at 0300, 29 April, Task Force George left Vicenza. Near the little town of Liseria on Highway 53 a fire fight occurred. While the task force was engaged in this fight, a company of enemy on bicycles, followed by a long column of troop-laden carts and trucks, approached the center of the task force from a side road. The entire task force opened fire on the column, and two tanks and one tank destroyer, protected by riflemen and heavy-machine-gun crews, moved down the road to fire into its rear. The entire enemy force was destroyed, including three artillery pieces, four trucks, and numerous carts and wagons. Eighty-seven prisoners were taken.

Three other fire fights occurred as the task force proceeded along Highway 53 toward the Brenta, but these barely slowed the rapid progress of the column. At 1100, 29 April, the first company was crossing the shallow river, which was made exceedingly hazardous by the large number of butterfly bombs[4] lying along the banks and approaches. Forty-five minutes later the town of Fontaniva had been secured. The perimeter defense of the bridgehead was quickly organized, and at 1200, 29 April, Task Force George had completed its mission.

Since leaving Orti at 1400, 27 April, the task force had: travelled 48 miles; destroyed one Tiger tank, 1 self-propelled gun and 69 vehicles; captured 972 prisoners; and killed an estimated 300 to 450 enemy. Most important, however, was the accomplishment of its missions, aiding in the capture of Vicenza and the seizure and consolidation of a brigehead across the Brenta River.

At the same time that Task Force George was making its dash from Adige to the Brenta, elements of the 361st Infantry were moving toward Vicenza by parallel routes. At noon, 27 April, the 3d Battalion had reached Zimella. At this point it was ordered to hold while the 1st Battalion continued the attack northeast in the direction of San Germano and the 2d Battalion attacked north to Lonigo.

The 2d Battalion advanced toward its objective against scattered resistance until 1730, when an enemy machine gun and small arms opened fire on the head of the column from the southern edge of Bagnolo, killing three Company F men. In the midst of the fire fight, the tank destroyers, which had been held up at the Adige, rejoined the battalion. With the added fire power of the tank destroyers' 3-inch rifles, the fire fight was brought to a sudden end, and the battalion continued on to Lonigo. It entered the southern outskirts at

[4]Self-activating antipersonnel bombs, dropped in clusters. Each bomb was about the size of a No. 8 tincan.

2200 and, after a short period of savage house-to-house fighting, cleared the town of resistance and assembled to rest for the night.

The 361st's 1st Battalion advanced with even less opposition. By 2200 it had reached Cas la Casetta, three and a half kilometers south of Lonigo. The 3d Battalion, following the 1st, had reached a point one kilometer south of Cas la Casetta. In these positions the three battalions halted until early the next morning.

While the 2d Battalion cleaned out pockets of resistance north and east of Lonigo, the 1st and 3d Battalions moved out toward Vicenza at 0500, 28 April. Numerous small fire fights slowed the column, but the most serious obstacle was a canal just south of Villa del Ferro, the bridge over which had been blown out by the retreating enemy. The infantry dismounted and continued the advance, while Company A, 316th Engineers, prepared a crossing for the armor and vehicles. By noon the infantrymen had reached Campolungo. All along the route the Powder River doughboys received joyous shouts of welcome from the happy natives and were showered with fruit and flowers. Soon after noon the vehicles and armor caught up with the troops again, and the motorized column headed for Vicenza. Light rains and hairpin curves in the rugged foothills of the sector made driving difficult, especially for the larger vehicles, but by 1645, 28 April, the 1st Battalion rolled into Vicenza. The 3d Battalion closed into the town a short while later, and by 2130 the 2d Battalion, which had left Lonigo at 1115 and joined the tail of the regimental column at Campolungo, had also closed in.

The 363d Infantry, which had reached Sossano the night of 27 April, also advanced to Vicenza on 28 April. Since, by 0450, the regiment's transportation, held up at the Adige, had not arrived, Colonel Magill ordered the regiment to start the move on foot. Ten minutes after the 1st Battalion had crossed the IP (initial point), the vehicles arrived, and, once again motorized, the battalion sped toward Vicenza. At the road junction east of Sossano, where the column was to turn north to Vicenza, the enemy had set up a roadblock, and a short, brisk fire fight ensued. It is characteristic of the fluid action of the last week of April that it was at this very point that Task Force George had had a fire fight twelve hours earlier. On neither occasion had the enemy roadblock halted the advance.

The troops of the 1st Battalion had entrucked again and resumed their progress northward when the rear of the column noted an enemy force approaching about 800 yards to the rear. The trucks were immediately pulled off to the side of the road, and the riflemen and machine gunners deployed in the fields on either side to engage the enemy. One 57mm antitank gun was placed in position alongside of the road so that it could fire down the length of the enemy column, while four heavy machine guns were set up in the field to the east and two on the west. The attached artillery—two tanks and the 81mm mortars—all moved into position to fire.

The battalion held its fire until the Germans, suddenly realizing the troops in front of them were Americans, started putting two 20mm AA guns into firing position. The forward gun was pushed 200 yards nearer the battalion's position and the crew was preparing to fire when Lieutenant Colonel Ralph N. Woods, battalion commander, gave the order for all guns to commence firing.

Two rifle companies, six heavy machine guns, one 57mm antitank gun, six 81mm mortars and twelve 105s opened fire on his command. The enemy's return was light, consisting of a few short bursts from machine guns and a few rounds from one 20mm gun. The 20mm gun which had been pushed up by the enemy never fired a round. It was destroyed, and its three crew members were killed instantly early in the fight.

When this battle had ended 153 prisoners had been taken, and 125 of the enemy had been killed. Three 88mm guns, at least fifteen trucks, three 20mm antiaircraft guns, several motorcycles, and about thirty horses had been destroyed. The battalion had not suffered a single casualty! By 1400 the action had ceased, and the battalion reorganized and once again entrucked for Vicenza. By 2345, 28 April, all three battalions of the 363d Infantry had reached the town of Debba, five kilometers southeast of Vicenza, where they spent the night. That night General Livesay received the following message from General Keyes:

CONGRATULATIONS ON TODAYS FINE WORK AND YOUR CONTINUED GOOD PROGRESS MY PARTICULAR COMPLIMENTS TO THE THREE SIX THREE.

By 2200, 28 April, General Livesay issued what was to be the last field order (No. 48) for combat operations in the Italian Campaign. It outlined the mission of the 91st Division as follows: Secure the crossing site on the Brenta River and then push forward rapidly to capture Bassano, Cornuda, and Treviso. To accomplish this mission General Livesay ordered Task Force George to move without delay to secure Fontaniva and assist the passage of the 361st Infantry through the bridgehead. The 361st was to move out after Task Force George along Highway 53 to secure Cittadella and there to prepare

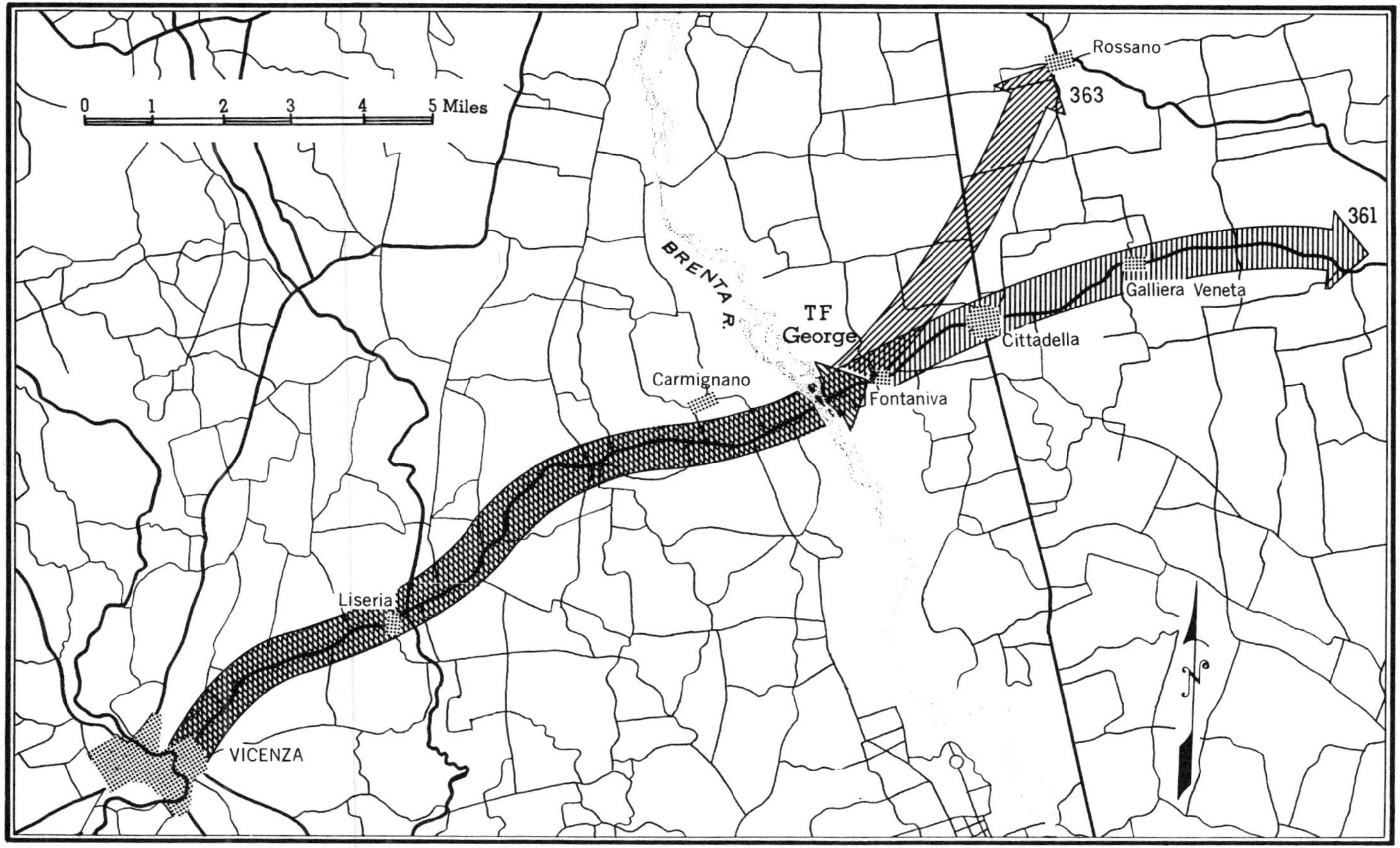
0 1 2 3 4 5 Miles
Rossano
363
361
BRENTA R.
TF George
Carmignano
Fontaniva
Cittadella
Galliera Veneta
Liseria
VICENZA
N

Map 30

to capture Bassano or to continue the attack to the northeast and east. The 363d was to move to the Brenta, and, having crossed it, move east to Treviso. The 362d, minus Task Force George, was to shuttle forward to the Brenta and assemble there, prepared to cross the river to secure Bassano or to attack through the 361st to secure Cornuda.

In accordance with these orders, the 361st's 3d Battalion headed out of Vicenza at 0500, 29 April. After Task Force George had secured the crossing over the Brenta, the 361st quickly crossed the river in DUKWs. Only twenty minutes were required to ferry all of the 1st Battalion troops across, and by 1430 the whole regiment had crossed. Lighter vehicles were then ferried over, while tanks and tank destroyers towed heavier vehicles to the east bank. The large number of unexploded butterfly bombs complicated the crossing. Several DUKWs, a bulldozer, and a jeep were damaged by these highly explosive bombs. Working rapidly and with unhesitating courage, members of the 316th Engineer Battalion cleared the crossing site of the bombs by pulling them with long ropes until they exploded.

Once across, troops of the 3d Battalion, 361st Infantry, advanced rapidly on foot, bypassing numerous pockets of resistance. By 1400 Cittadella had been seized against only light resistance. Shortly thereafter tanks and tank destroyers rejoined the column. Once again the foot-weary doughboys climbed aboard the armor, and, pushing relentlessly on through Castelfranco, the 3d Battalion reached Treviso by dusk of 29 April.

At the western edge of the town, where the head of the column had halted, the regimental commander himself directed the fire fight which ensued when an enemy personnel carrier arrived bringing men to set up a roadblock.

All during the night of 29-30 April fire fights occurred along Highway 53 between the Brenta and Treviso. It became apparent that the 361st Infantry had slashed through the enemy lines. Enemy columns were moving north and south across Highway 53, while elements of the 361st moved east and west along the route. The enemy, hard-pressed by the British Eighth Army to the south and unable to move during the daylight hours because of the overwhelming Allied air power, was making a last desperate effort to extricate himself from the ring of steel that was rapidly sealing off his only escape route. With the coming of darkness on 29 April, this enemy activity intensified. The 2d Battalion, 361st Infantry, following the 3d, became engaged in a fire fight at Galliera Veneta at 1700. When the enemy had been overcome, the battalion discovered that it had captured a large German field hospital intact.

While the 2d Battalion was mopping up Galliera Veneta and was reorganizing to continue the advance, the 361st's 1st Battalion passed through it and continued eastward. Two hundred yards west of Vedelago the head of the 1st Battalion column ran into the rear elements of the 3d, which had been cut off from their main force. This group reported that the Germans had set up a strong roadblock on the western edge of Vedelago and that a sizable motorized enemy column was moving north through the town. The infantrymen dismounted and deployed to engage this force. The two leading tanks in the column moved closer to the edge of town, and the lead tank was damaged by an enemy rocket. Casualties were sustained on both sides, but, by 0230, 30 April, before the enemy force could be enveloped and captured, it had withdrawn. Later it was discovered that it was the remnant of the entire German XIII Corps.

The next morning the 1st and 2d Battalions continued the move eastward along Highway 53 toward Treviso, and the regimental vehicles left the Brenta crossing site. All three groups moving along the highway were engaged by snipers at one time or another, and many Germans were taken prisoner, while many more were killed in running gunfights. By 1200, 30 April, however, the entire 361st Infantry had closed in Treviso, and the troops were patrolling the streets of the city to maintain order. During the forty-eight hour period, 29-30 April, the 361st had taken over 5,000 prisoners and had won its fourth commendation of the month from General Keyes. On 30 April he sent the following message to General Livesay:

CONGRATULATIONS ON THE SPLENDID RACE AND CAPTURE OF TREVISO PLEASE GIVE MY WARMEST CONGRATULATIONS TO THE THREE SIXTY FIRST.

In the meantime, the 363d Infantry, in accordance with Field Order No. 48, moved out from Debba early on the morning of 29 April in a column of battalions, the 2d Battalion leading and the 1st and 3d following in that order. The head of the column passed through Vicenza at 0620, cheered by the citizens who lined the streets, and proceeded toward the Brenta. Highway 53 was clogged with traffic, and the movement was considerably impeded. During the night of 29-30 April the 363d forded the river near a blown bridge four miles above the Division crossing, and at 0600, 30 April, the 1st and 2d Battalions had closed at Rossano. Two hours later the 2d Battalion, motorized, advanced from Rossano toward Riese. By the time the 3d Battalion, the last of the 363d to cross the Brenta, had reached

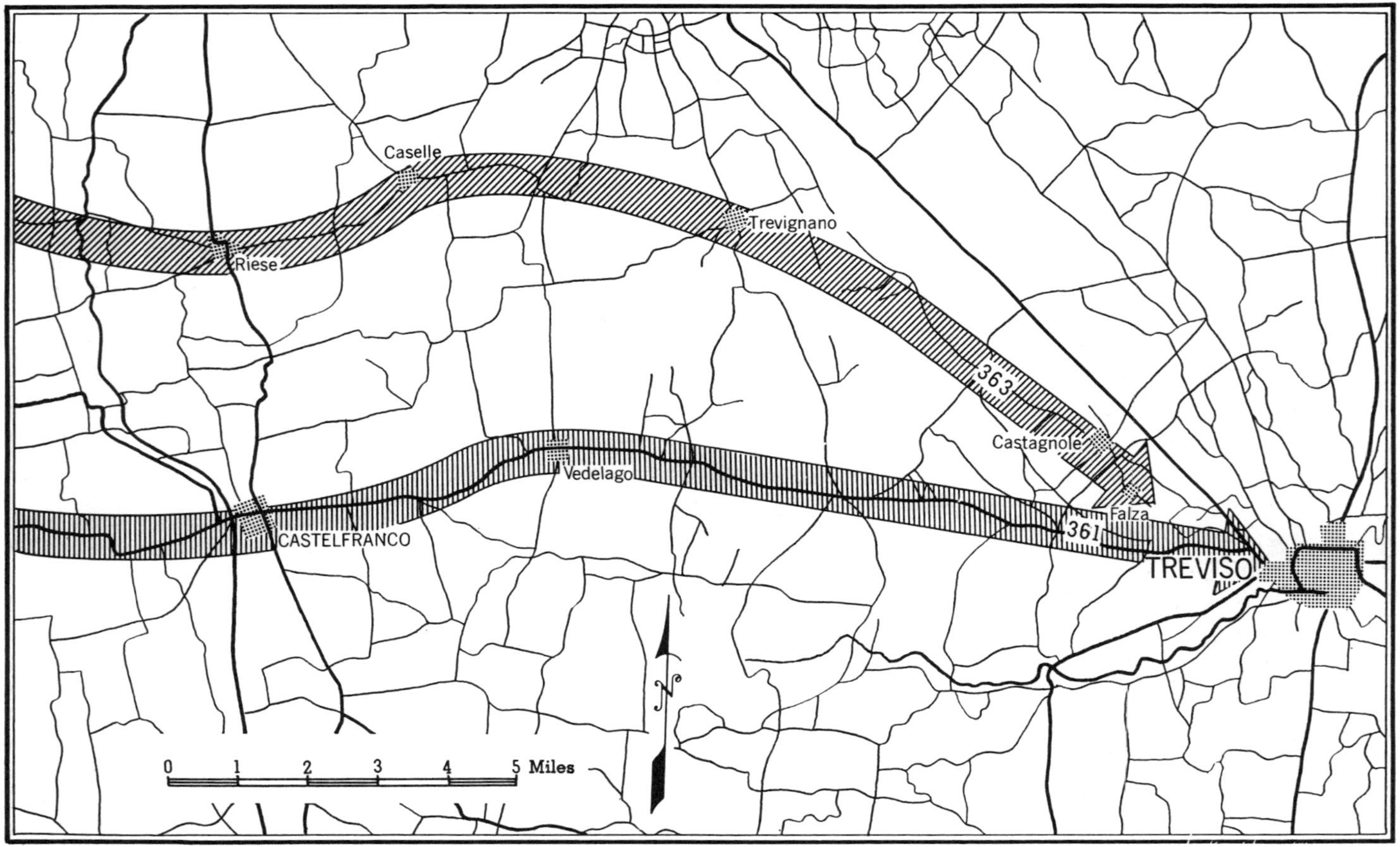
Caselle
Trevignano
Riese
363
Castagnole
Vedelago
Falza
361
CASTELFRANCO
TREVISO
0 1 2 3 4 5 Miles

Map 31

Rossano, the 2d had seized Riese and had captured 100 prisoners while cleaning out the town.

During the rest of the day the 363d's 1st Battalion captured Casella, and the 3d seized Trevignano. In each of these towns there were sharp fire fights, but the enemy was in both cases destroyed, captured, or put to flight. By 0800, 1 May, the 363d Infantry had assembled in areas in the vicinity of Treviso—the 3d Battalion at Falza, and the 1st and 2d Battalions at Castagnole.

After the 361st and the 363d Infantry Regiments had cleared the Brenta, the 362d, in mobile reserve, shuttled forward and went into an assembly position in the vicinity of Castelfranco.

Grouped in the vicinity of Treviso in II Corps reserve, the Division devoted 1 May to rest, reorganization, and the replenishment of its supplies and equipment. Small elements of each regiment and the 91st Reconnaissance Troop continued to mop up the area and round up the totally disorganized enemy forces. It was the 2d Platoon of the 91st Reconnaissance Troop which made the spectacular haul. The platoon was given the mission of cleaning out Montebelluna, and when it arrived, it was welcomed by a group of partisans who said there was a force of Germans who wished to surrender. Much to the surprise of everyone, the force turned out to be a Major General Zwad and his entire staff, together with 1,300 German officers and enlisted men.

Combat for the Powder River Division in the bitter Italian Campaign came to a joyous conclusion when, at 1900, 2 May, General Livesay announced that all German forces in Italy had surrendered unconditionally to Field Marshal Sir Harold L. Alexander, Supreme Commander, Mediterranean Theater of Operations. Satisfaction and pride in the share played by the Division in bringing about this crushing defeat filled every Powder River doughboy. The final and complete victory over Nazi tyranny could not be far off.

But the Division had new and important missions ahead of it, and immediately the energies of the men were bent toward preparing for its next assignment. New clothing was issued to all enlisted men, and every vehicle was inspected, overhauled, repaired if necessary, and washed or repainted.

The complete story of the 91st Division in the spring offensive of the Po River Valley Campaign may never be fully told. Actions and events moved with such startling rapidity as to preclude the recording of more than a tiny portion of the countless stories of individual valor and sacrifice. The cold statistics on the comparatively small number of casualties sustained do not begin to indicate the severity of the action. Almost every man in the Division has had at least one capture story to tell. The 347th Field Artillery Battalion took prisoners in numbers almost exactly equal to its strength—515 in all. Members of the 316th Engineer Battalion found themselves marching columns of prisoners to the rear. Even Division Headquarters men were saddled with prisoners as they moved forward from command post to command post. The total number of prisoners taken by the Division is somewhat confused. The official count of prisoners passing through the Division prisoner-of-war cage is 6,036 for the month of April, but units of the Division report having taken well over 10,000. Untold quantities of material of every sort were captured, and even more was destroyed or left behind. In fifteen days the Powder River Division had advanced more than 135 miles from its line of departure.

From Mt. Adone to Treviso, from the Apennines to the Alps and the Adriatic, the 91st had fought its last Italian campaign. It had conquered its enemies and helped bring about the complete and crushing defeat of the German armies in Italy.

PART THREE

AFTER VE-DAY

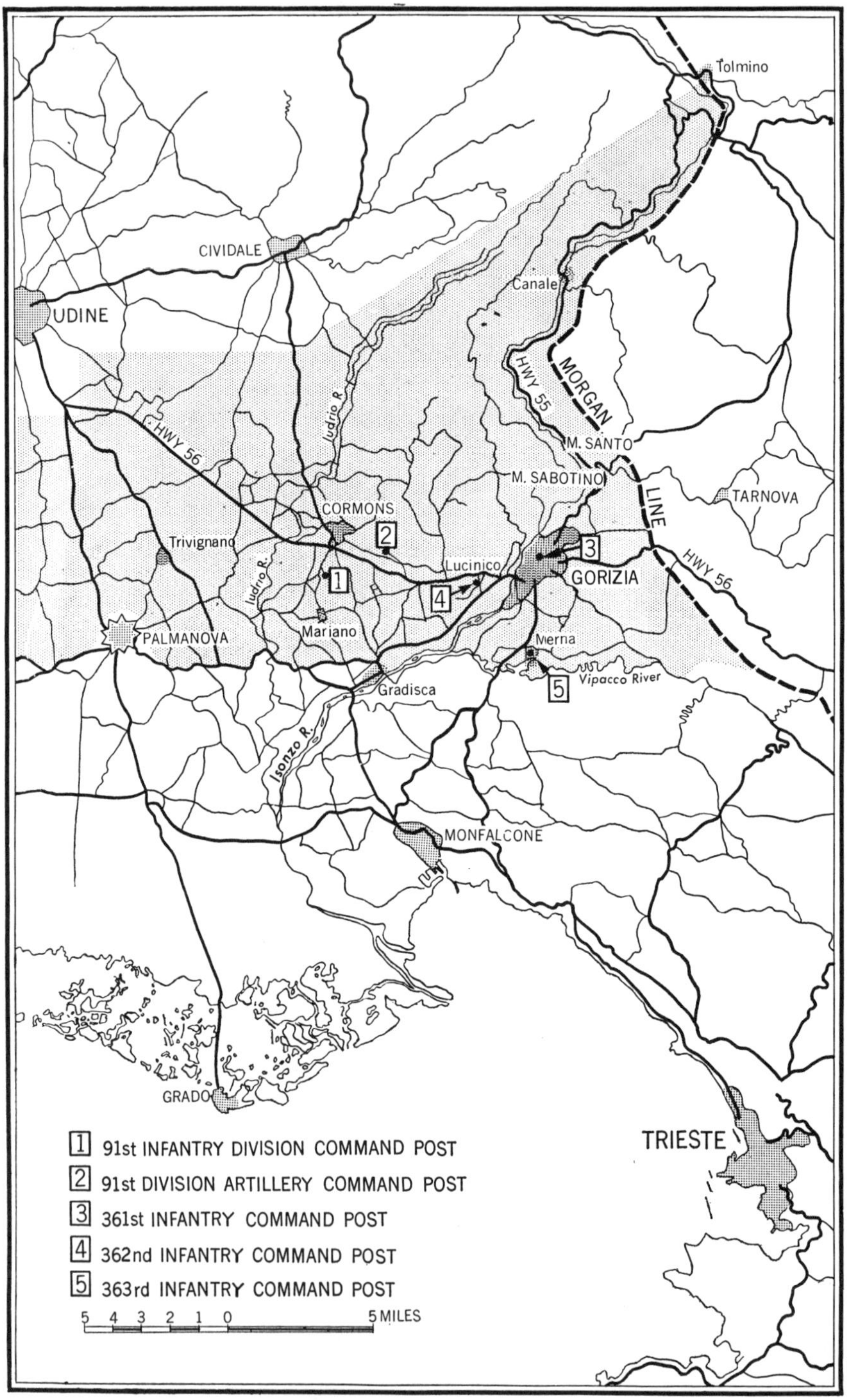

Map 32: Occupation Area

CHAPTER 13

GUARDING VENEZIA–GIULIA

WITH the unconditional surrender of all enemy forces in Italy on 2 May the long, bitter campaign of the Allied forces in the Mediterranean Theater of Operations came to a successful conclusion. At the moment of capitulation the Powder River Division, grouped in II Corps reserve in the vicinity of Treviso, was busily engaged in rounding up German prisoners and equipment. In addition it was bending every effort toward the rehabilitation of its troops and equipment. First- and second-echelon maintenance was quickly performed on all vehicles which needed attention after the swift, unremitting pursuit of the enemy across the Po Valley. Showers were set up for the troops, and each man received a complete new uniform. The Division rear echelon, as well as all the impedimenta, were brought up to Treviso. Special emphasis was placed on bringing the Division up to high garrison standards.

The day peace came to Italy General Livesay announced to the Division that it had been selected by General Truscott for a mission of the utmost importance and delicacy: the occupation of the western part of the province of Venezia Giulia, including Trieste—a region claimed by Marshal Tito for Yugoslavia.

Although in February 1944 Field Marshal Alexander and Marshal Tito had reached an agreement that Venezia Giulia should be administered jointly by the British and American military governments, the area had been conquered by Tito's forces, and a government controlled by him announced its intention of administering it by right of conquest. In support of our view that unilateral action should not be taken in an effort to present the world with a *fait accompli,* the 91st Division in conjunction with the British Eighth Army was ordered to the area to take up positions.

The prize was the excellent port of Trieste, southern gateway to Austria. Gorizia, the other large community in the region, had its chief military significance in the fact that it lay at the crossing of the major east-west highway, Highway 56, and the major north-south highway, Highway 55. The whole region, a flat coastal plain, was important from a military point of view, however, in that it is the traditional military avenue into northern Italy. It had been a battlefield for a long and bloody war of attrition between Italy and Austria during World War I, and as far back as Roman times Gorizia had been a fortress guarding the northeastern gateway to Italy.

To implement the occupation of Venezia Giulia General Livesay

Soldiers of the 1st Battalion, 363d Infantry, being welcomed by civilians in Trieste

issued Field Order No. 49. The Division was detached from II Corps and attached to British 13 Corps for operations only. While negotiations were in progress with the Yugoslav Army for the port of Trieste and the road and rail communications from Trieste to Austria, the Division was ordered to occupy and outpost the Gorizia–Merna–Segrado–Palmanova area. The 1st Battalion Combat Team of the 363d Infantry was attached to the 2d New Zealand Division and became part of the garrison of Trieste. One battalion of the Coldstream Guards was attached to the 361st Infantry in the vicinity of Gorizia.

General Livesay's instructions to the troops indicated the explosive nature of the situation. The following is quoted from the directive of 13 Corps:

> All troops, wherever located, will be tactically disposed, and will be immediately ready to defend themselves if attacked, or for immediate movement.
>
> If fighting between local factions breaks out in any part of the Corps area, the local commander will take steps to isolate the area in which fighting is taking place by roadblocks, preventing the entry into it of any persons bearing arms. On the conclusion of the fighting this headquarters will be notified by the fastest means. Any refusal to comply with such instructions will be reported to this headquarters before further action is taken. The guiding principle in all such matters will be to take all possible action *short of actually opening fire* to preserve law and order in the Corps.

General Williamson welcomes Field Marshal Alexander to the Division area

At 0740, 5 May, the 361st Regimental Combat Team, with colors and guidons flying, moved from Musano to Gorizia and completed the relief of elements of the 167th Brigade, British 65th Division, by 2000. On the arrival of the combat team in Gorizia, the 1st Battalion, Coldstream Guards, was attached to it. Immediately Colonel Broedlow deployed the 361st's 2d Battalion in an area 15 kilometers west of Gorizia to prevent any unfriendly force from penetrating beyond a line running generally north and south along the Iudrio River. The 361st's guards relieved British guards at the bridge across the Isonzo River in Gorizia, and one platoon of the 91st Reconnaissance Troop maintained patrols in Gorizia and vicinity.

The next morning, 6 May, the 1st Battalion Combat Team, 363d Infantry, under the command of Lieutenant Colonel Ralph N. Woods, moved to Trieste and, on arrival, was attached to the 2d New Zealand Division. The rest of the 363d Infantry took up new positions the same day south of Gorizia in the vicinity of Merna and Scariano. Meanwhile, the 362d Infantry remained in the vicinity of Treviso, carrying on light training. The Division Command Post was established near the town of Cormons. On 6 May Colonel Cotton, acting for the Division Commander, accepted the arms of 12 partisan brigades of a hundred men each. At the formal ceremony held in Treviso, Colonel Cotton commended the partisans for their accomplishments in helping to defeat the German armies in Italy. After his speech and one by the

In Trieste American and New Zealand soldiers discuss things over a cup of tea

Machine gunners of the 361st Infantry guard the bridge across the Isonzo River at Gorizia

head of the Italian Committee of Liberation, the brigades passed in review.

On 8 May the 362d Infantry moved to the area lying east of Palmanova, with the regimental command post established at Trivignano. By 10 May the Division was in place.

On 9 May, in Annex No. 2 to Field Order No. 49, more complete instructions concerning the nature of the Division's duties were issued:

> Negotiations are in progress to determine the boundaries and spheres of influence between Allied troops and Yugoslav troops. Until such time as a definite decision is handed down, it is necessary that higher headquarters be kept informed of activities of Yugoslav troops and partisan bands within the Division sector. Information must be collected unobtrusively and friendly relations maintained with Yugoslav troops and Italian partisans.

To gain this information each unit set up a system of observation posts within its area of responsibility. Bridge guards and traffic control posts, together with roving motorized patrols, observed concentrations or movements of Yugoslav and Italian partisans. One platoon of the 91st Reconnaissance Troop established an observation post on Mt. Sabotino, from which observation of the general area was possible. Primarily these observers sought to learn the location, strength, equipment, and identification of the Yugoslav troops and partisan bands; the movements of these groups; and the location and size of their supply stores. All these various sources, gathering information unobtrusively, reported to the Division G-2 Section, which consolidated the reports and forwarded the information to 13 Corps.

Extensive as this system of observation was, it occupied only a small portion of the personnel of the Division, so, also on 9 May, Training Memorandum No. 6 was issued, prescribing a training program for the Division. Military training occupied the morning hours, with special emphasis on the repair of equipment and light battle drill. A minimum of one battalion parade a week and two regimental reviews a month was scheduled. The principal aim of this training was to raise the conduct and appearance of the troops of the Division to the highest possible standards. It was pointed out that every member of the Division represented the United States of America, and every effort had to be made to impress the local populace. The afternoon hours were devoted to organized athletics to maintain the physical health and vigor of the men.

Until 20 May this was the pattern of activity for the Powder River Division. Check points and bridge guard posts were manned, and motorized patrols quietly went about their business of gathering information and keeping a close check on the local situation. Dances were held in the evenings, and both officers' and enlisted men's clubs were organized. Despite the fact that Trieste was the chief prize of the disputed area, life for the battalion combat team garrisoned there was devoid of incidents and very pleasant. However, the situation remained extremely delicate, and "eternal vigilance" was the watchword of the day.

A noteworthy incident occurred the night of 13 May, when, without previous notice, Yugoslav forces established roadblocks around Gorizia and refused to permit British or American vehicles to pass. Colonel Broedlow immediately ordered a general alert for the regiment. After a conference with Yugoslav authorities, it appeared that the Yugoslav forces were rounding up Italian Fascists, and by 2230 the general alert terminated, although a particularly strong guard was maintained in the city.

On 19 May Field Marshal Alexander announced that his negotiations with Marshal Tito had broken down:

> In this situation I tried my best to come to a friendly agreement with Marshal Tito, but did not succeed. The United States and British governments have therefore taken the matter up directly with Marshal Tito. The Soviet Government has been kept fully informed. We are now waiting to hear whether Marshal Tito is prepared to cooperate in accepting a peaceful settlement of his territorial claims or whether he will attempt to establish them by force.

The immediate result of this turn of events was a shift in the Division positions and the arrival in the area of more American troops. On 20 May the battalion combat team in Trieste was detached from the 2d New Zealand Division and returned to control of the 363d Infantry, closing into an area southeast of Gorizia at 1320, 20 May. By noon of 21 May the 1st Battalion, Coldstream Guards, having been detached from the 361st Infantry, had cleared Gorizia. On this date also elements of the 362d Infantry relieved elements of the 361st on Mt. Sabotino and at guard posts along the west banks of the Isonzo River in the vicinity of Gorizia. According to plan, the 2d Battalion, 362d Infantry, took over responsibility for guarding all bridge, road, ammunition, and supply-point positions in the zone of the Division, west of the Isonzo River. At 1200, 21 May the Division came under control of II Corps again, and simultaneously the 894th Tank Destroyer Battalion and the 757th Tank Battalion, minus Company D, were attached to it.

At 2000, 20 May, Field Order No. 52 was issued, recapitulating previous verbal orders for certain moves and ordering further deployment of the Division. Elements of the 361st Infantry were directed to move eastward in the Division zone and set up roadblocks at points leading into Tarnova from the east. The southern half of the Division sector was taken over by the 363d Infantry. The 91st Reconnaissance Troop was given the mission of securing Mt. Santo and establishing an observation post there. The units were cautioned to make these eastward movements in a peaceful manner. All individuals and units were warned not to fire on any Yugoslav or partisan troops unless first fired upon.

The tenseness of the situation may be judged from the fact that, in addition to the ground observation posts, from 20 May until the end of the month the Division Artillery air observation post flew observation patrols every two hours during daylight. All artillery batteries were laid and ready to fire in support of the infantrymen, should they be attacked in force. All members of the Division were ordered to carry individual weapons whenever they left their company or battery areas. Lastly, radio communication systems were set up for use in the event that ground wires were cut.

At first, it appeared that there might be difficulty as a result of this further deployment of the Division. At mid-afternoon of 23 May Colonel Grobar, Commanding Officer, Yugoslav 13th Division, and an officer of the Yugoslav Fourth Army Staff, arrived at the command post of the 361st Infantry to protest the presence of American troops east of the 80th Easting in the vicinity of Tarnova. He stated that unless the 361st's 1st Battalion withdrew from the area, the orders from the Commanding General, Yugoslav Fourth Army, were to eject it by force.

After ordering the 1st Battalion to maintain the highest state of alert, Colonel Broedlow notified Division Headquarters of the Yugo-

Easter church services, both Protestant and Catholic, were well attended

slav protest. General Williamson, Assistant Division Commander, met with the Yugoslav officers at the 361st Infantry command post. He informed Colonel Grobar that our troops had no offensive intentions and that they had been instructed to commit no hostile act which might bring about an armed conflict. General Williamson also informed the Yugoslav commander that the Division Commander had received orders not to withdraw.

The results of this conference were reported to the Division Commander, who in turn informed the II Corps commander. The latter immediately sent a letter to the Yugoslav 13th Division commander, informing him that the positions our troops were occupying would be held and reinforced as might be necessary to accomplish this purpose. Any conflict, he wrote, could result only from Yugoslav action, for his men had been instructed to commit no aggressive act unless fired upon.

Meanwhile, as this communication was being relayed, it was decided on closer survey that the two roadblocks east of Tarnova were tactically of little value, and they were moved to the high ground north and south of Tarnova. A new roadblock was also established on Highway 55, four and a half kilometers north of Gorizia. The strength of the old one west of Tarnova remained the same.

By 2200 the shifts in the roadblocks were accomplished, and the 361st Infantry remained alerted throughout the night, as over 1500 Yugoslav soldiers and some equipment continued to move west toward Tarnova. No unusual or alarming incidents occurred, however.

The next morning, 24 May, at 1100 a conference was held at the Division command post by General Keyes, General Livesay, Colonel Grobar, General Williamson, Colonel Donnovin, and other staff officers. At the meeting, which was marked by cordiality throughout, a satisfactory agreement was reached concerning the disputed area.

The same afternoon, at 1645, Colonel Grobar visited the commanding officer, 361st Infantry, in Gorizia and stated that his forces would commit no action to disturb the present status of all units concerned. At the same time he requested that our troops not be moved past the present positions and locations in the Tarnova area.

Thus the incident, which might easily have developed into more serious proportions, had been satisfactorily eased by conference. On the night of 24 May another brief period of tension occurred when Yugoslav troops began firing flares and machine-gun and small-arms ammunition with notable abandon. This, however, proved to be nothing more than a joyous celebration on the eve of Marshal Tito's birthday. The next morning Colonel Broedlow attended a review and celebration by the Yugoslav forces in Gorizia. And in the evening General Livesay, acting Commanding General, II Corps, General Williamson, acting Commanding General, 91st Division, and members of their staffs attended a banquet in Trieste given by the Yugoslav Army in honor of the day. On both occasions good will was exhibited by all parties concerned.

Minor events occurred almost daily, constantly reminding the men of the Division of the potentialities of the situation. No crises of prolonged or major importance arose however. Any accident involving the several nationalities in the area was promptly reported, as well

General Livesay and General Clark going over terrain features of the Trieste area

as the minor frictions which inevitably arose from the deep-rooted feelings of the Yugoslav and Italian military and civilians. The troublesome problem of Trieste was still smouldering, threatening to burst into flame again at the first opportunity.

After the original negotiations between Marshal Tito and Field Marshal Alexander had collapsed and the problem was placed in diplomatic channels, an agreement was finally reached and signed by the the Yugoslav Foreign Minister and the United States Ambassador in Belgrade on 9 June. This agreement was followed by another on 20 June between the Supreme Allied Commander, Mediterranean Theater of Operations, and the Supreme Commander of the Yugoslav Army, setting forth the areas of occupation for the Allies and the Yugoslavs and the regulations governing the actions of both parties. The boundary agreed on between the zones of occupation ran approximately three kilometers east of the Isonzo River and roughly followed the contour of the river. It was named the Morgan Line after Lieutenant General Sir Frederick Morgan, Chief of Staff of the Mediterranean Theater. The southern boundary of the sector was formed by the Vipacco and Isonzo rivers, while the northern boundary was a line running roughly three kilometers northeast of the Iudrio River to a point south of the city of Cividale and thence straight east.

This boundary, the long awaited result of negotiations between the Yugoslavs and Allies, clearly defined the zones of occupation between the two powers.

Within a short time after the announcement of this agreement, all Yugoslav forces west of the Morgan Line evacuated the area in a peaceful and orderly manner. A detachment of 2,000 Yugoslavs came by agreement under the command of British 13 Corps.

As time went on, the tension between Allied and Yugoslav troops eased, and as men from the two groups began to understand each other better, they became more friendly, attending the same dances and participating in sports contests.

On 4 June a report from Division G-2 sources indicated that hostile activity of the Yugoslavs on a rather large scale might be expected on 5 June. All units were immediately alerted. Guards were redoubled, mobile reserves were planned, and local security precautions were strengthened. However, when nothing happened, the alert status was lifted on 7 June, and troops resumed normal activities.

In another incident on 15 June several tanks from Company A, 752d Tank Battalion, attached to the 363d Infantry, moved slightly into the Yugoslav area so that they would be in the shade. A Yugoslav officer mistook this as a hostile move and immediately had his men surround the tanks, giving the tank commander fifteen minutes to get out. The incident was quickly settled, however, after the commanding officers had talked over the situation.

Again on 19 June several hundred unarmed civilians, mostly women and children, began to converge on Gorizia along the main highways after overrunning several of our roadblocks. Company A of the 361st Infantry was alerted and sent north to block a group moving south on Highway 55. Another large group was blocked on the eastern outskirts of Gorizia, near the 361st Infantry's Service Company, by trucks parked across the road. The purpose of these groups was to stage in the city square a parade and demonstration, strictly forbidden by Allied Military Government, in protest against two Fascists, who they believed were employed by the AMG in Gorizia. General Williamson and Colonel Broedlow held a conference with the leaders of the group, while the order forbidding the demonstration was enforced. The noise and appearance of our tanks, the use of yellow smoke grenades, and the presence of adequate guards were successful in breaking up the demonstrations. A few of the agitators were arrested. By early evening most of the groups had drifted out of the city, and peace and quiet were restored.

During those periods when they were not on alert, all units followed a planned training schedule, with military training in the morning

During the period of occupation, recreation included athletic competition among units of the Division. Shown here is one of the heats in the Division track meet.

and recreation during the afternoon. According to Training Memorandum No. 8, dated 19 June, the Division was placed in Category II for redeployment to the Pacific Theater of Operations. The training to be stressed during the redeployment period covered the following subjects: (1) Japanese tactics and techniques and Japanese matériel; (2) first aid and sanitation; (3) physical training and athletics; (4) chemical warfare; (5) maps and aerial photography; (6) military courtesy and discipline; (7) care and maintenance of arms, clothing, and equipment; (8) interior guard duty and local security; (9) unit training; (10) orientation, and (11) scouting and patrolling. The lessons that the men had learned in combat also played a big part in this training.

At all times adequate personnel remained on operations missions, fatigue, and other duty. Roadblocks were maintained, and all bridges, civilian installations, and military areas were strongly guarded. Roving patrols covered all roads in the sector and maintained contact between detachments. The War Department film, "On To Tokyo," was shown to the troops, reminding them that World War II was still in progress on the other side of the world. The afternoons and evenings were spent in organized athletics and recreation and Information and Education programs. The list of sports included, among others, softball, volley ball, tennis, track, boxing, and swimming.

The Division track and field meet was held 8-9 June with over a hundred men participating. The 363d Infantry track team won. Men placing first and second in each event were sent to Milan to represent the Division in the Fifth Army meet, which was held on 16 and 17 June. At Milan, two men, Corporal Luke M. Higgins, Company G,

General Livesay presents the track and field trophy to the 363d Infantry team, represented by its captain, Lieutenant Ross A. Notaro, Company A.

362d Infantry, and Private First Class George C. Willis, Company C, 362d Infantry, won first and third place respectively in the shot-put and 200-meter low hurdles, thus scoring the only points made by the Division in the meet.

In softball, company-level teams competed in unit leagues, and the championship teams from each of the seven units competed in a Division company-level, round-robin tournament which was held from 14 to 26 June. Antitank Company, 361st Infantry, won the championship. In the battalion All-Star Softball Tournament held in July, the 2d Battalion, 362d Infantry, emerged as victor and represented the Division in the Fifth Army tournament.

Volleyball teams were also very active down to company level. Tournaments were held in all the units, and championship teams of those units competed in a Division round-robin tournament, which began on 25 June and was completed on 7 July. The Medical Detachment of the 362d Infantry carried away the high honors in this sport. In addition, the Division staged a tennis tournament on the Grado tennis courts with 35 men participating.

Boxing shows put on by the 361st and 362d Infantry Regiments, the 316th Engineers, and Division Artillery drew record crowds. Primo Carnera, former world's heavyweight champion, gave several exhibition matches at these shows.

Col. George Fletcher gives instructions to Primo Carnera and Staff Sergeant Homer Blevins before an exhibition bout at the Gorizia stadium

Several USO shows, in addition to the regular movies, played for the Division—the Frank Sinatra Show, Duffy's Tavern, the Andrews Sisters, and "Pardon Me" were presented at eleven performances with a record attendance of over thirty thousand men. Soldier shows, produced with amateur talent in the battalions and regiments, drew large audiences. When possible, Italian performers were included in the soldier shows to add local color and a cosmopolitan air. The showing of regular movies was greatly improved during July by the addition of five new projectors and electric generators. Individual unit bands and orchestras were bolstered by the purchase of $4,000 worth of instruments through SSO funds. Parties and dances sponsored by officers' and enlisted men's clubs starred these GI orchestras with huge success and provided popular night spots for entertainment and relaxation.

USO shows. (1) The Andrews Sisters. (2) Archie of Duffy's Tavern. (3) "The Body"—Jinx Falkenberg

Red Cross clubs set up throughout the Division area rounded out the full off-duty program. The cheerful manner in which the Red Cross girls carried on their work gave a big boost to the morale of the troops. The girls themselves took pride in being assigned to the Powder River Division and wore their fir-tree insignia like veterans.

A generous quota of passes was issued to places such as Trieste,

Red Cross girls were always a welcome sight

Venice, Grado, the Italian Riviera, Alassio, and Austria. Visits to Fifth Army rest centers, many of which were internationally famous resorts, were a welcome relief from occupation duties. Once in the rest centers, the soldier was practically on his own—military regimentation was forgotten. It can be summed up in a few words: a good time was had by all. The schedule of passes was somewhat curtailed during July due to the lack of transportation after a large number of trucks had to be turned in. However, this shortage was alleviated to some extent when ten captured buses were requisitioned.

Unit schools were set up within the Division for men who wished to gain useful knowledge during their spare time. The Division's education program reached its climax during July with 180 organized classes meeting. Orientation and discussion reached a total of 1,107 meetings. In addition, a quota of men was sent to the University Training Center at Florence.

A forward step toward peacetime living for the Italians was taken on 24 June when the voluntary disarmament of all partisan bands and irregular police, requested by the AMG, took place. In Gorizia the formal disarming ceremony was held at 1400. After a brief explanation and congratulation by the 361st S-2, all arms were laid down. As a result of the splendid cooperation of all concerned, the request was carried out without incident. Company C, 361st Infantry, took over from partisans and Yugoslavs the guarding of foreign ammunition dumps and city installations. The latter included the communication systems, water works, city police building, municipal building, jail, railroad stations, and the like. Similar ceremonies were held in

Lt. Col. Ralph N. Woods

other cities and villages throughout the Division zone of occupation.

Two days later Operations Instruction No. 4 came out, establishing police zones to be patrolled by military personnel until such time as the AMG could set up a police force composed of Italian civilians.

Throughout the occupation period, many changes in command and personnel took place. The II Corps, under which the Powder River Division had served for so many months in Italy, received a new assignment with the occupation forces in Austria. The official order, dated 26 June, detached the 91st Division from II Corps command and attached it again to British 13 Corps. The loss of II Corps was felt by members of the Division like the loss of an old and trusted friend.

The redeployment plan, as set forth by the training film, "Two Down and One to Go," caused an appreciable change in personnel. Colonel Joseph P. Donnovin, Chief of Staff, departed on 25 June for the States on rotation. His duties were taken over by Colonel Woods.

The redeployment plan, or "point system," was developed by the War Department after very careful consideration and was based on the desires of soldiers as obtained from interviews of a representative

Reviewing the 363d Infantry at the presentation of the Medal of Honor to Sgt. Oscar G. Johnson of Company B, are: the Army group commander, General Clark, who presented the medal; Sergeant Johnson; Colonel Magill; General Livesay; General Keyes.

High-ranking officers visit the Division CP. Left to right: General Livesay, Lt. Gen. Sir Richard L. McCreery, General Williamson, General Hays, General McNarney.

cross-section of Army personnel. A scoring system was devised, giving men points for length of service in the Army, length of service overseas, decorations, and number of dependent children up to three. The critical score, announced by the War Department shortly after VE-day, was set at 85. Men with that many or more points were eligible for discharge. As a result of the point system, more than 500 enlisted men and officers left the Division for the States on 9 June after an appropriate ceremony at which General Livesay bade them farewell. Further changes in personnel took place when approximately 500 more high-point men of the Division were exchanged for an equal number of low-point men from the 85th and 88th Divisions.

The replacements, welcomed at a ceremony by General Livesay on 2 July, were assigned to vacancies in units and given training in their specific duties.

Division G-3 lost a very capable officer when Colonel George E. Fletcher departed on 5 July for a new assignment with Headquarters, II Corps.

The occupation of the disputed Venezia Giulia Province was highlighted by visits of high-ranking military officials, impressive ceremonies, and the celebration of important anniversaries. On 1 June Lieutenant General Richard L. McCreery, British Eighth Army commander, accompanied by General Livesay, then acting II Corps commander, and General Williamson, made an inspection tour of the troops of the Division. On the following day Field Marshal Alex-

Col. George Fletcher congratulates Sgt. Oscar G. Johnson after the latter was awarded the Medal of Honor

ander and his staff, accompanied by General Livesay, inspected the Trieste–Gorizia area. On 11 June General Joseph T. McNarney, Deputy Supreme Allied Commander, Mediterranean Theater of Operations, with General Livesay and General Williamson, reviewed troops in the Gorizia area. The Division was honored again on 23 July by a return visit of General McCreery, who was accompanied by Lieutenant General Sir John Harding, British 13 Corps commander.

The highest honor yet awarded to any man in the Powder River Division, the Medal of Honor, was presented to Sergeant Oscar G.

Johnson, Company B, 363d Infantry, by General Clark in a ceremony held on 25 June at the Gorizia airport. Sergeant Johnson received the medal for his heroic actions during the first phase of the Gothic Line campaign, when almost single-handedly he protected the left flank of his battalion in the fight for Monticelli Ridge in September 1944. General Clark's party, consisting of General Harding, General Keyes, General Livesay, and General Williamson, inspected and reviewed troops at the ceremony.

One of the most colorful and impressive military formations of the period occurred on 28 July, when a Division review was held at the Gorizia airport, with General Truscott as reviewing officer. The 3d Battalion, 363d Infantry, and the 3d Battalion, 361st Infantry, received the Distinguished Unit Citation for their achievements in battle—the 363d's 3d Battalion for its heroic fight for Monticelli and control of the Il Giogo Pass in the first phase of the Gothic Line campaign, and the 361st's 3d Battalion for its outstanding record on the Livergnano escarpment. At the same ceremony the Distinguished Service Cross was awarded to Technical Sergeant Johnny D. Lake, of Company F, 363d Infantry.

The 169th anniversary of the signing of the Declaration of Independence, 4 July 1945, was declared a day off for all personnel and was observed with short commemorative ceremonies, sports, social activities, and movies. It was the first time since Pearl Harbor that the majority of the Powder River men had had the opportunity to observe our oldest national holiday in this manner. In view of the experiences of the past year, the profound significance of this historic day had taken on a new meaning for every member of the Division.

Another date of particular importance to the men of the 91st, 12 July, marked the first anniversary of the entry of the Division into combat.

Although the mission of the Division was carried out without incident during the latter part of the occupation period and life on the surface seemed to be normal, there were many undercurrents in the local population that might have caused trouble. AMG, with the cooperation of the Division, attempted to uncover as much of this subversive activity as possible. Most of the activity was of a political nature, with pro-Yugoslavs threatening to seize control of Venezia Giulia after the departure of American troops. Girls seen with American soldiers were threatened by Slovene men and told to discontinue the association or they would have their hair cut off. Slovene anti-Allied

General Truscott decorates two battalions at a Division ceremony at Gorizia airport

General Truscott addressing the Division at Gorizia airport. With him are Technical Sergeant Johnny D. Lake and General Livesay.

propaganda held that the British Indian troops around Gorizia were in complete sympathy with Tito's men and, should there be trouble, would join the Yugoslav forces to express their desire to incorporate in a worldwide federation of Communism. Another point in the Slovene propaganda was that Stalin dominated the Big Three and directed things as he liked.

Numerous civilians brought in complaints of unfair distribution of food and failure to return to their homes of prisoners taken by the Yugoslavs when they occupied Gorizia. Reports also came in that the Slovenes intended to infiltrate into the area west of the Morgan Line through their demobilization plan. The idea was to increase the Slovene part of the population and add to their "information groups." Thus the battle of words and nerves continued, with constant patrolling and checking occupying the time of the troops.

On 13 July an order from British 13 Corps relieved the U.S. 10th Mountain Division of its mission of maintaining road posts along the Morgan Line and guarding ammunition dumps. This mission was temporarily taken over by the 91st Division, and by 1200, 14 July, the last relief of the 10th Mountain Division sector was completed.

Five days later the 34th Infantry Division assumed command of the former 10th Mountain Division sector and completed the relief of the 91st Division's temporary positions by 21 July.

At the same time Operations Instruction No. 5 changed the boundary between the 91st and 34th Divisions. The new boundary followed the Iudrio River south to the 13th Northing and from there east along the 13th Northing. The small occupation strip lying between the Iudrio River and the Morgan Line was bounded on the north by a line running northwest just south of the town of Tolmino.

Redeployment training continued according to schedule, with emphasis on subjects prescribed in Training Memorandum No. 8. To supplement the training, a combat course was constructed and made available to elements of the Division on 19 July. Parts of the course were the sniper, close combat, and crack-and-thump. Training Memorandum No. 10, dated 18 July, set forth the regulations governing the use of the course and the schedule of troops running the course.

As the occupation period progressed, speculation about the redeployment of the 91st was the chief topic of conversation wherever Powder River men gathered. Rumors were spiked by an article in *Stars and Stripes* stating that the Division would be redeployed to the Pacific via Camp Howze, Texas. General Livesay did not confirm this report and later issued the statement that the assembly station given by the War Department for the 91st Division was not Camp Howze. He reminded everyone that we could expect many changes and cautioned the men not to confuse facts with rumors.

More definite information on the redeployment of the Division came on 21 July in the form of a message from British 13 Corps, stating that the Division would revert to Fifth Army command on or about 1 August. At that time the 91st would begin movement to a training area to be announced later. Eight days later, on 29 July, an advance billeting party departed for the Volturno Redeployment Training Area, and at the same time a detachment, consisting of representatives of Division G-1, G-3, and G-4, left for Headquarters, Fifth Army, to receive detailed instructions concerning redeployment. Next day Field Order No. 53 was published, giving the details of the movement to the Volturno Redeployment Training Area between 1 and 16 August.

Thus, as July drew to a close, the occupation mission of the Powder River Division in the disputed Venezia Giulia Province terminated. In a delicate situation, which had required the utmost in diplomacy and firmness, the 91st, as in combat, had given a creditable account of itself. Despite minor incidents, the situation was at all times well in hand.

CHAPTER 14

RETURN AND INACTIVATION

FOR THE Powder River men, August 1945 was a month of mixed anticipation—grim when they thought of the combat with Japan which lay ahead of them; joyous when they thought of returning to the United States for a brief period of rest after their long, arduous months of fighting in Italy.

The journey westward began on 2 August, and on 9 August the last unit of the Division left the Venezia Giulia area on the 635-mile trip to the Volturno Redeployment Training Area near Caserta. Part of the Division travelled by train from Udine, while the rest went by truck convoy. The last unit closed in at Volturno on 13 August. On the movement down the boot, the Powder River doughboys saw, for the last time, many scenes which brought poignant memories.

Immediately after arrival at the staging area, often referred to as "The Farm," the Division set up a light training schedule, with close-order drill, calisthenics, orientation classes, and organized athletics as its main features. The packing and crating of all equipment was completed according to schedule by efficient crews of officers and men.

Pass quotas to such points of interest as Naples, the Isle of Capri, Pompeii, and Mondragone Beach were very liberal, and the men were well provided with healthful relaxation and entertainment. Additional recreation and entertainment at Volturno was furnished by an elaborate Red Cross Service Club and a cooperative Special Service.

The movement back to the United States began with the departure of the advance party from Naples on 13 August on the SS *Wakefield.* The party, consisting of 144 officers and men commanded by General Williamson, arrived at the Boston Port of Embarkation on 23 August. Members of the party were quickly sent home for thirty days of rest and recuperation so that they could arrive early at the Division's assembly point and prepare for the arrival of the rest of the Division.

While the Division was at Volturno, many hard-earned and well deserved awards were made to its members. The impressive list of decorations was topped by the posthumous award of the Medal of Honor, the second in the Division, to Sergeant Roy W. Harmon, of Company C, 362d Infantry, for action at Casaglia on 12 July 1944.

The first homeward movement of the main body of the 91st began on 14 August, when the *Kingston Victory* and the *Pauchaug Victory* left Naples with elements of the 361st and 363d Infantry Regiments and the 91st Signal Company on board.

Men of the Division take in historic sights of ancient Pompeii

Simultaneously, the long-hoped-for end of hostilities came—much sooner than anyone had dared to expect. Although it had been rumored for several days previously that Japan would surrender, the actual announcement was almost too good to be true. The official announcement came on 14 August. Immediately a storm of conjecture arose as to what disposition would be made of the Division in the light of the new situation. Regardless of what else the future held, however, the end of World War II meant no more shooting, and nothing could have pleased the veterans of the bitter Italian campaign more.

The date of embarkation for the remaining units of the 361st Infantry and additional elements of the 363d on the *Dominican Victory* was 19 August. Six days later, on 25 August, the *Philip De Neve,* a Liberty ship, left Naples with the 316th Medical Battalion on board.

The *La Crosse Victory,* which departed from Naples on 30 August, took with it the 347th Field Artillery Battalion, the 91st Quartermaster Company, and the 91st Military Police Platoon.

On the following day, 1 September, the last units of the Division left Naples on the SS *Mount Vernon.* These were the 362d Infantry, the remainder of the 363d, the 316th Engineer Battalion, the 91st Reconnaissance Troop, the 791st Ordnance Company, and Division Headquarters and Headquarters Company. So, at long last, after many

General Truscott chats with soldiers who were chosen as the most outstanding members of their respective divisions during a banquet at Fifth Army Headquarters. T/5 Rosario V. Lerma, of the 361st Infantry, represented the 91st.

weeks of impatient waiting, the Powder River Division bade a final *arrivederci* to the land of mud, mules, and mountains and headed for the Land of the Free . . . and home.

The return trip across the broad Atlantic was quiet and serene. Programs of religious services and entertainment were carried out by the committees in charge to the best of their ability with the facilities available. On 21 August, in a ceremony aboard the *Kingston Victory,* the regimental commander, 361st Infantry, presented to representative officers and enlisted men of the 361st's 2d Battalion the Distinguished Unit Citation for their participation in the battalion's heroic capture of Mt. Adone.

With the exception of the *La Crosse Victory* and the SS *Wakefield,* all ships transporting Division personnel docked at the Hampton Roads Port of Debarkation between 25 August and 10 September. From there the troops were quickly transported by train to Camp Patrick Henry, Virginia. The ships landing at Boston transferred their personnel to Camp Myles Standish, Massachusetts. From both of these camps the men were sent with dispatch and efficiency to reception stations, where they were speedily processed and started on their way home for a thirty-day period of rest, relaxation, and recup-

Aboard the SS Pauchaug Victory men of the Division wait for the U.S. shore line to come into sight

eration. A fifteen-day extension was granted later, giving the men a total of forty-five days at home. Men with point scores of 85 or more were authorized immediate separation from the service at their reception stations.

Shortly before the end of the rest period the Adjusted Service Rating Score was lowered to 60 points, and, as a result, most Division men were discharged when they returned to their reception stations after their 45-day vacation. Men who did not have a sufficient number of points for discharge went to Camp Rucker, Alabama, the assembly point of the Division. The last men returning to the Division arrived at Camp Rucker around the middle of November.

Since the major portion of the Division was separated at the reception stations, only a skeleton force remained. Many familiar faces that had been known throughout the Division for so many months were missing. Among the many who left before the final inactivation of the Division were General Livesay, who was transferred to Fort Leavenworth, Kansas, on temporary duty, and General Williamson, who was assigned to the Second Army at Memphis, Tennessee. On 7 November Brigadier General Neal C. Johnson arrived from the Infantry Replacement Training Center, Fort McClellan, Alabama, and assumed command of the Division. During World War II, General Johnson had served for two years in the Southwest Pacific Theater of Operations, where he commanded one of the first joint Army-Navy task forces.

General Johnson and staff

As his last official act before leaving the Powder River Division, General Livesay addressed the officers at Camp Rucker, declaring that command of the Division was "the greatest honor that has ever come to me." He expressed regret that a final review of the Division was impossible because of the large number of men already discharged.

The inactivation procedure was begun soon after the reassembling at Camp Rucker. Men with enough points for discharge were transferred to the Separation Point at the camp. The others were transferred to various units, among them the 35th Headquarters and Headquarters Detachment, Special Troops, First Army, Camp Rucker; the 4th Infantry Division, Camp Butner, North Carolina; and The Infantry School, Fort Benning, Georgia.

In recognition of the outstanding role played by the 91st Division in World War II, the following letter was written to General Livesay by the Commanding General, Army Ground Forces:

HEADQUARTERS ARMY GROUND FORCES
Office of the Commanding General
Washington 25, D.C.

20 October 1945

SUBJECT: Letter of Appreciation.
TO: Commanding General, 91st Infantry Division.

The officers and men of the Powder River Division, which is now being inactivated, will forever be honored and cherished by a grateful nation. Your

Men of the Division board ship at the harbor of Naples

determined efforts in the great battle against Nazi tyranny will stand through the years as a warning to those who would break the peace.

The 91st Infantry Division, after your training in Oregon, entered combat on the Italian front in the summer of 1944, and during the vicious compaigns which followed you fought brilliantly and decisively. At the Arno River, in the capture of Leghorn, during the terrific assault against the Gothic Line, and in the final drive across the Po Valley, your troops won for themselves the respect of the enemy and the tributes of the Allied world.

With your inactivation, many of your division will receive new assignments, and I am confident they will serve as loyally and devotedly as they did in battle, until the need has passed for a large miltary establishment in our country.

It is for me a deep privilege to have this opportunity to commend you, your officers, and your men for the superb part you played in the conquest of the enemy. The memory of your heroic deeds will never die, as long as men are willing to fight for freedom.

JACOB L. DEVERS
General, USA
Commanding

The inactivation of the Division was accomplished by inactivating the units individually as early as their records could be completed and closed. The first units to be inactivated were the 361st Infantry, on 13 November, followed by the 91st Signal Company on 14 November; the 91st Quartermaster Company, the 91st Military Police Platoon, and the Special Troops Medical Detachment on 15 November; the 791st Ordnance Company and the 91st Infantry Division Band on 16 November; the 316th Engineer Battalion on 17 November; the 91st Cavalry Reconnaissance Troop and the 316th Medical Battalion on 19 November; the 916th, 346th, 347th, and 348th Field Artillery Battalions, Headquarters and Headquarters Battery, Division Artillery, and the 363d Infantry on 27 November; and the 362d Infantry on 29 November. The Division itself was inactivated on 1 December, 1945, closing the record of its many glorious achievements in World War II.

The 91st Infantry Division Custodial Detail, composed of the personnel officers and enlisted details from each unit, and commanded by Lieutenant Colonel C. F. Primm, Division Adjutant General, remained at Camp Rucker until 7 January 1946 to close out the last administrative details involved in the inactivation of the division.

LEFT: The 361st Infantry's War Memorial at Livergnano.

BOTTOM: The 363d Infantry's War Memorial at Scarperia.

CONCLUSION

Fort Leavenworth, Kansas
1 December 1945

The Officers and Enlisted Men
91st Infantry Division
Camp Rucker, Alabama

Dear Comrades:

This day concludes the active participation of the 91st Infantry Division as a part of the Armed Forces of the United States. But by no means will the word "End" be written after the Division. It will live on, prepared to again spring to life ready and willing to carry on the splendid tradition of the Powder River spirit, so well established in World Wars I and II.

Upon reactivation, the World War II Division early developed a special spirit and pride of service that enabled it to carry on undaunted through all trials of service in training and combat in such a manner that it became outstanding among combat divisions. It emerged from the Italian Campaign a first-line division, battlewise, ready physically and spiritually, and willing for such other duty as it might be called upon to perform. Only the fortunate conclusion of the war "stopped" the Division.

To all of you who served, I express my grateful appreciation and pride. I wish you good fortune for the future, wherever you are or whatever you are doing. To those who paid the supreme sacrifice we all pay homage. Their sacrifice was not in vain, and we of the Division will long remember and respect them for what they gave. May God watch over you all.

Sincerely,

WM. G. LIVESAY
Major General, USA

PART FOUR

APPENDIX

HONOR ROLL

361ST INFANTRY

Adams, Dalvin J., Pvt
Edna Adams
Box 252, Galliaho, La.
Ace, Paul A., Pvt
Mable Ace
Marshalls, Creek, Pa.
Adkins, Clarence, Pfc
Harold Adkins
Sandy Hooks, Ky.
Alagna, August T., Sgt
Rose Alagna
4947 Beacon St., St. Louis, Mo.
Albert, Frank E., Pvt
Genevieve Albert
Rte. 3, New Brunswick, N. J.
Alexander, Thomas D., Pfc
Alice H. Alexander
Rte. 2, Box 81, Leeds, N. Dak.
Alexis, Oscar H., 2d Lt
Mrs. Oscar E. Alexis
2908 Everett St., Lincoln, Neb.
Alanzo, Marin, Pfc,
Leoner L. Alanzo
Box 523, San Juan, Tex.
Allen, Smith, Pvt
Lola Allen
Rte. 1, Rochester, Tex.
Alvarez, Cesario, Pvt
Cesario Alvarez, Sr.
1816 14th Ave., Tampa, Fla.
Anderson, Einar C., Pfc
Lars H. Anderson
184 Unbra Lane, Murray, Utah
Anderson, John W., 1st Lt
John W. Anderson, Sr.
1429 Aubrey Rd., Bend, Oreg.
Andrews, Samuel L., S/Sgt
Tillie Andrews
Rte. 1, Williamstown, N. C.
Anthony, William F., Pvt
Lizzie C. Anthony
Statesville, N. C.
Antoniello, Pasquale J., Pvt
Rose Antoniello
9544 115th St., Richmond Hill, N. Y.
Ardoin, Dominic, Pvt
Estea Ardoin
Oberlin, La.
Armstrong, Oscar E., T/5
Sarah Armstrong
Gen. Del., Nunnelly, Tenn.
Arroyo, Raymond, Pvt
Virginia Arroyo
56 South St., Hollister, Calif.
Ayers, L. D., Pfc
Love Ayers
Rte. 1, Box 52 A, Delhi, La.
Balanda, Clarence A., 1st Lt
Amelia Balanda
112 Sunnyside Ave., Waterbury, Conn.
Barker, Grady H., Pvt
Rosie Barker
Bryson City, N. C.
Barker, Joseph, Pfc
Pearl L. Barker
Rte. 9, Box 3585, Portland, Oreg.
Barnett, Guy V., Pvt
Bass, Arthur E., Pfc
Nellie M. Bass
Gonesville, Ia.
Bauer, Robert J., Pfc
Lola Bauer
4102 Field Ave., Detroit, Mich.
Beach, August T., Pvt
Mary C. Beach
265 Ege Ave., Jersey City, N. J.
Beardempl, Cecil, S/Sgt
Henry C. Beardemphl
422 Tacoma Ave., Tacoma, Wash.
Beckford, Joshua J., Pfc
Erma W. Beckford
Rte. 6, Findlay, O.
Behnke, Elson E., Pfc
Wilda Behnke
1805 Grand Ave., Ames, Ia.
Bell, Joseph, Jr., Pfc
Helen Hammel (Aunt)
7761 Conger, Detroit, Mich.
Bennett, Clyde E., Pvt
Ada Bennett
Rte. 1, Spencer, W. Va.
Benscotter, James, Pvt
James Crawford
Box 594, Chesney, Wash.
Bergdale, Arvid J., S/Sgt
Alice E. Bergdale
Rte. 5, Austin, Minn.
Bergstrom, Leroy, Pvt
Dunda P. Bergstrom
Rte. 1, Tremonton, Utah
Bertagna, John H., Pfc
Ernest L. Bertagna
5625 California St., San Francisco, Calif.
Bertao, Anthony, Pfc
Mrs. Rosie Bertao
Rte. 1, Box 70, Crows Landing, Calif.
Bertsch, Virgil, S/Sgt
Frederick Bertsch
Rte. 1, Hamilton, O.
Best, Stanley A., Pvt
Yolanda Best
1906 Broad St., Schenectady, N. Y.
Bevill, John T., Pvt
Fennie D. Bevill
Rte. 2, Yantow, Tex.
Bible, Ollie, Pvt
Mrs. Minnie Bible
Rte. 3, Box 10, Heflin, Ala.
Biehl, Vincent E., Pvt
Veronica Biehl
262 W. 107th St., New York, N. Y.
Birchfield, James, Pvt
Bitica, Konstantinos, S/Sgt
Grace Bitica
2061 E. Grand Blvd., Detroit, Mich.
Blaisdell, Joseph, Pvt
Marion Blaisdell
8 Barron St., Cottage City, Md.
Blake, Constantine, Pfc
Mary Blake
33 Sixth St., Auburn, Me.
Bolick, Bealer A., Pvt
Bealer A. Bolick, Sr.
119 Oil Ave., Lenoir, N. C.
Bolliger, Frederick C., Pfc
Mrs. Rosa Bolliger
1621 Clinton Ave., S. Plainfield, N. J.
Bowen, Grover E., S/Sgt
Grover C. Bowen
Seaton, Mo.
Breeden, James H., Pvt
Bertie Breeden
Rte. 4, Madisonville, Tenn.
Brewer, Paul, Pvt
Frank Brewer
Gilbert, W. Va.
Brindock, John, Pvt
Anna Brindock
Eberle, Pa.
Brinson, James W., Pvt
J. W. Brinson
418 E. Water St., Washington, N. C.
Broadrick, William, Pvt
Maggie C. Broadrick
Rte. 1, Lafayette, Ga.
Brodlo, Rudolph, Pfc
Mrs. Anna Brodlo
2859 W. 22d Pl., Chicago, Ill.

Brosmer, Frank R., Pvt
Mary Brosmer
122 Bidwell Pkwy., Buffalo, N. Y.
Brown, Harold R., Pvt
Lucille Brown
Lingleville, Tex.
Brown, James C., Pvt
Buchanan, Bennie D., Pfc
Allie J. Buchanan
Rte. 4, Pleasant Hill, N. C.
Bugaj, Joseph J., Sgt
Rose Bugaj
Box 425, Lansing, O.
Bunch, Charles H., Pvt
Estelle M. Bunch
2301 Buchanan St., Nashville, Tenn.
Burk, Woodrow W., Pfc
Mrs. Mindy Smith
4703 Averill St., Houston, Tex.
Burke, James J., Pvt
Madeline Burke
830 Amsterdam Ave., New. York, N. Y.
Burrows, Leroy, Pfc
Julia Ann Burrows
Rte. 2, Stratford, Okla.
Burton, Harvey, Pvt
Melvin Burton
Rte. 1, Steele, Mo.
Byron, Jack E., 2d Lt
E. C. Byron
Dayton, Tenn.
Bytner, John L., Pvt
Mary Bytner
41 Central Ave., Buffalo, N. Y.
Caristi, James J., Pfc
Irene Caristi
2137 E. 15th St., Brooklyn, N. Y.
Carlson, Edwin C., Pvt
Huld Carlson
1351 Hearst Ave., Berkeley, Calif.
Carlson, Erik O., Pfc
Alma Anderson
Kampingensala, Hensmala, Sweden
Carpenter, Israel R., Pfc
Israel Carpenter, Sr.
216 Gardner St., Plymouth, Pa.
Carpenter, Luther C., Pfc
Rosie McElroy
Box 332, Okemah, Okla.
Carter, James B., Pvt
Bernice Carter
Franklin Ave., Amherst, O.
Carver, Edmund T., Pfc
Ivy Carter
Gen. Del., Chelan, Wash.
Cassidy, John J., Pfc
Mary V. Cassidy
28 Sundview Ave., Huntington, N. Y.
Caudill, Kenneth, Pfc
Nelson C. Caudill
Sturgis, Ky.
Celiga, Steve A., Pvt
Anna Celiga
2906 Ambler St., Cleveland, O.
Cepull, August, Pfc
Margaret Cepull
Box 83, Delmont, Pa.
Chambers, Ben J., Pvt
Elizabeth Chambers
108 Temple St., Aliquippa, Pa.
Chernko, John, Pvt
Margaret Chernko
6116 Woodlawn Ave., Chicago, Ill.
Chiapusio, John, S/Sgt
Madeline Chiapusio
429 5th St., Iron Mountain, Mich.
Churchich, Milan, 1st Lt
Mrs. Leona Churchich
119 W. 103d St., Chicago, Ill.
Cioban, Walter S., Pvt
Joseph Cioban
2943 N. Fairfield Ave., Chicago, Ill.
Clark, King, Pvt
Mrs. Alice Hensley
4715 N. E., 18th Ave., Portland, Oreg.
Clarke, Donald, Pvt
Mabel Clarke
698 W. Grand St., Muskegon, Mich.
Coble, Guy E., Pfc
Mary Fox
327 W. 2d St., Port Angeles, Wash.
Cockrell, Forrest, Pfc
Julia E. Cockrell
41 College St., Mooresville, N. C.
Coffman, Franklin, Pvt
Mabel Coffman
E. Davisville, Mo.
Cole, John W., Pfc
Hazel Hutchins
Fryburg, Me.
Colina, Kenneth B., Pvt
Elizabeth Colina
2486 Sheridan Ave., Detroit, Mich.
Collins, Chester, Sgt
Beatrice L. Collins
Gen. Del., Blackburn, Okla.
Connell, James M., Pvt,
Vera Mae Connell
572 4th Ct., Mar. Home, Tampa, Fla.
Conrade, Raymond H., Sgt
Jake Conrade
Rte. 2, Olpe, Kan.
Constantinople, Tommy, Pvt
Amelia Constantinople
105 S. Garden Ave., Benseville, Ill.
Cooley, Malcolm V., Pfc
William T. Cooley
Rte. 1, Manchester, Tenn.
Cooper, Arhcie B., Pfc
Marion Cooper
Mortons Gap, Ky.
Copeland, Glen, Pfc
Dorothy I. Copeland
Rte. 2, Marion, Ind.
Corlin, Robert, Pfc
Isabel Stankin
Hillside Ave., Kingston, Pa.
Corta, James, Pfc
Maria Corta
Jordon Valley, Oreg.
Crego, Russell F., 2d Lt
George Crego
283 E. Aston, Depew, N. Y.
Crook, William D., Pfc
Lucille Crook
Rte. 2, Doniphan, Mo.
Crosby, Jack M., Pfc
J. C. Crosby
Rte. 1, Jonesboro, Tex.
Cross, Millard, 1st Lt
Edna P. Cross
Rte. 1, Oneida, Tenn.
Cruz, Candido F., Pfc
Jacob Cruz
Chacon, N. Mex.
Cukerman, Abram, Pfc
Rykalaja Okierman
2215 8th St., Bklyn, N. Y.
Durole, Louis J., Pfc
Maude Curole
Box 10b, Raceland, La.
Curran, Richard, Pfc
Ellen M. Curran
7 Concord Ave., Charlestown, Mass.
Currier, Donald V., Pvt
Leila Currier
66 Wren St., W. Roxbury, Mass.
Curchet, Francis, S/Sgt
Grace Curchet
Rte. 1, Box 201, Chow Chilla, Calif.
Czubak, Leo J., Pfc
Anna Czubak
1600½ S. Logan St., Lansing, Mich.

Czubek, Raymond, Sgt
Stella Czubek
3247 S. 14th St., Milwaukee, Wis.
Dalies, Carl J., Pfc,
Alice S. Dalies
6836 N. Kenton, Lincolnwood, Ill.
D'Amico, Phillip P., Sgt
Josephine D'Amico
61 Theresa St., San Francisco, Calif.
Darragh, Ralph F., Pfc
Margaret Darragh
Box 253, New St., Fords, N. J.
Davis, Glen C., Pfc
Dorthy M. Davis
921 S. Fairmont St., Sioux City, Ia.
Davis, Robert G., Pvt
Mary E. Davis
311 Leona St., Fairmont, W. Va.
Davis, Willie L., Cpl
Norman J. Davis
Shaffer & Walker Ave., Ontario, Calif.
Dawe, George R., 2d Lt
Deal, Bernice L., Pfc
Lucy R. Deal
1069 Southern Ave., Fayetteville, N. C.
Decareaux, Leo A., Pvt
Evelyn Decareaux
800 Southern Ave., Algiers, La.
DeCarlo, Charles, Pvt
Helen DeCarlo
4432 Ellery St., Detroit, Mich.
Delgado, Santiago, Pvt
Lupe L. Delgado
13th St., Silver City, N. Mex.
Dellinger, Charles, Sgt
Ella P. Dellinger
Rte. 1, Cary, N. C.
Demas, George, Pvt
Mrs. Anna Demas
80 Linden Ave., Springfield, Mass.
DeFuglio, Frank A., Pvt
Carolina De Puglio
Marfa Airfield, Marfa, Tex.
Derickson, Robert, Pvt
Kathleen Derickson
601 Washington St., Monroe, Mich.
Derouen, David P., S/Sgt
Emonie Derouen
Hockley, Tex.
Diamond, Seymour, 2d Lt
S. Dora Diamond
504 Grand St., New York, N. Y.
Dixon, James, Pfc
Cora E. Dixon
RFD 1, Box 196, Atkins, Va.
Donahue, Raymond L., Pvt
Mary Donahue
RFD 1, Newcastle, Neb.
Done, Cyril W., Maj
Julia A. Done
c/o J. W. Fisher, Cotton Valley, La.
Dougan, Jack, Pfc
Mabel Dougan
Rural Rte., Oakland City, Ind.
Drake, William L., Pfc
Cordie Drake
Rte. 2, Box 5b, Richmond, Calif.
Dufore, Harold L., Pvt
Pearl D. Dufee
Belgrade, Neb.
Dugger, Peter, Pfc
Elsie Malcoll
Rte. 2, Box 768, Watsonville, Calif.
Dukes, Franklin, Pfc,
Margaret Dukes
205 Terrace Blvd., Lewistown, Pa.
Dunch, Victor J., Pvt
Mary Dunch
Box 235, Atlasburg, Pa.
Elbert, George H., Sgt
Beulah Watson
250 Tabor Rd., Philadelphia, Pa.
Edd, Jess A., S/Sgt
Mrs. Minnie Edd
817 5th Ave., Williston, N. Dak.
Edwards, Delbert L., S/Sgt
Bonnabell Edwards
Box 268, Lynnen, Ind.
Eisensmith, Murray, Pfc
David Eisensmith
108 McKibbin St., Brooklyn, N. Y.
Elliott, Ray O., Pvt
George Elliott
Fourmile, Ky.
Eng, Glen, Pfc
Ida Eng
Box 104, Boaz, Wis.
Erickson, Emmett, Pfc
Louise Erickson
Edmore, N. Dak.
Espinoza, Jesus P., S/Sgt
Reguio Espinoza
124 S. Convent, Tucson, Ariz.
Faircloth, Fred J., Pvt
Mae Faircloth
Whitetop, Va.
Fanchin, Lino J., Pfc
Regina Fanchin
2536 S. E. 14th St., Portland, Oreg.
Fasanella, Thomas J., Pvt
Phyllis Fasanella
3379 12th Ave., Bklyn., N. Y.
Fayant, Adolphus, Pfc
Victor Fayant
Box 645, Lewistown, Mont.
Felich, Christopher J., Pfc
Sarah Felich
573 W. Ave., Buffalo, N. Y.
Feneis, Ernest J., Pvt
Ferrill, Richard W., Sgt
Mrs. Edna Ferrill
8198 Nightingale Ave., Dearborn, Mich.
Fickel, Curtis J., S/Sgt
Bella Blackman
232 Williams St., Marlin, Tex.
Fieberling, Ernest, 1st Lt
Mrs. Anna Fieberling
877 Alma Ave., Oakland, Calif.
Findlay, David, Pfc
Mary Findlay
RFD 3, Milford, Mich.
Fischer, Paul N., Pfc
Alice Fischer
403 Iowa St., Mineral Point, Wis.
Fisher, Leon, Pvt
Eva Fisher
Gen. Del., Bogota, Tenn.
Fitzgerald, Russell E., Pfc
Waty Fitzgerald
Ellington, Mo.
Fitzsimmons, Clifford, S/Sgt
Agnes Fitzsimmons
Gen. Del., Sallisaw, Okla.
Fletcher, James J., 1st Lt
Julia Fletcher
1328 Lee St., Charleston, W. Va.
Foisy, Thomas, Pvt
Anita Foisy
27 Vale St., Pawtucket, R. I.
Fonaas, Vernon, Pfc
Ruth E. Fonaas
1500 E. Oklahoma Ave., Milwaukee, Wis.
Frankovsky, George, Pvt
Virginia Frankovsky
47 Hays St., Newark, N. J.
Franks, Harley, Pfc
Olive P. Franks
Gen. Del., Union, S. C.
Frazier, Clyde, Cpl
Frazier, George N., Pfc
Martha Frazier
4018 Olive St., Kansas City, Mo.

Froelich, Leonard, S/Sgt
William Froelich
Rte. 2, Cambria, Wis.
Fykson, Robert, Pfc
Martin Fykson
326 Gilead St., Duluth, Minn.
Galea, Elmer L., Pfc
Mrs. Etta May Gallea
10 N. 3d St., Lebanon, Oreg.
Gallo, Joseph M., Pvt
Mary Gallo
2444 E. Tremont Ave., Bronx, N. Y.
Galm, Lawrence W., 2d Lt
George P. Galm
1885 1st Ave., New York, N. Y.
Garcia, Transcito, T/5
Victoria Garcia
Box 161, Karnes City, Tex.
Garmon, Kelley, Pfc
Betty Garmon
Beaumont, Ky.
Garven, Clarence T., Cpl
Anna Garven
509 Meinecke Ave., Milwaukee, Wis.
Gass, Clifton B., T/Sgt
Sara J. Bird
Rt. 2, Marion, Ky.
Gathwright, Robert, Pfc
Ruth Gathwright
5237 Theodosia Ave., St. Louis, Mo.
Gellert, Jack H., 2d Lt
Gentry, William E., Pfc
Sarah Gentry
421 Fairview Hgts., Shelbyville, Tenn.
Giantomasi, Gelard, Pfc
Dominica Giantomasi
Box 391, Elkland, Pa.
Gilton, Elwood N., Pfc
Mary Gilton
6042 Wissahickon Ave., Philadelphia, Pa.
Girvan, Thomas, Pvt
Elenor Girvan
575 W. 177th St., New York, N. Y.
Givens, John M., Capt
Mary Givens
253 Summer St., Ruffleville, Ky.
Glee, Paul W., Sgt
Kathrine Stewart
1219 2d Ave., Rochelle, Ill.
Glidewell, Leo, Pvt
Nina P. Glidewell
510 Catlett St., Hickman, Ky.
Goelz, Robert A., Pvt
Bernice W. Goelz
4214 N. Wolcott, Chicago, Ill.
Goldberg, Avery, Pvt
Bertha Goldberg
2455 Clairmont Ave., Detroit, Mich.
Goldstein, Albert, Pvt
Anna Goldstein
2116 Crotona Parkway, New York, N. Y.
Gomez, Frank, Pfc
Anita Gomez
Rte. 2, Box 458, Hollister, Calif.
Gonzales, Benjamin, Pfc
Josefa Gonzales
143 4th St., Chino, Calif.
Gonzales, Camil, Sgt
Manuel Uribe
Box 59, Raymondville, Tex.
Gonzales, Jose, Pvt
Guillermina Gonzales
2024½ 4th Ave., Tampa, Fla.
Goodkin, Bernard, Pvt
Harry Goodkin
173 Mansigh St., Poughkeepsie, N. Y.
Grandbois, Zenophile, S/Sgt
Olga Grandbois
Rte. 1, Fargo, N. Dak.
Granger, Donald, Pvt
Gladys Granger
443 Cherry St., Milton, Pa.
Grant Frank D., Jr.,
Frank D. Grant, Sr.
410 Blackstone Blvd., Cleveland, O.
Grassano, Nicholas, Pvt
Mary Grassano
685 Grand St., Brooklyn, N. Y.
Green, Clyde, Cpl
Flossie Green
Rte. 1, Randlett, Okla.
Greer, Cecil, Pvt
Juanita Greer
322 S. 15th St., St. Joseph, Mo.
Gressman, Herbert, Pvt
Gustav Gressman
2758 Cruger Ave., Bronx, N. Y.
Groves, Kenneth, Pvt
Helen C. Groves
128 Fremont St., Gloversvile, N. Y.
Gruby, Charles, Pvt
Mrs. Cora Gruby
P. O. Box 12, Saunders Farms, Rome, N. Y.
Guidros, Adam, Sgt
Lucia Guidros
Mix, La.
Gulgas, Paul, Pfc
Fannie Gulgas
2005 E. 33d St., Kansas City, Mo.
Gurule, Frank, Pfc
Mary S. Gurule
921 Elm St., Trinidad, N. Mex.
Gutierrez, Jesus J., Pfc
Tonicia Cizneros
3025 Cypress St., El Paso, Tex.
Haag, Joseph H., Pvt
Henry Haag
514 Belleville St., New Orleans, La.
Hagins, Jack P., Sgt
Walter M. Hagins
940 Henry Ave., Columbus, Ga.
Haley, Bradley E., T/Sgt
Ada Haley
Oakwood, Tex.
Hall, Harold J., Pvt
Barbara J. Hall
7 S. Main St., Pittsfield, N. H.
Hamilton, James, Pvt
Mrs. Hattie Chase
Rte. 3, Hanford, Calif.
Hamm, Fred E., Sgt
Irene Hamm
Gen. Del., Sullivan, Mo.
Hammond, Herbert L., Pvt
Betty L. Hammond
Rte. 2, Friendship, Tenn.
Hammons, John H., Pvt
George Hammons
Rte. 2, Maryville, Tenn.
Hanks, Charlie W., S/Sgt
Roosevelt P. Hanks
Benham, N. C.
Hanna, Lee A., Pvt
Virginia Hanna
1013 Darthmouth Ave., Canton, O.
Hargrove, Douglas, Pvt
Lillia Hargrove
Rte. 1, Hamer, S. C.
Hari, Ben R., Pfc
Sam Hari
Rte. 3, Silverton, Oreg.
Harrison, Theron, Sgt
Harrison, Tyrus H., Pvt
Carl Harrison
116232 Washburn St., Detroit, Mich.
Hassler, Clarence W., Pvt
Gladys Hassler
Glenville, Minn.
Hatfield, Herman, Pvt
Maggie Hatfield
Black, Ala.

Haukvik, Floyd, Sgt
Mildred Haukvik
Rte. 1, Hanska, Minn.
Haworth, Albert, Pfc
Ethel Haworth
2114 Clearfield St., E. Philadelphia, Pa
Hayes, Rhuel, S/Sgt
Mrs. Ethel Mason
1421 Cornell Ave., Berkeley, Calif.
Hayes, Robert L., Pvt
Ella M. Hayes
Rte. 1, Lakewood, N. Y.
Haynes, Willie L., Pfc
Ida H. Haynes
Mertie, N. C.
Hearn, Roy G., Pvt
Margaret Hearn
Rte. 1, Wingate Run, O.
Hegel, Harold, Sgt
Martha Hegel
450 E. St., Lincoln, Neb.
Heidt, Roy, Pfc
Steve Heidt
312 S. 13th St., Bismark, N. Dak.
Heiss, Clarence, Pfc
Frances Heiss
Rte. 2, Carlisle, Ark.
Hernandez, John Y., Pvt
Mrs. Micholase Hernandez
14835 Calvert St., Van Nuys, Calif.
Hernandez, Salvador T., Pfc
Geneyeva Roycroft
2320 Grant St., Laredo, Tex.
Hickey, Gaylord O., Pvt
Myrtle Hickey
9018 Escanaba Ave., Chicago, Ill.
Hill, Thomas E., Pvt
Marjorie H. Hill
1207 Andrew Ave., N. E., Massillon, O.
Hobson, Dixon, Pfc
Hoffman, Leo E. J., Sgt
Carmela Rooney
625 Lincoln St., Watsonville, Calif.
Holman, Eather J., Pvt
May Holman
Rte. 4, Weatherford, Tex.
Honeycut, James, S/Sgt
Mary B. Honeycutt
Sioux, N. C.
Hood, Willa, Pvt
Mr. Eldrige Hood
Gen. Del., Crossville, Tenn.
Hopkins, William B., Pvt
Mary Hopkins
Methodist Orphans Home, Versailles, Ky.
Howe, Earl B., Sgt
Ida Samuelson
3519 Reta St., Chicago, Ill.
Huber, Arthur J., S/Sgt
Margaret Huber
217 Elden St., Lodi, Calif.
Huchette, Leon, S/Sgt
Angelique Clement
Rt. 1, Box 196, Lutz, Fla.
Huckle, Ivan B. B., Pfc
Vallie B. Huckle
901 E. Barcloft St., Toledo, O.
Hughes, Teal F., Pvt
Homer Hughes
Gen. Del., Opal, Ark.
Hull, James, Pvt
Elizabeth Hull
Rte. 1, New Florence, Pa.
Humphrey, Donald, Pfc
Juliette Humphrey
320 Vine St., Albert Lea, Minn.
Hurley, Robert E., Pvt
Arkie Hurley
Stantonville, Tenn.
Hurry, Obie, Pfc
Frances B. Hurry
Rte. 1, Eaton, Ind.
Hurt, Deaain E., Pfc
Mrs. Connie Hurt
Rte. 1, Swords Creek, Va.
Hutzell, Delbert, Pfc
Amanda Hutzell
217 David St., Johnstown, Pa.
Idzikowski, George A., S/Sgt
Josephine Idzikowski
2459 N. Cramer, Milwaukee, Wis.
Janiec, Walter, Pfc
Mary Jennings
44 Hathoway St., Wallington, N. J.
Jarman, Ervin T., T/Sgt
Beulah Jarman
Rte. 1, Comfort, N. C.
Jefcoat, Billie D., Pvt
Jennings, Alvin J., Pvt
Catherine Jennings
592 E. Grant St., Lebanon, Oreg.
Jocelyn, Edward J., Pvt
Lydia Jocelyn
70 Summit St., E. Providence, R. I.
Johnson, Charles, Pfc
Helen Johnson
Star Terrace, Northport, Wash.
Johnson, George L., Pfc
Alice Johnson
2036 Wynant St., Portland, Oreg.
Johnson, Theodore, Pvt
Cheryl Johnson
3956 Van Buren St., N. E., Minneapolis, Minn.
Johnson, Carless, Pvt
Pearl Pruner
Asco, W. Va.
Jolly, Ross, Pfc
Mrs. Flora Jolly
Rte. 1, Glade Valley, N. C.
Jones, Samuel, Pfc
Clayton Jones
306 Shasta Dr., Toledo, O.
Joseph, Cyrus, Pvt
Anna Joseph
970 Elm St., Zanesville, O.
Joss, Robert O., S/Sgt
Oscar J. Joss
503 Payton St., Ft. Des Moines, Ia.
Julien, George J., Pfc
George F. Julien
436 6th Ave., Troy, N. Y.
Justice, Carol, Pfc
Mary M. Justice
Barnardsville, N. C.
Kalina, Robert J., Sgt
Mary Kalina
3929 Hyde Ave., Cleveland, O.
Karwoski, Theodore A., Pvt
Laura Karwoski
2520 Carson St., Pittsburgh, Pa.
Kaucic, Ladimer, Pfc
Mary Kaucic
1111 E. 71st St., Cleveland, O.
Keegan, John F., Pfc
Elizabeth Keegan
15 Naylor St., San Francisco, Calif.
Kelleher, Joseph D., Pvt
Kelly, Delmer P., Pfc
Nettie Kelly
Blandville, W. Va.
Kennedy, James S., Pfc
Lillie Whisenhunt
Gen. Del., Ringold, Okla.
Kenney, Wilmer C., Pfc
Mildred Kenney
424 S. Park St., Medford, Wis.
Keyes, John L., Pfc
Richard Keyes
Rte. 2, Titusville, Pa.
Kiltles, John H., Pfc
L. P. Kiltles
Box 3, Ferndale, Wash.

King, Ralph T., Pfc
Alice Harrison
Box 14, Columbia, Tex.
Kirscher, Frank, Pfc
Klase, John E., T/5
Mamie Klase
RFD 3, Middleburg, Pa.
Klein, Walter O., Pfc
Martha Klein
Rte. 1, Bark River, Mich.
Kmets, Frank
Mary Kmets
95 Irene St., Buffalo, N. Y.
Knaus, Albin, S/Sgt
Mary Knaus
Box 13, Tranunik, Mich.
Knuckles, Theodore, Pfc
Avery Knuckles
New Madrid, Mo.
Koballa, Charles W., Pvt
Mary Koballa
11122 Woodland St., Cleveland, O.
Kracht, Gerald W., 1st Lt
Elizabeth Kracht
7164 Coolidge Ave., Centerline, Mich.
Krotz, Ernest L., Pvt
Mary J. Krotz
648 E. 131st St., Cleveland, O.
Kuczynski, Theodore J., Pvt
Dorothy Kuczynski
25 N. 25th Ave., W., Duluth, Minn.
Kudasz, Joseph L., Pvt
Catherine Kudasz
211 India St., Brooklyn, N. Y.
LaBarge, Francis M., Pvt
Laure LaBarge
Ellenburg, Deport, N. Y.
Lackey, William J., Pfc
Charles Lackey
17 W. Vine St., Oxford, O.
LaCrosse, Carl E., Pfc
Almo B. LaCrosse
319 Buena Vista St., Nevada City, Calif.
Ladenforf, William C., Pfc
Daisy Dubord
1857 Circle Ave., DesPlaines, Ill.
LaFace, Floren, Pfc
Mrs. Cosma LaFace
99 Jefferson St., Norwalk, O.
Lamp, Henry, Pvt
Frances Lamp
525 E. Nebr., St., Blair, Neb.
Lanigan, Edward T., Pfc
Helen Lanigan
220 Zeralda St., Phila., Pa.
Laslie, Jefferson, Pfc
Grace Laslie
1415 4th St., Lawrenceville, Ill.
Latham, Horace E., Pfc
Minnie H. Latham
Kings Mt., N. C.
Laube, Elmer C., Pfc
Annie Laube
Rte. 2, Morgan, Minn.
Leary, Joseph E., Pvt
Mary Leary
2141 Tasker Ave., Philadelphia, Pa.
Leathers, Leo C., S/Sgt
Ida May Leathers
Breedsville, Mich.
Lentz, William A., Pfc
Mrs. Dessa Lentz
17411 Russell St., Detroit. Mich.
Levinson, Lewis M., Pvt
Samuel Levinson
852 W. 177th St., New York, N. Y.
Lenz, Carl H., Pvt
Ella Lenz
2472½ Auburn St., Toledo, O.
Lewis, Dilmus L., Pvt
Homer Lewis
Rte. 3, Box 351, Lakeland, Fla.
Lex, Weinard E., Pvt
Doris Lex
2206 N. 21st St., Milwaukee, Wis.
Lindsay, Melvin H.,
Myrtle F. Lindsay
428 Virginia Ave., Essex, N. Dak.
Livas, Santos, Pfc
Mr. Euenis Livas
Alamo, Tex.
Livingston, George D., Pvt
Elivera Livingston
8335 S. Kerfoot, Chicago, Ill.
Loden, Nim R., Pvt
Leslie G. Loden
Rte. 1, Dorsey, Miss.
Loe, Joseph D., Pfc
Edwin Loe
RFD 5, Coffeyville, Kan.
Loeb, Harold S., Sgt
Mayetta Loeb
Manatawny, Pa.
Long, Richard E., Pvt
Lonzetta, Michael, Sgt
Mary Lonzetta
844 N. Locust St., Hazelton, Pa.
Lopez, Jesus C., S/Sgt
Cresencia Castro
Box 242, Rte. 1, Atwater, Calif.
Lound, Ray S., S/Sgt
Anna Lound
Rte. 2, Brainard, Minn.
Lourick, Nick, Pvt
Lowrie, Brannon E., Pfc
Nora M. Lowrie
Rte. 4, Terrell, Tex.
Ludovina, Frank S., Pfc
Mary S. Ludovina
Box 832, Lincoln, Calif.
Lulas, Andres G., Pfc
Lois M. Lulas
1061 Main St., N., Jamestown, N. Y.
Lundy, Wayne L., Sgt
Viola A. Lundy
1523 S. W., 49th St., Portland, Oreg.
Lynwood, Philip A., Pvt
Joyce Lynwood
533 Terrace Ave., Marietta, Wis.
Lyth, Robert W., Pvt
Jane Lyth
Rte. 2, Sharpsville, Pa.
Mabe, Boyd B., Pvt
Floyd L. Mabe
Turtle Creek, W. Va.
Macias, Samuel, Pvt
Trinidad Macias
1108 S. Ashland Blvd., Chicago, Ill.
Mack, Harry S., Pvt
Winifred G. Mack
1617 2d Ave., S. Minneapolis, Minn.
Malisch, Howard J., Pvt
Cora Malisch
117 River St., Portage, Wis.
Malszycki, John J., Pfc
Mary Malszycki
Dodge, Wis.
Marinko, Jack J., Sgt
John Marinko
924 S. 5th St., Milwaukee, Wis.
Marquez, Julio, Cpl
Mary Marquez
1870 17th St., Santa Monica, Calif.
Martin, Forest E., Pvt
Alice Martin
Rte. 2, Oklee, Minn.
Martin, George D., Pvt
Sadie Martin
1901 Florida Ave., Richmond, Calif.
Matthews, Milton R., Pfc
Mary Matthews
Rte. 1, Fackler, Ala.
Meyer, Walter A., Pfc

Miller, Richard P., Pfc
Raymond Miller
630 Foraker St., Sidney, O.
Miller, Thomas A., Pvt
Mrs. Jewel Miller
Rte. 2, c/o A. T. Jenkins, Athens, Tex.
Morris, Everett C., Pfc
Ida Morris
Gen. Del., Mustang, Okla.
Mowry, Chas. A. T/5
Lester Mowry
Box 292, Scottsdale, Ariz.
Mulitsch, Harry F., S/Sgt
Mary Mulitsch
3079 N. 19th St., Milwaukee, Wis.
McTiernan, John J., Capt
Mrs. John McTiernan
353 1st St., Mineola, L. I., N. Y.
Newell, Lester G., S/Sgt
Norris, Warren A., Pfc
Pauline Norris
Ord St., Salisbury, Pa.
Olson, Earl W., Pfc
Ellen Olson
Rte. 1, Stephen, Minn.
Oppenheimer, Eugene, Pfc
Lizzie Woods
Allen, Ky.
Palonen, Jorma, Pvt
Alice Palonen
Massachusetts Ave., Fitchburg, Mass.
Parker, Orbie, Pvt
James A. Parker
Gen. Del., Pixley, Calif.
Pergler Charles, Pvt
Mildred J. Pergler
2442 S. Homan Ave., Chicago, Ill.
Plata, Natalie, Sgt
Lucy Plata
Sevaign Liege, Belgium
Polzin, Ralph M., Pvt
Barbara Polzin
Lindsay, Neb.
Porter, Raymond A., Pvt
Florence Porter
E. Homer, N. Y.
Prokopia, George W., Pfc
Timothea H. Prokopia
9 Handly St., Jamestown, N. Y.
Salyards, Douglas W., 1st Lt
Mrs. Mary Salyards
702 N. Burt St., Omaha, Neb.
Sander, Vernon R., Pvt
Paul B. Sander
Rte. 2, Banner, Okla.
Taylor, Milton, Pfc
John Taylor
Rte. 2, Columbia, N. C.
Tucker, Clarence T., Pfc
Florence Borek
8820 Georgia, Detroit, Mich.
Verke, Clayton E., Pfc
Carolina Verke
Rte. 1, Fairdale, N. Dak.
Verner, Abraham S., 2d Lt
Mrs. Flora Ossoff
18 Martinack, Peabody, Mass.
Vernon, George R., S/Sgt
Hilda E. Vernon
Rte. 2, West Iron River, Mich.
Vires, Roscoe, Pfc
Nancy Byrant
Lone, Ky.
Wahlbrink, Karl R., Pvt
Walka, Harold C., Pfc
Christ D. Walka
Baldwin, Mo.
Walker, Robert C., Pfc
Ruth E. Walker
Rte. 1, Ravenna, Tex.
Walker, Wayne K., Pfc
Minnie Walker
P. O. Box 109, Carbon Cliff, Ill.
Warran, John, 2d Lt
Wanholm, Alvin H., Pvt
Arthur Wanholm
14838 Wilfred St., Detroit, Mich.
Way, Clifford L., S/Sgt
Dorothy R. Way
4506 W. 16th Pl., Los Angeles, Calif.
Wayman, Marcus C., Sgt
Thelma B. Wayman
360 Harrison Ave., Salt Lake City, Utah
Weaver, John C., Pvt
Anna M. Kopera
Rte. 1, Defiance, O.
Wehlan, Paul N., Pfc
Sadie Wehlan
3504 Lewis St., Blue Island, Ill.
Welsh, Donald M., T/Sgt
Jessie L. Welsh
1615 Atlanta Ave., Long Beach, Calif.
White, Robert W., Pvt
Mary White
Brownlee, Neb.
Whitman, Willie L., Pfc
Annie L. Whitman
Rte. 1, Eclectic, Ala.
Wilder, James D., Pvt
Mrs. Matilda J. Wilder
Rte. 2, Whitesburg, Tenn.
Williams, Billy C., Pfc
Mary L. Williams
Rte. 1. Exeter, Mo.
Williams, Charles R., Pvt
Mrs. Elizabeth B. Williams
Begler, Pa.
Williams, Marion R., Pfc
Loura A. Williams
2345 Albion Pl., St. Louis, Mo.
Williams, Stephen R., Pvt
Mrs. Jennie Williams
297 E. Main St., Amsterdam, N. Y.
Wilson, George G., 1st Lt
Hilda Wilson
1415 Willamette St., Eugene, Oreg.
Wilson, Raymond L., S/Sgt
Annie S. Wilson
Oteen, N. C.
Winkles, Joseph N., Pvt
Ida Winkles
337 F St., Thoristore, Ga.
Winstead, Kermit M., Pvt
Mathew Winstead
Rte. 2, Little Rock, Miss.
Winter, Gerald H., Pfc
Ann M. Winter
Wahpeton, Ia.
Wolf, Kenneth D., Pvt
Geneva M. Wolf
1420 N. Van Buren, Topeka, Kan.
Wood, Earl F., Pvt
Edith F. Wood
34 Frances St., Cranston, Ill.
Yanowski, Joseph J., Pvt
Yarborough, Fred T., Pvt
Elizabeth Yarborough
Timmonsville, S. C.
Yarbrough, Wiley W., Pvt
Anne Yarbrough
Gen. Del., Canadian, Okla.
Yates, Oda A., Pfc
Delia E. Yates
501 N. 5th St., Phoenix, Ariz.
Zachrisson, Arthur N., S/Sgt
Martha Zachrisson
425 N. E., 2d St., Camas, Wash
Zaffuto, Anthony P., Pfc
Josephine Zaffuto
2110 Arthur Ave., New York City, N. Y.
Zamora, Ventura J., Pfc
Ventura Zamora
1225 3d St., Sacramento, Calif.
Zamores, Andres V., Pvt
Zielinski, Andrew, Pvt

362ND INFANTRY

Acuna, Raymond M., Pfc
Lupe Acuna
P. O. Box 176, San Juan Bautista, Calif.
Adkins, Willie L., Pvt
Lorena Adkins
Rte. 1, Box 92, Lavalette, W. Va.
Alvarado, Reyes R., Pfc
Santos Alvarado
Rte. 2, Big Spring, Tex.
Ambrose, Francis G., S/Sgt
Lucy A. Ambrose
94 W. 3d N., American Fork, Utah
Anderson, Varges, Pvt
Clara Anderson
Gen Del., Lakeport, Calif.
Arismendiz, Felix, Pfc
Ingnacia Arismendiz
1026 S. Dacotah St., Los Angeles, Calif.
Armstrong, Earl, Pfc
Laura Armstrong
Gauge, Ky.
Baker, Harold F., Pvt
Mrs. Josephine H. Baker
3114 Pleasant Ave., Hamilton, O.
Barker, James T., Jr.,
James T. Barker, Sr.
RFD 1, Lake Lynn, Pa.
Barger, Vernon G., Sgt
Lector Barger
Cooter, Mo.
Barton, Andrew J., S/Sgt
David Barton
Gen. Del., Box 438, Brentwood, Calif.
Beers, Leonard A., Pvt
Archie B. Beers
2430 S. 18th St., St. Joseph, Mo.
Bell, Kenneth C., Pvt
Anna M. Bell
Railroad St., Barton, Md.
Berry, James A., Pfc
Mary E. Berry
Jack, Mo.
Bertino, Frank J., Pvt
Catherine Bertino
Worcester, Mass.
Blakeman, Francis J., Pfc
Frank Blakeman
1927 Caleb Ave., Syracuse, N. Y.
Bleclic, Carl, Pfc
Helen Bleclic
Rte. 3, Willard, O.
Bogenschutz, Arnold G., Pfc
Elizabeth Bogenschutz
101 Charles St., Albert Lea, Minn.
Boren, Oscar V., T/Sgt
Mrs. W. B. Boren
Box 777, Blythe, Calif.
Borseth, Alvin M., Pfc
Joseph Borseth
Rte. 2, Galesville, Wis.
Borst, Donald F., Pfc
37314175
Bradshaw, Claude H., Pvt
John W. Bradshaw
1325 W. Colorado, Arkansas City, Kan.
Breitling, Vernon D., T/5
Jake G. Breitling
Rte. 1, New Plymouth, Ida.
Brieseacher, Harold G., Pfc
Bertha Brieseacher
1107 Bell St., Mt. Vernon, Ill.
Brockway, Frederick, Pfc
Wealthy Brockway
Howell, Mich.
Brott, Kenneth C., T/5
Bertha Mae Brott
R. R. 3, Mountain Grove, Mo.
Brown, L. D., T/Sgt
Marie M. Brown
3285 S. Acoma St., Englewood, Colo.
Brown, David S., Pfc
Virginia Clay Brown
Gen. Del., Achilles, Okla.
Brown, George W., Lt
Mrs. Dorothy M. Brown
565 Elmwood Dr., N. E., Atlanta, Ga.
Bruno, James L., S/Sgt
James L. Bruno, So.
524 N. Main St., Helper, Utah
Bruns, Bloyce L., S/Sgt
Margaret Bruns
RFD 1, Galva, Ia.
Bryant, Leon, Pfc
Mrs. Alda Roberta Bryant
400 S. Clinton, Collinsville, Ill.
Buchanan, Eugene L., Sgt
Maude Viola Crump
Rte. 1, Twin Falls, Ida.
Buchanan, Vernon L., Pfc, NBC
John Buchanan
Stillwater, Ky.
Buckley, Wadie, Pvt
Gussie Buckley
Rte. 2, Wetumka, Okla.
Burgess, Roy, Pvt
Elizabeth H. Burgess
Gen. Del., Overton, Nev.
Burnett, Marion R., Pfc
Fred A. Burnett
Box 502, Cushing, Okla.
Caballero, Marcelo F., Pfc
Genoveva Caballero
Box 304, Whitsett, Tex.
Caccia, Edward, Pvt
Charles Caccia
Bonita St., San Francisco, Calif.
Calabria, Carlo, Pfc
Beatrice Calabria
RD 1, Bensenville, Ill.'
Cambell, Robert, Pfc
Fred Cambell
Grace, Ida.
Campbell, Franklin D., Pvt
Laura E. Campbell
River St., Fair Haven, Vt.
Cannon, John R., Pvt
Margaret Cannon
257 Lenox Ave., Providence, R. I.
Carter, Earl, Pvt
Henry Pearl Carter
Box 72, Haines, Oreg.
Carlson, Franklin E., T/5
Hulda Melving Carlson
908 B St., Coeur D'Alene, Ida.
Carter, Walter D., Sgt
William Dave Carter
Oran, Mo.
Castaneda, Caralampio, Pfc
Josefina M. Castenada
Gen. Del., Rio Grande City, Tex.
Castro, Frank J., M/Sgt, NBC
James Castro
347 Haight Ave., Alameda, Calif.
Cavallaro, Bernard J.,
Fort Cavallaro
1312 Lodi St., Syracuse, N. Y.
Chacon, Ezequiel, Pfc
Ezequiel Chacon, Sr.,
429 Charles St., El Paso, Tex.
Chandler, Lonnie, Pfc
Margie S. Chandler
RFD 1, Dry Prong, La.
Christie, Calvin C., Pvt
Mrs. Nancy Christie
Rochester, Ky.
Cieslik, Julius H., Pvt
Marcella Cieslik
Box 75, Pelham Hills, N. H.

Clark, Archie T., Pfc
Bessie Clark
Clinton, Mo.
Coates, Robert E., Pvt
Norman O. Stratton
1922 Pierce Ave., Camden, N. J.
Coe, Joseph B., Jr., Pfc
Alice J. Coe
Rte. 1, Lyons, Ga.
Cofield, Rudolph, S/Sgt
Ethel M. Cofield
Rte. 3, Phil Campbell, Ala.
Collins, Thomas Jr., Pfc
Thomas Collins
Box 301, Wayland, N. Y.
Columbus, Ernest H., Pfc
William Columbus
361 Halladay St., Seattle, Wash.
Connelly, Jack J., Pvt
Loretta Connelly
631 Bakewell St., Covington, Ky.
Contreras, Harvey J., Pfc
Carlos Contreras
1455 Eastern Ave., S., Los Angeles, Calif.
Conway, John A., Pfc
Thelma H. Conway
Gen. Del., Benton, Calif.
Cornwell, Lester H., Pfc
Dora Cornwell
45 Ogden St., Hammond, Ind.
Courtright, Lewis M., Pfc
Mary E. Cotta
634 Gem St., Tulare, Calif.
Czinki, John, Pfc
Mary Czinki
561 White St., Lincoln Park, Mich.
Dale, James M., Pfc
John W. Dale
McQueen Blvd., St. Albans, W. Va.
Dalke, Emil U., Pvt
Sarah Dalke
Newport, Wash.
Daniels, John, Pvt
Agnes Daniels
Colerbrook, Conn.
Danna, Jesse P., Pfc
Mrs. Tom Marino
3901 10th Ave., N., Birmingham, Ala.
Dano, Maxwell E., Pvt
Louine Dano
Chaumont St., Depauville, N. Y.
Daugherty, Leon, Pfc
Lessie Duke Daugherty
Rte. 2, Delta, Ala.
Dawson, Wallace, Pfc
Florence P. Dawson
Rte. 2, Donaldsonville, Ga.
Dechiara, Donald D., Pvt
Lucille Dechiara
1313 S. E. 12th St., Canton, O.
De John, Frank, Pfc
Constance De John
Rte. 1, Box 128-A, Liberty, Tex.
Del Vicario, Joseph A., Pvt
Joseph Del Vicario
2074 Ogden Ave., Chicago, Ill.
Detmer, James A., Pvt
Frances Detmer
28 Grand St., Poughkeepsie, N. Y.
Devlin, Gerald J., T/Sgt
Bernard F. Devlin
Gen. Del., Laird, Colo.
Diefendorf, Lewis E., Pvt
Edith J. Diefendorf
Yates City, Ill.
Di Giovanna, Kelly C., Pvt
Josephine Di Giovanna
516 Palist St., Harvey, La.
Dollar, Raymond, Pvt
Ella Dollar
Deer, Ark.
Dominquez, Joaquin R., Pfc
Mrs. Escolastica R. Dominquez
Box 81, Peru, Calif.
Downing, Robert D., Pfc
Mary Pierce
1023 No. Mayfield Ave., Chicago, Ill.
Drake, Dale I., Pvt
Mrs. Virginia Drake
1240 Bensch St., Lansing, Mich.
Drawyer, Delbert E., Pfc
Lena Drawyer
Jasper, Oreg.
Duncan, Donald A., Pvt
Mrs. Margie Duncan
Rte. 1, Thurmont, Md.
Dunn, Hollis E., Pvt
Neva M. Dunn
Rte. 1, Kodak, Tenn.
Eagon, Willard, Pfc
Elwood Eagon
RFD 1, Buffalo, Mo.
Edwards, Jack, Pfc
Tove H. Edwards
Rte. 3, Greenville, N. C.
Eggers, Carroll D., Sgt
Lona F. Eggers
RFD 1, Vilas, N. C.
Ekis, Archie R., Pfc
Zeffie Ekis
Rte. 2, Moatsville, W. Va.
Elder, Inue E., Pvt
Ola Hanson Elder
Rte. 3, Jefferson, Ga.
Eldridge, Wallace K., Pvt
Alden B. Eldridge
Rte. 1, Box 507, Edmunds, Wash.
Elliott, Leslie H., Lt
0515453
Ellis, Bert R., Pfc
Lula Ellis
RFD 1, Essex, Mo.
Enlow, Roy, S/Sgt
Mildred Enlow
1120 State Ave., Olympia, Wash.
Farr, Woodrow, Pvt
Saddie Farr
109 Elm Ave., Takoma Park, Md.
Faunt, George L., Pfc
Florence Faunt
238 S. 3d St., Philadelphia, Pa.
Fawver, Forest B., Pfc
Elmer C. Fawver
Gen. Del., Logan, Kan.
Finn, Sheldon, S/Sgt
Gertrude Baker
4544 Dockweiler St., Los Angeles, Calif.
Firsty, Morris, Pvt
Mrs. Freda Firsty
1545 Archer Rd., New York City, N. Y.
Foley, Thomas P., S/Sgt
Ellen Foley
5230 Marion St., Germantown, Pa.
Forpanek, James W., Pfc
Emily Forpanek
4860 W. Cornelia Ave., Chicago, Ill.
Foster, Chester M., Pfc
Florence Foster
Rte. 1, Dekalb, Tex.
Francis, Simon A., Pfc
Minnie Francis
Gen. Del., Lemont, Calif.
Frank, Stanley, Lt
Franks, Rupert A., Pfc
Florence Franks
Box 117, Midland, Tex.
French, Hugh S., Capt
Mrs. Christine L. French
Rte. 4, Versailles, Ky.
Futch, Merle H., Sgt
Nancy Futch
Gen. Del., Dickens, Tex.

Galaviz, Fidencio C., Pvt
Ivana C. Galaviz
Gen. Del., Tornillo, Tex.
Galimore, Weldon, Pfc
Bertha Nelson Galimore
Rte. 2, Elba, Ala.
Garcia, Alfonzo, Pfc
Mrs. Lupe Garcia
Rte. 7, Box 3400, Pioneer Dr., Bakersfield, Calif.
George, Lloyd J., Pfc
Audrey S. George
Rte. 1, Shelby, N. C.
Gerber, Emil E., Pfc
Stella Hensley
546 Doris St., Wichita, Kan.
Gerety, William P., Pvt
Kathryne S. Gerety
716 Carroll St., Brooklyn, N. Y.
Gerhardt, Richard H., Pvt
William H. Gerhardt
1870 Knox St., Cincinnati, O.
Gerry, Frank C., Pvt
Mrs. Ada C. Gerry
10611 Spaulding Ave., Chicago, Ill.
Gette, John T., Pvt
Kathryn Gette
2843 N. Commercial, Portland, Oreg.
Gibbs, Lloyd F., S/Sgt
Bill Samuel Gibbs
RFD 4, Salem, Mo.
Gifford, Hughie D., Pfc
Fannie Mae Gifford
Box 151, Westwood, Calif.
Gilbert, Joseph M., Pfc
Mary Gilbert
2053 W. 65th St., Cleveland, O.
Gilbert, William J., T/3
Mrs. Theresa Gilbert
1489 E. Colorado St., Pasadena, Calif.
Gillett, David L., Pfc
Glenn D. Gillett
635 Ocean Ave., New London, Conn.
Ginsberg, Isidore, Pvt
Alex Ginsberg
1977 Gleason Ave., New York City, N. Y.
Glasser, Joseph S., Pfc
Ann Glasser
Box 75, Carrolltown, Pa.
Goldblum, Leon, Pfc
Anna Collins
3145 Rochambeau Ave., Bronx, N. Y.
Gomez, Jesus M., Pfc
Matilde Alvarez
Higley, Ariz.
Gonzales, Esiquio, Pvt
Mercedes Rodriguez
1000 Sabine St., Austin, Tex.
Goodbird, Stanley, Pvt
Mary Z. Goodbird
Sisseton, S. Dak.
Goreham, Vernon W., Pfc
Nora May Goreham
Gen. Del., Alpaugh, Calif.
Goren, Isadore W., Pfc
Mrs. Clara Goren
2809 W. Oakdale St., Philadelphia, Pa.
Gorkowski, Lewis G., Pfc
Stanley Gorkowski
Milwaukee, Wisc.
Gower, James H., Pfc
Hazel M. Gower
4 Water St., Oakland, Md.
Grant, Horace G., Pfc
Ruby V. Grant
Rte. 2, Pomona, Mo.
Grant, Jack, Pfc
Dorothy O. Hamilton
Segno, Tex.
Graves, Clarence C., Pfc
Mary Graves
1003 Hope St., Austin, Minn.
Gray, Earl D., Pfc
Ethel Jane Gray
RFD 5, Salem, Mo.
Greczek, Alex S., S/Sgt
Bernice Greczek
5004 S. Kedvale Ave., Chicago, Ill.
Green, John T., Pvt
Susie I. Green
Rte. 3, Buchanan, Tenn.
Griffin, Willie R., Pfc
Myrtle May Griffin Guyman
Chicago, Ill.
Groe, Orin T., Pfc
Lillian G. Groe
107 6th St., Northwood, Ia.
Gruen, John H., Lt
Mrs. Betty Jane Gruen
1025 Main St., Wellsville, O.
Guiliano, Vincent, Pvt
Lillian Guiliano
233 Parkway, Rochester, N. Y.
Gulley, Willard C., Pfc
Mrs. T. B. Gulley
Rte. 5, Lafayette, Tenn.
Gussler, Milton M., Pvt
Ugie Ethel Gussler
Adams, Ky.
Guzinski, Sylvester T., S/Sgt
Mary Guzinski
3046 N. Buffum St., Milwaukee, Wis.
Haimes, Henry D., Pfc
Mary Ila Haimes
Rte. 2, Box 66, Tyronza, Ark.
Hajduk, Alfred M., Pfc
Michael Hajduk
46 Gorski St., Buffalo, N. Y.
Hale, Elbert S., Pfc
Florence Hale
4 Dorsey St., Saranac Lake, N. Y.
Halpern, Jerome B., Pvt
Esther Halpern
20 Cabrini Blvd., New York City, N. Y.
Hancock, Hubert V., Sgt
Florence Hancock
Royce City, Tex.
Hanley, Jean D., T/Sgt
Ethel Hanley
Plevna, Mont.
Hansen, Bernard J., Pvt
Katherine M. Hansen
Dallas, S. Dak.
Harmon, Paul B., Pvt
Maggie M. Harmon
501 Margaret St., Charleston, W. Va.
Harmon, Roy W., Sgt
R. E. Harmon
Box 698, Pixley, Calif.
Harper, Cecil R., S/Sgt
Willie W. Harper
102 10th Pl. N. E., Hampton, Ia.
Harris, Elmer L., T/Sgt
Earl L. Harris
624 E. Austin, Nevada, Mo.
Harris, James L., Pfc, NBC
Essie S. Harris
Rte. 4, Box 11-A, Greenville, N. C.
Having, Lawrence H., Pfc
Mrs. Mary E. Having
210 E. 19th St., Covington, Ky.
Heisinger, Ralph W., Lt
Mrs. Jeannette H. Heisinger
615 Adams St., Jefferson City, Me.
Henning, Harold, Pfc
Emalia Henning
3360 N. 21st St., Milwaukee, Wis.
Henson, Howard L., S/Sgt
Nora Ford Henson
Dover Mill, Shelby, N. C.
Hernandez, Santos C., Pfc
Alvira Hernandez
437 Silver Ave., Mt. View, Calif.

Herrera, Benjamin W., Pvt
Raymonsita Herrera
509 Porter St., Vallejo, Calif.
Hibben, Mike, Pfc
Matt Hibben
806 24th Ave., S. Minneapolis, Minn.
Hickman, Clifford E., Pvt
Ethel Hickman
325 Bernard St., Denton, Tex.
Hicks, Houston G., Pfc
Ezell R. Hicks
Fallston, N. C.
Higgins, Burl S., Sgt
Bessie Reynolds Higgins
Rte. 7, Shelbyville, Ill.
Hinkel, Raymond F., Pfc
George John Hinkel
RFD 2, O'Fallon, Mo.
Hodge, Otto, Pfc
Carrie Hodge
Trinidad, Calif.
Hoffman, Oakleigh M., Pfc
Hubert F. Hoffman
Glendenin, W. Va.
Holcomb, Rupert R., Jr., S/Sgt
Dorothy D. Holcomb
Castle Rock, Wash.
Holland, Arthur N., Pfc
Cora B. Holland
Vowells Mill, La.
Holmes, Willie, Pfc
Clara Holmes
Gen. Del., Lakeport, Calif.
Hoover, Samuel V., Jr., Pfc
Edna Hoover
1709 Windsor Ave., Lancaster, Pa.
Howdy, Mervin T., Sgt
Catherine Alcolde
Ione, Calif.
Hughes, Lamer L., Pfc
Robert T. Hughes
Crab Orchard, W. Va.
Hughes, Orville L., Pvt
Annie Sarah Hughes
Rte. 2, Delaware, Okla.
Hulett, David E., Pfc
Evelyn H. Hulett
Rte. 1, Milan, Ga.
Hunting, Edmond V., Pfc
George E. Hunting
Rigby, Ida.
Huter, Theodore F., T/4
Martin Huter
Raymond, Wash.
Hyland, Kenneth L., Lt
Mrs. Mary R. Hyland
521 N. 12, Pocatello, Ida.
Ilaszcat, William, Pfc
Henrietta Ilaszcat
219 E. 69th St., New York City, N. Y.
Imboden, Guy V., T/5
Elsie May Imboden
Vienna, Mo.
Imondi, Basil, Pvt
Fanny Imondi
56 Clemece St., Cranston, R. I.
Jackaway, Lester B., Pvt
Lena Jackaway
8307 Harford Rd., Parkville, Mo.
James, Cecil, Pvt
Willis James
RFD 3, Box 151, Provo, Utah
Jarzambowski, Donald P., Pfc
Mary Jarzambowski
5300 Morgan St., Detroit, Mich.
Jenkins, Benjamin H., Lt
Jerpbak, Jas E., Pvt
Inga Jerpbak
514 2d St., Devils Lake, N. Dak.
Johnson, Buck, Pfc
Fannie Johnson
Gen. Del., Leon Junction, Tex.
Jones, Albert A., Pvt
Mabel Jones
P O Box 542, Naugatuck, Conn.
Jones, Maurice E., Pvt
Margaret Collins
Bald Hill Trailer Park, Natick, R. I.
Karna, Wesley, Pvt
Mrs. Ida Karna
Box 457, Painesdale, Mich.
Kemp, Benjamin F., Jr.
Benjamin F. Kemp Sr.
320 Cedar Ave., Takoma Park, Md.
King, Fred M., Pfc
P. C. King, Sr.
P O Box 155, Ft. Gaines, Ga.
Kirstine, Andrew N., Pfc
Mary Kirstine
527 Oak St., San Francisco, Calif.
Klaphaak, John R., Sgt
Clara Klaphaak
539 4th St., Ann Arbor, Mich.
Klein, Aaron, Pvt
Anna Klein
986 Fox St., Bronx, N. Y.
4 Water St., Oakland, Md.
Knaus, Henry R., Pvt
Francis Knaus
Box 27, Traunik, Mich.
Knight, Wendell A., S/Sgt
Harry A. Knight
Rte. 2, Centerville, Kan.
Koch, Earl G., Pfc
Bertha Koch
RFD 1, Chelsea, Mich.
Koch, Paul W., Pfc
Mildred M. Koch
1707 Cardwell St., Baltimore, Md.
Kolak, Rad J., S/Sgt
John Kolak
Box 271, Ripon, Calif.
Koogler, Jack R., Pvt
Mrs. Myrtle E. Koogler
1968 Dresden Rd., Zanesville, O.
Kunkel, Howard R., Lt
Mrs. Chalcie M. Kunkel
Tioga, Tex.
Kunz, Joseph C., Pvt
Margaret S. Kunz
1221 Myrtle Ave., Brooklyn, N. Y.
Ladd, James E., Cpl
Lottie Ladd
Gen. Del., Hornbrook, Calif.
Lager, John R., Sgt
Alfreda N. Lager
163 Chautauqua St., Jamestown, N. Y.
Lalonde, Roland A., Pvt
Laura Lalonde
10 Washington Ave., Old Orchard Beach, Me.
Lamb, Carlton A., Jr., Pfc
Carlton A. Lamb, Sr.
3705 S. Garfield St., St. Louis, Mo.
Land, Alvis B., Pvt
Annie I. Land
1326 Fremont St., Delano, Calif.
Landreth, Earl E., Pvt
Sarah G. Landreth
Rte. 5, No. Chattanooga, Tenn.
Landry, Lawrence, Pfc
Laura Landry
2916 Knodell So., Houston, Tex.
Lanier, Haines H., Cpl
Benny L. Lanier
2222 Hershall St., Jacksonville, Fla.
Lara, Jesus R., Pvt
Joseph Lara
Kickapoo St., No. 901, Hiawatha, Kan.
Lavendoski, Gerald P., Pvt
Martha Lavendoski
Rte. 2, Cable, Wis.

Lewis, Norman, Lt
Mrs. Rose Lewis
1131 Hollywood St., Far Rockaway, N. Y.
Levitt, George M., Pvt
Bessie Jordon
Rte. 1, Palestine, Ark.
Lindberg, George R., Pvt
Edla Lindberg
205 No. Court St., Gladstone, Mich.
Lines, Howard S., Pfc
Mildred Lines
RFD 1, Knoxville, Pa.
Linstedt, Fred G., Jr., Lt
Mrs. Fred G. Linstedt Jr.
4836 Julian St., Denver, Colo.
Lucas, Russell S., Pfc
Mrs. Sadie A. Lucas
42 Hawthorne Ave., Bend, Oreg.
Lumpkin, Chester T., Jr., Pvt
Ila Lumpkin
Gen. Del., Canton, Tex.
Lynes, Floyd W., Pfc
Edith E. Lynes
946 E. 9th St., Upland, Calif.
Lux, Alfred E., Pvt
Mabel M. Lux
1105 E. 13th St., Wilmington, Del.
Lyons, Richard L., Pfc
Irene Lyons
Box 364 B, Kelso, Wash.
Macchio, Andrew A., Pfc
Anna Macchio
332 E. 109th St., New York City, N. Y.
Maciolek, John A., Pvt
Antoinette Region
97 Tariff St., Thompsonville, Conn.
MacDonald, George M., Pvt
William MacDonald
Winslow, Me.
Mahoney, Wade E., Pvt
Ella Ann Mahoney
Rte. 1, Box 55, Gilbert, La.
Mann, William F., Pfc
Mary E. Mann
13 Middle St., Watertown, Mass.
Mansor, Mansor J., Lt
Marcille, Adelard A., Pfc
Aurelia Marcille
30 Meadow St., Bristol, Conn.
Marcyan, Edward J., Pvt
Josephine Marcyan
7114 Hartwell St., Dearborn, Mich.
Maret, George H., T/5
Mabel Stogner
77A W. 9th St., Tracy, Calif.
Marker, Watson, Pfc
Luther Marker
500 W. 5th St., Wilmington, Del.
Martineau, Gerard A., Pvt
Mrs. Lorraine Martineau
10 Highland St., Lowell, Mass.
Martinez, Ernest, Pfc
Delia Martinez
Box 393, Del Norte, Colo.
Masso, Louis A., Pfc
Mary J. Masso
12 Orchard Ave., N. Providence, R. I.
Masters, Harry, S/Sgt
June E. Woodward
Rte. 6, Box 577, Salem, Oreg.
Mathis, T. W., Pfc
Ruby A. Mathis
Rte. 2, Paducah, Tex.
Medran, Joseph R., Pvt
Marta Bates
Tyrone, N. Mex.
Melott, Theo., J., Pfc
Roger Melott
922 Stark St., Utica, N. Y.
Merritt, Thos., D., Pfc
Mrs. Helen Sutter
2811 F St., Liberty Boro., McKeesport, Pa.
Merritt, Wm. H., Jr., Pfc
Nora Merritt
137 Rectory St., Port Chester, N. Y.
Meyer, Lawrence E., Pvt
Mary Meyer
Fullerton, Neb.
Meyerhoefer, LaVerne L., Pvt
Mrs. Leo Meyerhoefer
781 Blenwood Av., Buffalo, N. Y.
Miller, Robt., P., Pvt
Anna L. G. Miller
715 Brown St., LaFayette, Ind.
Miller, Theodore R., Pfc
George W. Miller
1106 W. 37th Dr., Los Angeles, Calif.
Milligan, Verl J., Pvt
Lula Milligan
Crabtree, Ore.
Mills, Leonard D., T/Sgt
Roy S. Mills
Pampa, Tex.
Milstead, Amos A., Pfc
Rosie Bane
Rte. 2, Henderson, Tex.
Mitchell, James L., Pvt
Flora L. Mitchell
Box 3, Falkner, Miss.
Modrak, Michael, Pfc
Mary Modrak
20 Hurley St., Miner's Mill, Pa.
Monahan, Thomas F., Pfc
Margaret Monahan
124 Maple St., Holyoke, Mass.
Moor, John F., Pfc
Mrs. Hermina Casapo
559 Almon Ave., Woodbridge, N. J.
Moore, Ellis W., Pvt
Edith Moore
169 Walnut St., Ravenna, O.
Moore, Mervin B., Pfc
Hettie H. Moore
Rte. 3, Walstonburg, N. C.
Moraine, Russel E., Jr., Lt
Albert T. Moraine
2703 W. Brood St., Columbus, O.
Murry, Samuel D., Pfc
Katherine S. Murry
Cotton Plant, Miss.
Murry, Warren D., Pvt
Mrs. Sarah Murry
Wagoner, Okla.
McCarthy, Edward A., Pvt
Edw. M. McCarthy
436 Gladston St., Sheridan, Wyo.
McClure, William N., Pfc
May C. Cady
Joplin, Mont.
McCusker, Robert E., Pvt
Susan McCusker
134 Tremont St., Cambridge, Mass.
McDonald, Donald, Pvt
Mrs. Lily Dietrick
Gen. Del., Stockton, Calif.
McEldowney, Charles T., Pvt
Lula M. McEldowney
Stricklett, Ky.
McGee, John R., Sgt
Iva Burghdoff
Ismay, Mont.
McIntosh, Earl N., S/Sgt
Della McIntosh
Star Rte., Norwood, Mo.
McNamara, Patrick J., Pvt
Mary McNamara
514 W. Main St., Susquehana, Pa.

McNichols, Murrel E., Pfc
Viola McNichols
317 E. Grant St., Osceola, Ia.
McQuage, Arthur C., Pfc
Jane H. McQuage
Rte. 2, Box 303, Wadesboro, N. C.
Nance, Frances D., Pvt
Henry G. Nance
Salem, Va.
Nanneman, Aloysius, S/Sgt
Fred Nanneman
Rte. 1, Bynumville, Mo.
Neide, George R., Pfc
Catherine C. Neide
208 Chestnut St., Audubon, N. J.
Nelson, Frank, Pfc
Jno. Dingman
Palm Ave., 10th St., Palm City, Calif.
New, Orville F., Pfc
Mrs. Melba New
Hopkins, Mo.
Nicholson, George A., Pfc
Mrs. Carrie L. Nicholson
Rte. 2, Lafayette, Ga.
Nielson, Russel A., Pfc
Axel A. Nielsen
509 17th St., NW, Minot, N. D.
Norton, Leroy F., Lt
Mrs. Lucille M. Teasdall
5752 Lincoln Ave., Hollydale, Calif.
Nunez, Antonio, Pvt
Mrs. Manuela N. Fernano
315 Hicks St., Brooklyn, N. Y.
O'Brien, Nelson E., Pvt
Rose E. Gladue
126 Bath St., Providence, R. I.
Ochal, Joseph, Pvt
Adam Ochal
165 Carson St., Phila., Pa.
O'Donnell, Charles W., Lt
Mrs. Evelyn O'Donnell
25 School Lane, Ardmore, Pa.
O'Donnell, James J., Pvt
Elizabeth O'Donnell
724 E. 156th St., Bronx, N. Y.
Oelberg, Andrew J., Pvt
Emma A. Oelberg
Rte. 2, Postville, Ia.
Ohman, Roy C., Pfc
John A. Ohman
1217 W. Maxwell Ave., Spokane, Wash.
Ojeda, Richard M., Pfc
Chas. P. Ojeda
P.O. Box 16, Paso Robles, Calif.
Olsen, Eldon A., Pvt
Myrtle Olsen
Monalla, Ore.
Orr, George C., Pvt
Daisey A. Orr
Rte. 3, Yukon, Okla.
Orsak, Alfons J., Pvt
May Orsak
Rte. 5, Schulenberg, Tex.
Osgood, Richard E., Pfc
Emma Osgood
112 W. Minnehaha Pkwy., Minneapolis, Minn.
Ostaszewski, Stanley, Pvt
Valentina Ostaszewski
Imperial, Pa.
Parodi, Arthur, Pfc
Theresa Parodi
519 14th St., Modesto, Calif.
Parrish, Doyle M., Pvt
Ruthie M. Parrish
Rte. 1, Boydsville, Ark.
Passarino, Samotor, Pfc
Senator Passarino
52 Ward St., Healpsburg, Calif.
Paulson, Chester S., Pvt
Helen M. Pellikan
Rte. 2, Patton, Mo.
Pearson, James L., Pvt
Mrs. Maude Pearson
Box 523, Wink, Tex.
Pease, William H., Sgt
Katy Pease
Box 750, Tacoma, Wash.
Pelletier, Romeal A., Pfc
Laura Pelletier
RFD 3, Ft. Kent, Me.
Pemberton, George E., Pvt
Elizabeth Pemberton
Rte. 2, Bambridge, O.
Perry, James J., Pfc
Susie W. Perry
Baxley, Ga.
Perry, Mason, Pfc
Ella Perry
Huntington State Hospital, Huntington, W. Va.
Petersen, Billy E., Pvt
Wm. H. Petersen
5500 Garrettson Ave., Sioux City, Ia.
Peterson, Harry L., Pfc, **NBC**
Fritz Peterson
42 E. Elizabeth St., St. Paul, Minn.
Peterson, Orville H., Pvt
Lena K. Peterson
Astoria, S. D.
Petrick, Elmer M., Pfc
Louise Petrick
Rte. 4, Appleton, Minn.
Pickle, Henry, Pvt
Magdaline Pickle
Rte. 2, Filion, Mich.
Pierce, James V., Sgt
Frances Pierce
Doniphan, Mo.
Pincus, Max, Pvt
Sarah S. Moscowitz
3349 W. Roosevelt Rd., Chicago, Ill.
Pineda, Felipe, Pfc
Liandra Pineda
1114 E. 7th St., El Paso, Tex.
Piscitelli, Mario A., Pvt
Madeline Piscitelli
32 Murray Ave., Paterson, N. J
Pitcher, Harry B., Sgt
Bernadine Pitcher
Rte. 3, Box 1, Mt. Clemens, Mich.
Plate, Arthur, Pvt
Gertrude Plate
28 Stewart Ave., Hampstead, N. Y.
Plona, Casimer, Pvt
Anne Plona
Avon, Conn.
Poley, Henry A., Lt
Pool, James W., Pvt
Audrey K. Pool
Rte. 1, Brookneal, Va.
Porter, Charles W., Pfc
Marie Porter
Gen. Del., Macon, Ill.
Porter, Joseph S., Pvt
Ida Porter
Gen. Del., Milfay, Okla.
Poulopoulos, Peter, Pvt
Florence Poulopoulos
810 N. Evergreen, Los Angeles, Calif.
Powell, Harvey L., Pvt
Annie Powell
Rte. 2, Brookland, Tex.
Prall, Robert W., Cpl
Mildred Prall
1627 N. 7th St., Phoenix, Ariz.
Prawdzik, Melvin A., Pvt
Helen Prawdzik
295 Pulaski St., Bridgeport, Conn.

Prebonich, Frank J., Pfc
Frank J. Prebonich, Sr.
Rte. 1, Nahwauk, Minn.
Price, Earnest C., Pvt
Arvie J. Price
Gen. Del., Jamestown, Tenn.
Price, Glenn, Pfc
Hayden Price
Rte. 1, Clyde, N. C.
Proffer, Lloyd E., Pfc
Alice Susan Proffer
Cardwell, Mo.
Provencio, Raymond T., Pfc
Felix Provencio
Gen. Del., Gonzales, Calif.
Pyle, Woodrow W., S/Sgt
Gallie Belle Pyle
Rte. 9, Box 547, Oklahoma City, Okla.
Quinton, Levi, Pvt
Dave Quinton
Gen. Del., Stilwell, Okla.
Radin, Samuel, Pvt
Beckie Radin
104 Division Ave., Brooklyn, N. Y.
Ramirez, Inez, Pvt
Isabel Ramirez
916 Judson St., Los Angeles, Calif.
Ramirez, Lupe R., Pvt
Santoes Ramirez
Humboldt, Kan.
Ramp, Francis E., Pfc
Edith Ramp
14866 Fairfield, Detroit, Mich.
Raney, John L., Pvt
Sarah E. Raney
Rte. 1, Boydsville, Ark.
Red Eagle, Moses M., Pvt
Mary Deartrack
Rte. 2, Ralston, Okla.
Reinhardt, August L., Pfc
Bertha L. Reinhardt
Gen. Del., Jasper, Tex.
Regier, Henry H., Sgt
Abraham J. Regier
Rte. 1, Box 32, Reedley, Calif.
Renfro, Orvel R., Pvt
Stella Renfro
Rte. 1, Pottsboro, Tex.
Renda, Michael, Pfc
Angeline Renda
3801 Sullivan St., St. Louis, Mo.
Reyes, John G., Pvt
Josephine Reyes
1417 No. 12th St., Saginaw, Mich.
Richardson, Bruce A., Pvt
Eunice M. Richardson
c/o Deer Park Lodge, Manitowish, Wis.
Rickabaugh, Everett E., Pvt
Goldie Rickabaugh
Gen. Del., Medford, Okla.
Rico, Louis J., Sgt
Maria P. Rico
210 Stevens Ave., Bellflower, Calif.
Ristine, Tyrus C., Sgt
Maggie C. Ristine
5346 Harlan Dr., Klamath Falls, Oreg.
Robinson, Warren, Pfc
Emma Robinson
Lamero, Ky.
Robinson, Wilbur H., Pvt
Annie B. Robinson
Rte. 2, Abernathy, Tex.
Roeser, Clifford R., Pfc
William Roeser
RFD 2, Junction City, Kan.
Rogers, Charles E., Pfc
Will Rogers
28 Jackson St., Joplin, Mo.
Romano, Mario U., Pvt
Mrs. Frank N. Romano
2918 Butler St., Elmhurst, L. I., N. Y.
Rowzee, Oscar B., Jr., Pvt
Iva Rowzee
Points, W. Va.
Roy, Ferris G., Pvt
Pearl M. Roy
Rte. 1, Hedley, Tex.
Rykena, Albert C., Jr., Pfc, **NBC**
Alberta Rykena
6127 Traymore, Brooklyn Village, O.
Saenz, Humberto, Pfc
Pilar Saenz
210 No. Granna St., El Paso, Tex.
Salisbury, Patrick H., Cpl
Emma E. Salisbury
5724 1/2 Fayette St., Los Angeles, Calif.
Salomon, Ernest F., Pvt
Louise Salomon
26 Meadway-Southgate, London, England
Sanchez, Antonio F., Pvt
Rosa Sanchez
Box 332, Junction, Tex.
Sarle, John L., Lt
Mrs. Maretta H. Sarle
212 W. Pine St., Midland, Mich.
Sauter, Byron, Pvt
Nellie Sauter
RFD 1, 314, Temperance, Mich.
Savage, Patrick J., S/Sgt
Mrs. Hilda M. Savage
2538 28th Ave., San Francisco, Calif.
Schaeffer, Arlan M., Pvt
Mabel Schaeffer
Bernville, Pa.
Schreir, William A., Pvt
Louise Schreir
27 McClure St., Elmont, N. Y.
Schuettpelz, Burton W., Pvt
Henrietta Schuettpelz
Suring, Wis.
Schultz, Kedrick W., S/Sgt
Mamie I. Schultz
6130 Burnwood Ave., Los Angeles, Calif.
Scott, Roy E., Pvt
Nancy W. Scott
811 22d St., Mt. Vernon, Ill.
Selinsky, George J., Pvt
Monica Selinsky
79 7th St., NE., Minneapolis, Minn.
Sipes, Charles B., Pvt
Elsie Sipes
Rte. 2, Olney, Tex.
Skipper, Andrew J., Pfc
Howard L. Skipper
Rte. 2, Conway, S. C.
Smiley, Richard E., Pvt
Mildred E. Smiley
315 E. Pearl St., Lebanon, Ind.
Smith, Archie C., Pvt
Floyd Smith
Ogle, Ky.
Smith, Dorvin L., Sgt
Nellie Smith
834 E. Johnston St., Madison, Wis.
Smith, J. T., S/Sgt
Victoria T. Smith
Clay Springs, Ariz.
Smith, John G., Pfc
Mrs. Nannie J. Smith
Rte. 1, Byrdstown, Tenn.
Smith, Moses L., Pfc
Mary K. Smith
Elk Rapids, Mich.
Smith, Omer L., Pfc
Eunice Smith
Rte. 1, Townville, S. C.
Smith, Robert D., Pvt
George C. Smith
Morgan, Utah

Smith, William A., Pvt
Herbert R. Smith, Sr.
414 Franklin St., Gainesville, Fla.
Smith, William C., Pfc
Mrs. Philips Smith
Rte. 5, Cookeville, Tenn.
Soujak, Joseph, Jr., S/Sgt
Julia Soujak
2245 N. Fremont St., Chicago, Ill.
Sousa, Wesley J., Pfc
May Sousa
2501 1st Ave., Sacramento, Calif.
Southern, Lowell E., Lt
Spatz, Ralph R., Pvt
Matie Spatz
Rte. 1, Bernville, Pa.
Spence, George K., S/Sgt
Isabella Spence
7281 Standard Ave., Centerline, Mich.
Stahosky, Deval B., Pvt
Virginia Stahosky
1029 N. Chatsworth St., St. Paul, Minn.
Starling, Paul R., Pvt
Milfred Starling
Coats, N. C.
Steckling, Howard W., Sgt
William Steckling
1417 Neenan St., Watertown, Wis.
Steele, Charles R., Pvt
June Steele
824 W. 10th St., Texarkana, Tex.
Steele, Jack C., Pfc
Cleve Steele
Rte. 2, Joplin, Mo.
Stirewalt, Ernest B., Pfc
Mrs. Ruby C. Stirewalt
Rte. 1, Chinagrove, N. C.
Stock, George C., Sgt
Lillian Morgan
c/o Court House, Grants Pass, Oreg.
Stocklos, Stacey F., T/5
Frank Stocklos
Princeton, Mo.
Striefer, Louis, Pfc
Harry Streifer
1838 Bryant Ave., Bronx, N. Y.
Stubbs, Gene M., Pvt
Glen J. Stubbs
10 E. 29th St., Kearney, Neb.
Suey, Michael, Pvt
Anna Suey
736 Virginia Ave., Johnstown, Pa.
Suttle, John D., Pvt
Ethel M. Suttle
Rte. 1, Morgantown, N. C.
Swift, Howard W., S/Sgt, NBC
Frances Swift
103 College St., Oriskany Falls, N. Y.
Tank, Arthur J., Pvt
Charles Tank
Rte. 2, Chagrin Falls, O.
Thomas, Tyree C., Jr., Pvt
Vedda Thomas
Box 54, Bivins, Tex.
Throne, John H., S/Sgt
Gertrude Stymus
22 Danforth St., Rochester, N. Y.
Trawick, Thomas E., Lt
True, Luther L., Pvt
Evie N. True
Gen. Del., Sweetwater, Tenn.
Turner, Honor R., Jr., Pfc
Felix J. Turner
Chunchula, Ala.
Utley, William R., Pvt
Clara E. Utley
312 No. Ave. 62, Los Angeles, Calif.
Van Der Vaart, Edwin F., Pvt
Louise Van Der Vaart
1317 No. 13th St., Sheboygan, Wis.

Van Lierop, Carolus P., S/Sgt
Peter Van Lierop
5125 Loran Ave. S., Minneapolis, Minn.
Van Osdol, Joseph O., Pfc
Paul De Witt Van Osdol
S. High St., Jefferson City, Mo.
Vaughan, Joseph B.
Madge Vaughan
Independence, Va.
Villalogos, John H., Pfc
Mrs. Damiana Garcia
94 Spruce St., Houston, Tex.
Vincent, Homer R., Pvt
Alta Lorene Vincent
Rte. 2, Anton, Tex.
Visintainer, Rudolph, Lt
Waechter, Harry H., Pfc
Mamie Waechter
643 Jefferson St., Quincy, Ill.
Walker, Eura I., Pfc
Dorothy I. Walker
Box 191, Wild Horse, Colo.
Wall, Elbie W., Pvt
Ida Wall
Rte. 3, Quitman, Ark.
Walsh, Robert J., T/Sgt
Frank D. Walsh
103 S. Hayworth Ave., Los Angeles, Calif.
Wamsley, Carl D., Pfc
Lauthberry Wamsley
Huttonville, W. Va.
Webb, Sydner J., Sgt
Jess L. Coleman
Dalton, Mo.
Weldon, Raymond O., Pvt
Mrs. Olive E. Weldon
Rte. 2, New Bloomfield, Pa.
Wendt, Edwin O., Pfc
Reinhold Wendt
609 Cody St., Watertown, Wis.
Wertz, Edward R., Pfc
Katherine R. Wertz
104 M. Street, Lincoln, Neb.
Wetzel, Raymond A., Pfc
White, Jacob A., Pvt
Mrs. Maude G. White
Rte. 2, Chuckey, Tenn.
Whiteley, Arther J., Pfc
James H. Whiteley
Gen. Del., Telico, Tex.
Whittington, Virgil J., Pfc
Emma R. Whittington
Eureka, Ill.
Whitworth, Hector J., Pfc
Bill W. Whitworth
1012 Tournament St., Ft. Meyers, Fla.
Wickman, Chester H., Sgt
Louise Wickman
Star Rte. 1, Iron Mt., Mich.
Wight, Harry W., Cpl, NBC
Wilk, Stanley, Pvt
Mrs. Agnes Wilk
River St., Thorndike, Mass.
Wilkof, Sanford, S/Sgt
Mrs. Eva Wilkof
321 Arlington Ave., Canton, O.
Wilson, John L., Pfc
Estelle Driggers
Rte. 2, Box 345 A, Charleston, S. C.
Wishow, Clifford A., Pvt
Bertha A. Wishon
902 Queen St., Johnson City, Tenn.
Witherspoon, James C., Pvt
Bessie Witherspoon
Rte. 2, Jones, Okla.
Wood, Clarence E., Sgt
Veda Wood
Rte. 2, Loganville, Wis.
Wood, Emmet L., Pvt
Cora L. Wood
Dobson, N. C.

Worley, William C., Pfc
William H. Worley
Rte. 2, Lancaster, Tex.
Wray, Rufus R., Pfc
Carrie B. Wray
RFD 1, Mayoden, N. C.
Wright, Jewel D., Pfc
Gideon Wright
Rte. 3, Marcelline, Me.
Young, Waldron D., Pfc
Florence Young
Lookout, Pa.
Zarunski, John J., Pvt
John Zarunski
447 W. 42d Pl., Chicago, Ill.
Zenty, Emeryk V., Pvt
Mary Zenty
1063 Parmalt St., Dickson City, Pa.
Zimmer, Arthur P., Pvt
Florence Zimmer
150 Wetmore Park, Rochester, N. Y.

363D INFANTRY

Abowitz, Herbert M., Sgt
Molly Abowitz
Arkansas City, Ark.
Ackley, Richard W., Pvt
Charles Ackley, Sr.
Odanah, Wisc.
Acosta, George C., Pfc
Mary Acosta
1329 J St., Reedley, Calif.
Acosta, Tony M., Pfc
Julia Acosta
1801 E 41st St., Los Angeles, Calif.
Adams, Jay W., Pfc
Buster C. Adams
161 North St., Tulare, Calif.
Akers, Alfred, Sgt
Ruth Clemins,
Iron Gates, Joplin, Mo.
Alvarez, Julian T., Pfc
Jose Alvarez
Santa Anna, Sonora, Mexico
Anderson, George A., Pfc
Dorothy Anderson
6032 McClellan, Detroit, Mich.
Anderson, Robert F., Pvt
Edward Anderson
Rte. 1, Convick, Minn.
Anderson, Robert W., 1st Lt
Alice M. Anderson,
3711 7th Ave., Sioux City, Iowa
Antill, Wayne H., Pfc
Edna Antill
Rte. 1, Elba, Ohio
Arendt, Anthony J., Sgt
Anthony H. Arendt
4345 S. Frisco, Chicago, Ill.
Arthur, Orbin L., Pfc
Martha Ann Arthur
Box 633, O'Donnell, Texas
Ashmore, James L., Sgt
Florence J. Ashmore
16321 Orchard, Los Angeles, Calif.
Bacciglieri, Louis T., Pfc
Anna Baccigliere
Rte. 1, Martinez, Calif.
Bahner, Leo O., Pvt
Joseph L. Bahner
1007 S. Harrison, Sedalia, Mo.
Baitz, Anthony M., Pfc
Betty Baitz,
4022 W. Liberty St., Cincinnati, Ohio
Baker, William N., Pfc
Maggie M. Baker
Rte. 3, Dickson, Tenn.
Bakowski, Felix J., Pfc
Sophie Bakowski
1020 2nd Ave., Schenectady, N. Y.
Banks, Mathew V., 1st Lt
Frank J. Banks, Sr.
628 E. Sheridan, Ely, Minn.
Barber, Lenzly R., Pvt
Mrs. Ivey D. Barber
Rte. # 1, Hartsville, Tenn.
Barbiero, Fausto, Pvt
Felice Barbiero
Box 35A, Mount Harris, Colo.
Barlow, Graham J., Pfc
Lula Barlow
Box 26, Vernon, Mich.
Barnes, Clarence L., Jr., Pvt
Olive Barnes
8448 Prairie Ave., Chicago, Ill.
Barnes, Cletus, Pfc
Mary E. Barnes
Rte. # 2, Selmer, Tenn.
Barrios, Stanley T., Pfc
Helen M. Barrios
56 Seville, St., San Francisco, Calif
Bauer, Wm. D., Lt
Mrs. O. W. Bauer
3998 Sherman, Englewood, Colo.
Baum, John S., Pvt
Alice Baum
335 W. James St., Lancaster, Penna.
Bazidlo, Peter E., Pfc
Mr. Stanley Bazidlo
126 Fairview, Syracuse, N. Y.
Beal, Charles E., S/Sgt
Jean Beal
311 E. Commercial St., Anaconda, Mont.
Beer, Paul L., Pvt
Hiram K. Beer
Williams St., Easton, Penna.
Beland, George L., Pvt
Mary Beland
119 Margin St., Lawrence, Mass.
Beltran, Jose D., Pvt
Amelia N. Beltran
Box 11, Lincoln, N. M.
Bender, George C., Pfc
Matilda Bender
1510 McKinley Ave., Yakima, Wash.
Benge, Herman E., T/4
Alpha Bell Benge,
Box 584, Central Point, Ore.
Benshoff, Alvin W., Pvt
Goldie Benshoff
340 Fair St., Bloomsburg, Penna.
Bixler, Henry H., Pvt
Elizabeth Bixler
2424 N. Leithgow St., Philadelphia, Penna.
Bloedorn, Elroy H., Pfc
Shirley J. Bloedorn
3119 N. 24th St., Milwaukee, Wisc.
Blunt, Roger W., Sgt
Alice Blunt
Box 32 F, S. Milwaukee, Wisc.
Bonvicino, John, S/Sgt
Christine Bonvicino
723 Adams St., Albany, Calif.
Booth, Coy, T/Sgt
Ima Faye Booth
Ceres, Calif.
Bourey, Donald J., Pvt
Wilfred S. Bourey
Dannemora, N. Y.
Boyles, Alfred J., Pvt
Effie May Boyles
Gen. Del., Fitzhugh, Okla.
Brannon, Cleo, Pvt
Kate Brannon
Rte. # 2, Headland, Ala.

Braun, Robert C., Pvt
Francis Guber Braun
100305 Charles St., Chicago, Ill.
Brookman, Edward H., Pfc
Lena Russel
Marion, Va.
Brown, Charles E., Pvt
Neeley Brown
Rte. #1, Hickory, Ky.
Brown, Jessie W., Pfc
Mary P. Brown
Rte. #2, Blue Ridge, Ga.
Brown, Raymond A., Pfc
Josephine Brown
157 Crimeo St., Ventura, Calif.
Brown, Warren E., Cpl
Virginia Brown
253 Olmstead St., Winona, Minn.
Bryan, James O., Pfc
Jenny W. Bryan
1200 Forrest Ave., Nashville, Tenn.
Bryant, Miller J., Pfc
Nettie Bryant
Gen. Del., Blue Sulphur Springs, W. Va.
Brzozowski, Stanley J., Pfc
Marcy Anna Brzozowski
1021 Butterworth St., S.W., Grand Rapids, Mich.
Buchanan, Jack E., Pfc
Ibbia D. Buchanan
Elk Park, N. C.
Buie, Duie E., Pfc
Lessie M. Buie
Rte. #2, Centerville, Tenn.
Bunce, Leon M., Jr., Sgt
Helen E. Bunce
340 B St., Yuba City, Calif.
Bunker, John W., Pfc
Iva Bunker
420 Elm St., Sidney, Ohio
Bunyard, Elmer V., Pfc
Angeline Bunyard
Rte. #2, Colfax, Iowa
Burkhalter, Ralph H., Pfc
Audrey J. Burkhalter
121 D St., Lindale, Ga.
Burklow, John R., Pfc
Effie May Burklow
Rte. #2, Hermileigh, Texas
Byars, Ralph J., T/5
May Byars
RFD #3, Mexico, Mo.
Byrd, Calvin L., Pfc
Elizabeth Byrd
Star Rte., Box 162, Shelton, Wash.
Cain, Gaither H., Pfc
Minnie S. Cain
Rte. #5, Fayetteville, N. C.
Campbell, Robert L., Pfc
Violet D. Campbell
2200 Potter St., Lincoln, Nebr.
Canavan, Austin P., S/Sgt,
Mae Canavan
11263 Virginia Ave., Lynwood, Calif.
Carlos, Ventura B., Pvt
Elena Louise Carlos
Rte. #3, Tucson, Ariz.
Carneal, Patrick H., Pfc
Mabel B. Carneal
521 Main St., Fredericksburg, Va.
Carter, Morrill L., Pfc
Edith I. Carter
Rte. #1, Stark City, Mo.
Caruso, Frank S., Pvt
Miss Anna R. Caruso
422 S. 11th St. Newark, N. J.
Case, Ulysses, Pvt
Fred Case
Selina, Ark.
Catt, Clyde B., Pfc
Clyde Catt
Willow Hill, Ill.
Caywood, Theodore C., Pfc
Cleo T. Caywood
1302 S. Cincinnati, Tulsa, Okla.
Cervini, Joseph P., Pvt
Louise Cervini
769 E. 224th St., Bronx, N. Y.
Chavez, Raul A., Pfc
Victoria Chavez
3616 Eagle St., Los Angeles, Calif.
Cheatham, Ermy W., Jr., Pfc
Ermy W. Cheatham, Sr.
515 Granite, Pawnee, Okla.
Chittick, Robert V., Pfc
Essie A. Chittick
Riddle, Ore.
Christiano, Vincent P., S/Sgt
Angela Christiano
15 Winchester Ave., Buffalo, N. Y.
Choff, Ralph M., Pvt
Theda V. Choff
Rte. #1, Grafton, W. Va.
Cobb, Murl L., T/Sgt
Mayme Cobb
Rte. #1, Benronville, Mo.
Coberly, Hayden E., Pfc
Pearl Coberly
Gen. Del., Pittsburgh, Okla.
Cole, Alvin C., Pfc
Bertina Cole
Rte. #2, Poynette, Wisc.
Collins, Edward S., Pfc
Anna Collins
Rte. #2, Dover, Del.
Collins, Vernon J., S/Sgt
Ida M. Collins
Box 182, Cupertino, Calif.
Colunga, Felipe B., Pvt
Virginia Colunga
207 S. San Saba St., San Antonio, Texas
Connelly, James P., Pfc
Amanda H. Connelly
Rte. #1, Sabin, Minn.
Cook, Elmer L., Sgt
Anna Cook
1323 W. 90th Place, Los Angeles, Calif.
Cooley, James E., Pvt
Bessie Mae Cooley
Rte. #1, Fountain Inn, S. C.
Corton, Edwin E., Pfc
Cornelia Corton
Manston, Wisc.
Couch, George W., Pfc
Viola C. Couch
Rte. #2, Adairville, Ga.
Cozad, Karl W., T/Sgt
Ella D. Cozad
2705 Hope St., Huntington Park, Calif.
Craig, Edwin E., T/Sgt
Mary C. Craig
Monticello, Iowa
Crews, Robert F., S/Sgt
Opal Crews
Rte. #3, Lawrenceville, Ill.
Criner, Quention, Pfc
Wonza Fay Criner
1250 University Ave., Berkeley, Calif.
Cullar, Lewis B., Sgt
Evelyn Cullar
245 Monroe St., Eugene, Ore.
Curto, Charles C., Pfc
Salvatore Curto
159 River Rd., Cos Cob, Conn.
Cvijanovich, Sam L., T/Sgt
Louis Cvijanovich
Jerome, Ariz.
Daniels, Claude, Sgt
Thomas F. Daniels
Star Rte., Neelyville, Mo.

Davis, Robert R., Pfc
Elizabeth Davis
1562 Nye Ave., Pomeroy, Ohio
DeBord, Charles B., Jr., Cpl.
Goldie DeBord
Rte. #3, Marion, Va.
Degerlomoe, Tony, Pfc
Evelyn Degerlomoe
Box 783, Klamath Falls, Ore.
Dell, Franklin, Pfc
Blanche E. Dell
624 N. Fair St., Allentown, Penna.
Depew, Clair H., Pfc
Mrs. W. Lloyd Putnam
Hammondsport, N. Y.
Des Rosiers, Aime J., Pvt
Jacqueline Des Rosiers
49 Midland, Rte. #2, Highland Park, Mich.
Diner, Irving, S/Sgt
Cynthia Diner
1025 Boynton Ave., Bronx, N. Y.
Doble, Harry A., Cpl
Emma A. Doble
Box 84, Rexford, Mont.
Dodd, Denver H., Cpl
Frieda L. Dodd
Box 472, Rigby, Idaho
Dover, Robert A., Pfc
Susie H. Dover
Rte. #1, Hull, Ga.
Driscoll, Thomas L., Pvt
Daniel Driscoll
7 Summit Ave., Pittsfield, Mass.
Dubla, Carl D., Pfc
Michael F. Dubla
1416 E. Russell Ave., Milwaukee, Wisc.
Dugas, Alfred H., Pfc
Hortense Dugas
282 Gaulin Ave., Woonsocket, R. I.
Duncan, Howard E., Pfc
Florence Duncan
Rte. #1, Server, Penna.
Duncan, Marvin M., Pfc
Eliza C. Duncan
Chimney Rock Rd., Hendersonville, N. C.
Dunne, Thomas L., Pvt
Rose G. Dunne
5759 Hazeltine Ave., Van Nuys, Calif.
Durham, Floyd, Pvt
Pearl Durham
Sand Gap, Ky.
Dzik, Casimir J., 1st Lt
Kathrine R. Dzik
288 Fisk St., Pittsburgh, Penna.
Eberhart, Raymond A., S/Sgt
George C. Eberhart
1046 Penna. Ave., Lawrence, Kan.
Eberle, William, Jr., Pvt
Wilma F. Eberle
Pawnee City, Nebr.
Eells, Edwin C., 2d Lt
Rev. Edwin Eells
2134 W. Pierce Ave., Chicago, Ill.
Egert, Frey D., Pfc
Katharine Egert
Rice St., Elmore, Ohio
Eggert, Albert A., Pfc
Augusta Eggert
232 Benziger Ave., N. Brighton, L.I., N.Y.
Elliott, John J., Pvt
Agnes M. Elliott
Box 53, Calhoun, Tenn.
Elser, Earl H., Pvt
Grace Elser
82 E. Fulton St., Ephrata, Penna.
Elwood, Harry R., Pfc
William Elwood
Baraboo, Wisc.
Emig, Robert E., S/Sgt
Mildred M. E. Emig
Rte. #1, Midland Rd. Bay City, Mich.
Emlinger, Henry C., Pfc
Della Emlinger
Box 122, Kalida, Ohio
Eschallier, Jean M., 1st Lt
Florence M. Eschallier
1652 Mentor St., Philadelphia, Penna.
Esquibel, Daniel, Jr., Pvt
Virginia G. Esquibel
Roberts, N. M.
Fain, Alvis M., Pvt
Maude A. Fain
Rte. #1, Lepanto, Ark.
Farnham, Gerald E., Pfc
Floyd Farnham
Rte. #2, Westfield, N. Y.
Fast, Walter V., Pfc
Audrey Fast
Rte. #1, Box 58, Finleyville, Penna.
Feilbach, Paul T., Pfc
Bessie L. Feilbach
4335 Vermaas Ave., Toledo, Ohio
Ferrimani, Vincent D., Pvt
Romige Ferrimani
427 West Ave., E. Rochester, N. Y.
Figueroa, Damen S., T/Sgt
Maria S. Figueroa
Rte. #2, Tucson, Ariz.
Fisher, Phillip, Pvt
Gertrude Fisher
133 Powell St., Brooklyn, N. Y.
Fleharty, Lester C., S/Sgt
Pearl Tobin
Winifred, Mont.
Flores, Joaquin C., Pfc
Angela G. Flores
201 S. 22nd St., Phoenix, Ariz.
Flud, Haldeman J., Pfc
Charles Flud
814 Lincoln Ave., Oildale, Calif.
Forrest, Byron, T/Sgt
Winnie Forrest
Rte. #3, Hartford, Ala.
Fortune, Ernest L., Pfc
Laura E. Fortune
Rte. #1, Killbuck, Ohio
Franciskovich, William, S/Sgt
Anna M. Franciskovich
2424 N.W. Overton St., Portland, Ore.
Franklin, Harley E., 1st Sgt
Emma Franklin
Rte. #1, Box 151A, Big Springs, Texas
Fratarcangeli, Joseph G., Pvt
Antoinett Fratarcangeli
2438 Fuller St., Bronx, N. Y.
Friedrich, Joseph F., Cpl
Mrs. Joseph Friedrich
2175 Wash. St., Dubuque, Iowa
Fullwood, Stanley A., S/Sgt
Amanda B. Fullwood
Box 337, Orland, Calif.
Gabhart, Marvin D., Pvt
Hallie Divine Gabhart
Rte. #4, Jacksonville, Ill.
Gallegos, Herman, Pfc
Eduardo Gallegos
Belen, N. M.
Gallo, Alfred, Pfc
Jennie Gallo
2431 S. 9th St., Philadelphia, Penna.
Gallus, Lester A., Cpl
Mary A. Gallus
627 S. David St., Casper, Wyo.
Ganner, Allan B., Pvt
Ina Lois Ganner
2232 Florence, Tulsa, Okla.
Garcia, Marcelo M., Jr., S/Sgt
Mrs. Belinda G. Sarola
128 N. Colo. St., Mercedes, Texas
Garner, Clarence C., Pfc
Alvin Garner
Rte. #3, Box 100, Sherman, Texas

Garrett, Paul H., Pfc
Sarah E. Garrett
37 W. Gates St., Columbus, Ohio
Garrison, Robert F., Pvt
Frances R. Garrison
131 N. Fla. St., Mobile, Ala.
Gayhart, Claude W., Pvt
Clotine C. Gayhart
3752 Gallio St., New Boston, Ohio
Giallombardo, Frank J., Pfc
Carl Giallombardo
129 Cherry St., New York, N. Y.
Giddings, Carl R., 2d Lt
Jessie R. Giddings
Rte. #1, Jerome, Mich.
Gifford, Harry F., S/Sgt
George Gifford
Rte. #2, Steele, Mo.
Gilhooley, Thomas A., Pfc
Virginia Gilhooley
3213 M St., N.W., Washington D. C.
Gilliland,, Lee M., Pfc
Isabille Gilliland
228 E. Southern St., Springfield, Ohio
Gleason, Edmund M., T/Sgt
John J. Gleason
3180 43rd St., Astoria, L.I. N.Y.
Goldsmith, Sidney W., Pvt
Anna Goldsmith
1171 Morrison Ave., Bronx, N. Y.
Gomez, John C., Pfc
Mrs. Sacramento Gomez
2616 E. Wash. St., Stockton, Calif.
Gonzalez, Claudio, Pvt
Marcelo Gonzalez
Box 771, Pearsall, Texas
Gorath, Richard, Pfc
George A. Gorath
Rte. #1, Yuba City, Calif.
Gordner, William A., Pvt
Myrtle Gordner
Drums, Penna.
Graham, Keith B., Pvt
Mattie M. Graham
Box 346, Aurora, Minn.
Grecsek, Edward R., Sgt
Sophia Mae Grecsek
Box 386, Newport, Ore.
Grandokken, Andrew, Pvt
Margaret Grandokken
Haywood, Minn.
Green, Oscar W., Pfc
Ruth A. Green
Keytesville, Mo.
Greig, Alexander M., S/Sgt
Selma Greig
201 Hillside Ave., Stoughton, Wisc.
Grimes, Wilbur E., Pvt
Mary M. Grimes
Rte. #5, Cadiz, Ohio
Grittini, Joseph J., Pfc
Josephine Grittini
2625 Jos. Campau, Detroit, Mich.
Grodt, Earl K., Pfc
Mrs. Peter G. Grodt
Rte. #2, Bonetrail, N. D.
Grossman, Abraham, Pvt
Sylvia Grossman
404 Girard Ave., Philadelphia, Penna.
Grosso, Michael A., Pfc
Jennie D. Grosso
402 Vincelette St., Bridgeport, Conn.
Gutches, Ralph J., Pvt
Vilma Karpes
83 Trenton Ave., Newark, N. J.
Hageman, Robert W., Pfc
Mario Hageman
365 Elmwood Ave., S. Fargo, N. D.
Hamann, Leo E., Pvt
Anna L. Hamann
Green Valley, Ill.
Hanke, Laverne G., Pfc
Mrs. Laverne G. Hanke
211 Chippewa St., Eau Claire, Wisc.
Harder, Albert E., T/5
Mary E. Harder
820 S. Brook St., Louisville, Ky.
Harris, Harold T., Pfc
Sara Harris
300 Hamilton Ave., Paterson, N. J.
Harrison, Louie D., Pfc
Viola H. Harrison
501 E. Oak St., Lodi, Calif.
Hatcher, Robert C., 1st Lt
C. K. Hatcher
Box 608, Concrete, Wash.
Hartline, Roy C., Pfc
Mary Hoke Hartline
Rte. #1, Box 301, Statesville, N. C.
Heeg, Joseph T., Pvt
Marie B. Heeg
9108 175th St., Jamaica, N. Y.
Hemphill, Harvey L., Pvt
Mildred I. Hemphill
Ramseur, N. C.
Hertel, Norbert F., Sgt
Elizabeth Hertel
Rte. #1, Newton Manitowoc, Wisc.
Hester, William E., Sgt
Verna B. Hester
Rte. #1, Cottontown, Tenn.
Higdon, Joseph D., Jr., S/Sgt
Mildred L. Higdon
715 N. 13th St., Waco, Texas
Higgins, Berlyn, Pvt
Irene Higgins
1132 S. 16th St., Manitowoc, Wisc.
Himmelstein, Saul S., Pvt
Martha Himmelstein
1089 Fox St., Bronx, N. Y.
Holan, Charles W., 1st Sgt
Dorothy J. Holan
Sunset St., Rte. #1, Albert Lea, Minn.
Hollinghead, Ernest I., Pfc
Emerson G. Hollinghead
Rte. #1, Flint, Mich.
Hoober, Bertram E., Pfc
Alta S. Hoober
6625 Wayne Ave., Philadelphia, Penna.
Hood, George C., Pfc
Emma A. Hood
Mound City, Mo.
Hopkins, Pharoah I., Pvt
Mandy B. Hopkins
Rte. #3, Baily, N. C.
Houltram, John, Pvt
Ruth Eades
Hillcrest Ave., Inglewood, Calif.
Howell, Claude H., Pvt
Violet Howell
1605 W. 3rd St., Perry, Iowa
Hubbard, Vernon, S/Sgt
Flora Hubbard
Rte. #2, Coleman, Texas
Hughes, Ottis D., S/Sgt
Goldie T. Lawson
1303 Swan St., Ogden, Utah
Hulsizer, Harold L., Pvt
Margie E. Hulsizer
4176 Meade Ave., San Diego 5, Calif.
Hunter, Garland M., Pfc
Ella S. Hunter
Rte. #1, Mt. Airy, N. C.
Hurlbert, Robert L., Pfc
Ezra Hurlbert
1510 Clarimont Ave., Antigo, Wisc.
Hutchinson, Henry A., Pfc
Peter Hutchinson
Rte. #3, Box 48, Hammond, La.
Hyland, Frank T., Capt
Marie Hyland
6346 S. Sangamon St., Chicago, Ill.

Ide, Walter J., Pvt
Julia A. Ide
44 Maverick St., Fitchburg, Mass.
Jagodzinski, Frederick J., Pvt
Mary Jagodzinski
91 Oakland St., Brooklyn, N. Y.
Jasie, Milton, Pvt
Eva H. Jasie
15 N. 10th Ave., Mt. Vernon, N. Y.
Jeffrie, Harvey H., Sgt
Minnie Jeffrie
Nora Springs, Iowa
Johnson, Elvin, Pfc
Sena Johnson
Rte. 4, Clear Brook, Minn.
Johnson, George A., Sgt
Mrs. Bertha Johnson
Rte. 2, Delsea Dr., Millville, N. J.
Johnson, Gerald L., Pvt
Delia M. Johnson
Box 2, Pontoosuc, Ill.
Johnson, Herbert C., Pfc
Laura Johnson
738 E. Sutherland St., Ironwood, Mich.
Johnsen, Russell J., Pfc
Olga Johnsen
4552 33rd Ave., S. Minneapolis, Minn.
Jolley, William, Jr., Sgt
Mrs. William Jolley, Jr.
150 S.E. 4th St., Miami, Fla.
Jones, George C., Pvt
Sally Jones
Box 157, Rte. 2, S. Boston, Va.
Jones, Lonnie M., Pfc
Ada F. Jones
Rte. 1, Seagrove, N. C.
Jones, Rosco E., Pfc
John Jones
Alvin, Texas
Jopke, Henry C., Pfc
Lillian Jopke
2033 E. Johnson St., Madison, Wisc.
Judd, Lloyd, Pfc
(Next of Kin Unknown)
Kalck, LawrenceW., Jr., Pfc
Elizabeth Kalck
2053 Roscoe St., Chicago, Ill.
Keahne, Frank, Pvt
Bituse Keahne
Steamboat Canyon, Ganado, Ariz.
Kelch, Clarence J., T/5
Dorothy Kelch
8264 Brenner, Overland, Mo.
Kentner, Lewis K., Pfc
Mrs. Capitola H. Kentner
348 Chicago Ave., Savanna, Ill.
Keyes, Charles M., Pvt
Bonnie B. Keyes
Fillmore, Ill.
Kimrey, Boyd R., Pvt
Annie K. Kimrey
Rte. 1, Liberty, N. C.
Kjellerson, Ernest W., Pfc
Maxine Kjellerson
Perth, N. D.
Kleinert, Frederik C., Pfc
Ruth E. Kleinert,
711 McKoskry, Saginaw, Mich.
Kline, Hugh D., T/5
Florence D. Kline
11701 Longview Ave., Detroit, Mich.
Klos, Nicholas J., Pfc
Anna Klos
1859 Madison St., Ridgewood, N. Y.
Knight, Jack H., Pvt
Eula G. Knight
Western Ave., Hemet, Calif.
Kohler, Telford C., T/Sgt
Kathryn J. Kohler
Wescoeville & Mine Rd., Rte. #2, Allentown, Penna.
Koivisto, Edwin R., Pfc
Einar W. Koivisto
Box 25, Esko, Minn.
Kor, Warren H., Pfc
W. E. Kor
155 O St., Dimuba, Calif.
Koranda, Joseph W., Pvt
Marie Koranda
2814 S. Springfield Ave., Chicago, Ill.
Koski, Toivo W., Pfc
Onni G. Koski
Box 654, Old San Diego Sta., San Diego, Calif.
Kozinski, Edward P., Pfc
Magdaline Koziara
4803 Trenton, Detroit, Mich.
Kozlowski, Bernard E., Pvt
Julia Kozlowski
5044 S. Justine St., Chicago, Ill.
Krajenta, Wallace, Pvt
Lillian Krajenta
4200 Dodge Ave., Detroit, Mich.
Kramer, William, Pvt
Emma Kramer
2019 L Ave., Galveston, Texas
Kriens, Roy N., Pfc
Mrs. Anne M. Henne
134 Franklin Ave., Hasbrouck Hgts., N. J.
Kulakowski, William W., Pvt
Antoinette Kulakowski
8066 Curt, Detroit, Mich.
Kwiatkowski, Edwin, Pfc
Rose Kwiatkowski
2485 S. 9th Pl., Milwaukee, Wisc.
LaFranchi, Clarence A., Pfc
Elvira LaFranchi
619 43rd St., Oakland, Calif.
Laganowski, Frank A., Pfc
Stella Laganowski
2759 Breman St., Milwaukee, Wisc.
Lambrix, Irving R., Pfc
Sophie Lambrix
18275 Riverview, Wyandotte, Mich.
Lampi, Reino, W., Pfc
Lempi Lampi
328 10th St., Cloquet, Minn.
Land, Larry J., S/Sgt
Catherine J. Land
Abernathy, Texas
Layland, Gerald I., Pfc
Grace O. Layland
112 Pacific Ave., Audubon, Iowa
Leath, Don C., S/Sgt
Naomi Leath
Rte. #3, Lafayette, Tenn.
Lee, Fred W., Pfc
Fred A. Lee
3825 E. Bruce St., Wichita, Kan.
Leehman, Walter H., Pfc
Marie Leehman
Box 74, Minden, Nev.
Lennox, Sidney A., Pfc
Esther Lennox
1112 Ferris St., Alma, Mich.
Leonard, Harold E., Sgt
Gordon A. Leonard
58 W. 5th St., S. Price, Utah
Leoni, Nando D., Pfc
Achille Leoni
114 Downer Ave., Uniontown, Penna.
Leos, Porfirio V., Pfc
Pedro Leos
819 S. Picos St., San Antonio, Texas
Leslie, Edward A., Pvt
Betty G. Leslie
498 Stryker Ave., St. Paul, Minn.
Levy, Albert M., Pvt
Mrs. Ethel Thomas
1440 Clay St., San Francisco, Calif.

Likins, George A., Pfc
Lucy Thurman
1025 W. 12th St., Owensboro, Ky.
Lohmeyer, Edgar G., Pfc
Ann Varie Lohmeyer
405 Clark St., Florissant, Mo.
Lomonaco, Vincent A., Pfc
Catherine Lomonaco
174 Canal St., Canastota, N. Y.
Long, Edgar, Pvt
Crate Long
Hiram, Ga.
Lowry, Clifford A., S/Sgt
Flora Lowry
Rte. #1, Lyman, Nebr.
Lukowitz, George M., Sgt
Francis Lukowitz
2539 N. Richards, Milwaukee, Wisc.
Lucas, Samuel E., Pfc
Alta Lucas
402 Theo Ave., San Antonio, Texas
Luna, Rito, Pfc
Miguel Luna
Box 273, Sedgwick, Kan.
Lunders, Howard E., Pvt
Sylvester J. Lunders
Craigmont, Idaho
Ludwig, James A., Pvt
Rose Ludwig
5025 N. Santa Monica Blvd., Milwaukee, Wisc.
Lynch, Claude A., S/Sgt
Edward Lynch
479 Wilson Ave., Morgantown, W. Va.
Mackowey, Arthur W., Pfc
Emma Mackowey
Rte. #1, Abbotsford, Wisc.
Mahaffey, James C., Pfc
Agnes Mahaffey
106 George St., Johnstown, Penna.
Malone, Cleve T., Pfc
Susie B. Malone
Rte. #5, Boonville, Miss.
Malone, Savadore J., Pvt
Rita Malone
8734 Plum St., New Orleans, La.
Marano, Anthony, Pfc
Daniel Marano
1532 73rd St., Brooklyn, N. Y.
Marcum, Wade H., Pfc
Mahala K. Marcum
Dunlow, W. Va.
Marshall, Max M., Pvt
Mary Almeda Marshall
114 N.E. 7th St., Oklahoma City, Okla.
Martin, Harold, Pvt
Lillian Martin
36 3rd St. Brooklyn, N. Y.
Martin, William R., T/Sgt
Mrs. Augusta T. Lopes
1050 81st Ave., Oakland, Calif.
Martinez, Domingo R., Pvt
Sylviana V. Martinez
Cebolla, N. M.
Martinez, Pete A., S/Sgt
Antoinette M. Martinez
710 W. 1st St., Salida, Colo.
Mason, Thomas W., Pvt
Mae E. Mason
Sunset, S. C.
Massetti, Marino, Pvt
Mary Massetti
5630 Neckel St., Dearborn, Mich.
Mathes, Edwin M., Sgt
Mrs. Roxie J. Mathes
Rte. #2, Centerville, Iowa
Meir, Vernon A., Pvt
Alma Louis Meir
2603 Louis Ave., Brentwood, Mo.
Meinardus, Edward E., 2d Lt
Eloise Meinardus
1274 N. Ardmore Ave., Hollywood, Calif.
Melchiori, Clarence, Pfc
Yvonne Melchiori
549 Belmont Ave., Chicago, Ill.
Mendez, Frank, Pvt
Olga Mendez
1725 Park Ave., New York, N. Y.
Merx, Wilbert I., Sgt
Elizabeth R. Merx
901 Madison St., St. Charles, Mo.
Metzger, Edward F., Pvt
Bertha Metzger
Box 15, Shavertown, Penna.
Meyners, Ferdinand H., T/5
Augusta E. Meyners
Rte. #1, Boyden, Iowa
Miles Clifford E., Pfc
Pearl B. Miles
Rte. #1, Pocatello, Idaho
Miller, Bernard W., Pfc
Ethel Miller
404 W. Grand, Muskegon, Mich.
Miller, Charles H., Pfc
May Miller
Box 683, Miles City, Mont.
Miller, John J., Pfc
Letha P. Miller
Versailles, Mo.
Miller, Robert E., Pvt
Luella M. Miller
Warren, Ind.
Mitchell, Jack, Lt
Charles J. Mitchell
3418 Gillham Rd., Kansas City, Mo.
Moles, Floyd H., Pfc
Lizzie Moles
Keystone, W. Va.
Moore, Thomas K., Pfc
Tom L. Moore
Rte. #2, Marshville, N. C.
Moore, William T., Pfc
John F. Trombley
834 E. Coast Hgwy., Long Beach, Calif.
Moreno, James P., Pvt
Margaret Moreno
1926 E. 1st St., Los Angeles, Calif.
Morris, William L., Pfc
Lille Morris
Barterville, Ky.
Mosbeck, Vernon E., T/5
Beatrice Mosbeck
418 N. State Ave., Theif River Falls, Minn.
Muckerheide, Lowell J., Pfc
Eva C. Muckerheide
2129 N. 45th St., Milwaukee, Wisc.
Mullen, Chester L., Pvt
Annis D. C. Mullen
Rte. #1, Box 200, Belzoni, Miss.
Mullen, Edward T., 2d Lt
Julia Mullen
16 Ogden St., Newark, N. J.
Mumford, Geddes, Pfc
Sophia Mumford
Amenia, N. Y.
Murphy, Travis J., Pvt
Walter R. Murphy
Rte. #1, Hodges, Ala.
Murray, James M., Pfc
Viola A. Murray
526 S. Lehigh St., Baltimore, Md.
Myers, Oscar M., Sgt
Kate S. Myers
Rte. #1, Carbon Hill, Ala.
McAllister, William V., Pvt
Viola A. McAllister
26 Grove Ave., Pt. Richmond, N. Y.
McBurnett, James B., Pvt
Rosa M. McBurnett
1111 Arlington Ave., Atlanta, Ga.

McCarley, Jack C., Pvt
Flossie McCarley
Rte. #2, Temple, Okla.
McCauley, Delbert A., Pvt
Mamie A. McCauley
Rte. #2, Box 55, Orlando, W. Va.
McClendon, J. W., Pvt
Mary A. McClendon
3620 Pickett St., Greenville, Texas
McComas, Jennings, H., Pvt
Watt McComas
Amherstdale, W. Va.
McCormick, John W., Pvt
Helen L. McCormick
Box 306, Old Fort, N. C.
McDonald, William B., Pfc
Verla McDonald
Rte. #2, Mystic, Iowa
McElravy, Clardean L., Pvt
Mina M. McElravy
131 W. Idaho St., Weiser, Idaho
McGinn, William P., Pfc
James McGinn
1741 Arapahoe St., Boulder, Colo.
McIlhargey, James D., Cpl
Montigue L. McIlhargey
Box 128, Whitefish, Mont.
McKinney, Richard C., S/Sgt
Virgie O. McKinney
Lake Tozaway, N. C.
McLaughlin, John J., Sgt
Charolotta Soyre
421½ W. Virginia, Charleston, W. Va.
Nagel, Clifford V., Pfc
Mable Nagel
Rte. #2, Arlington, Minn.
Nanos, John, Pfc
Stella Nanos
118 S. Hoyne St., Chicago, Ill.
Neal, Eugene B., Pfc
Ernest M. Neal
Rte. #1, Chase, Kansas
Nehoitewa, Owen R., Pfc
Lillian Nehoitewa
Oraibi, Ariz.
Neider, Charles R., Pfc
Mariam Neider
Temple, Penna.
Nelson, Edwin R., Pfc
Ethel N. Nelson
Fennon, Utah
Niemiste, Emanuel E., Jr., Pfc
William M. Niemiste
Iron, Minn.
Nix, James F., T/5
Mary Davis Nix
Rte. #5, Rome, Ga.
Norris, Lloyd O., Pvt
Anna L. Norris
Rte #1, Bristow, Ky.
Norton, Sylvester, Pfc
Sara Jane Norton
Rte. #5, Chuckey, Tenn.
Nye, Robert D., Pfc
Eunice M. Nye
120 High, Carey, Ohio
Ochoa, Arthur O., Pfc
Beatrice Ochoa
1052 N. Herbert, Los Angeles, Calif.
O'Leary, Gerard E., Pvt
Patrick O'Leary
529 E. 105th St., Bronx, N. Y.
Olmstead, Clifton B., T/4
Bessie A. Olmstead
Rte. #2, Malta Bend, Mo.
Olson, Walter A., Pfc
Katherine Olson
1324 Norwood, Red Wing, Minn.
Olszewski, Joseph, Jr., Pvt
Helen Olszewski
4429 Tillman, Detroit, Mich.
Ona, Arthur, Sgt
Ludvik J. Ona
Rte. #5, Thief Falls, Minn.
O'Rear, Levi W., Jr., S/Sgt
Levi W. O'Rear, Sr.
Rte. #2, Willington, Texas
Ortega, Juan J., Jr., Pvt
Mary E. Ortega
Rte. #1, Trinidad, Colo.
Owen, Ralph L., Sgt
Grace O. Owen
Hallsville, Mo.
Palola, George E., Pfc
Irene Palola
2424 10th St., Berkeley, Calif.
Parinella, Joseph J., T/3
Joseph Parinella
1102 Filbert, Oakland, Calif.
Pearce, Oscar E., Pvt
Lucile B. Pearce
Folkston, Ga.
Pederson, Peter, Pvt
Rogna Pederson
124 Calif. St., McCloud, Calif.
Penning, Wm. J., Pvt
Helen Penning
1014 Westnedge, Kalamazoo, Mich.
Pereau, William P., Pvt
Myrtle Pereau
914 N. Avon, Burbank, Calif.
Perez, Florenzo, Pfc
Lupe Perez
601½ Studebaker Rd., Norwalk, Calif.
Peterson, Waldo, Pvt
Emma Peterson
Rte. #1, Meyers Mill, S. C.
Petter, Alfons F., Pfc
Matilda Petter
Rte. #2, West, Texas
Petty, Gene W., Pvt
Annie Petty
Rte. #4, Waverly, Tenn.
Phillips, Lee E., Pfc
Viola Phillips
301 Spruce, Rosewood PO, Eureka, Calif.
Phillips, Newell E., Pfc
Delbert Phillips
578 Grand Blvd., Detroit, Mich.
Pierce, James T., Pvt
Mary N. Pierce
Willow, Ark.
Piller, Frank J., Pvt
Betty A. Piller
Box 497, Washougal, Wash.
Pillow, Clyde E., Pvt
Ophelia R. Pillow
Atwood, Tenn.
Pizzati, John, Pvt
Evelyn Pizzati
2031 W. 12th St., Brooklyn, N. Y.
Pojanowski, Raymond T., Pfc
Harriet Ann Pojanowski
Rte. #4, Anoka, Minn.
Pokorni, Herbert A., Pfc
Josephine Pokorni
2149 Irving Pk. Rd., Chicago, Ill.
Post, Gerald D., Pvt
Bernice Post
Rte. #3, New Castle, Penna.
Provencher, Wm. H., Pvt
Pauline E. Provencher
320 Lowell, Manchester, N. H.
Ptomey, Jas. E., Pfc
Benj. F. Ptomey
Rte. #2, Troy, Kan.
Purcell, Walter L., Pfc
Paul Purcell
1124 Jackson St., Owensboro, Ky.
Raine, Thomas W., 1st Lt
Mrs. J. W. Raine
Alderson, W. Va.

Ray, William G., T/Sgt
George G. Ray
Rte. 1, Franklin Rd., Birmingham, Mich.
Redmond, Cecil R., Pfc
Luthasha R. Redmond
Rte. #1, New Hope, N. C.
Reece, Charles L., Jr., Sgt
Opal T. Reece, Jr.
539 Hudson St., Gainesville, Ga.
Reed, Jim N., Pfc
Marvin R. Hughett
Langley Rt., Glenwood, Ark.
Reed, Ollie W., Jr., 1st Lt
Mrs. Ollie W. Reed, Jr.
Manhattan, Kansas
Rempel, Dietrich B., Pfc
Dietrich B. Rempel, Sr.
Rte. #1, Peabody, Kansas
Reyes, Samuel G., Pvt
Juan Reyes
Rte. #3, San Benito, Texas
Reynolds, Fred R., T/4
Ida Mae Reynolds
Rte. #3, Jefferson, Texas
Rhiner, Gilbert F., Pfc
Bela B. Rhiner
Rte. #1, Godwin, N. C.
Rice, John W., S/Sgt
Mrs. Verna Ulum
760 6th Ave., San Francisco, Calif.
Richter, Frederick M., Jr., Pvt
Eleanor M. Richter, Jr.
57 Central Ave., Albany, N. Y.
Riley, Arthur E., Pfc
June Riley
21 2nd Ave., N. Troy, N. Y.
Riley, Kenneth, Pfc
Nellie Riley
Auberry, Calif.
Roberts, James E., Pfc
John T. H. Roberts
Rte. #6, Hopkinsville, Ky.
Robinson, John B., Pvt
Jean Robinson
817 Clifton Ave., Clifton, N. J.
Rodriguez, Alonzo, Pfc
Simon Rodriguez
321 E. Fay St., Edinburgh, Texas
Rogers, Victor M., Sgt
Gloria L. Rogers
Elm St., Holly Hill, Fla.
Rosellini, Bruno R., 1st Lt
R. H. Nahood
1286 Gilamn Ave., San Francisco, Calif.
Royster, James C., Pfc
Justine C. Royster
Edgefield, S. C.
Rubinstein, Isidore, Pfc
Mary Rubinstein
2555 Grand Concourse, New York, N. Y.
Ruccio, George W., Pvt
Rose A. Ruccio
Darragh, Penna.
Russell, Howard M., Cpl
Florence D. Russell
2549 41st Astoria, L.I. N.Y.
Russomano, Arthur, Pfc
Fannie Russomano
63 S. 5th Ave., Long Branch, N. J.
Sadler, Bertram H., 2nd Lt
Sara R. Sadler
727 Locust Ave., Charlottesville, Va.
Sams, Lyle, J., Pfc
Ethel E. Sams
479-A 14th St., San Francisco, Calif.
Sanchez, Luciano, Pvt
Rose G. Sanchez
1501 N. 8th St., Albuquerque, N. M.
Sanford, Albert L., S/Sgt
Emma J. Sanford
3646 S. Hope St., Los Angeles, Calif.
Sansing, David L., Pfc
Wilma J. Sansing
Box 28, Peniel, Texas
Saponiere, Wm., Pvt
Anna Saponiere
366 Vermont St., Brooklyn, N. Y.
Saxon, William B., S/Sgt
Helen Saxon
Rio Grande City, Texas
Saylor, Daniel D., Sgt
Rena Saylor
Brussett, Mont.
Schank, Orland W., S/Sgt
Margaret Pauline Schank
111 E. Waldo, Independence, Mo.
Schaper, Vernon W., Pvt
Hilda Schaper
201 Brady St., Portage, Wisc.
Scheiner, Henry, Pvt
Ana Katz
626 Longdale Ave., Toronto, Canada
Schiano, Thomas L., Pfc
Pedro Schiano
1645 Grant Ave., San Francisco, Calif.
Schirmer, Walter H., Pfc
Hazel Schirmer
2805 E. 26th St., Minneapolis, Minn.
Schmidt, Loren F., Pfc
Minnie Schmidt
Ellendale, Minn.
Schneider, Robt. E., Pfc
Viola E. Schneider
Day, Calif.
Schronce, William H., Pfc
Eunice C. Schronce
Rte. #1, Maiden, N. C.
Scott, Herbert, Pfc
Celestia H. Scott
2629 Abbe St., Oakland, Calif.
Shaffer, Arnold D., Pfc
Scott A. Shaffer
Rte. #1, Gridley, Kan.
Shaffer, Henry P., Sgt
Christina Shaffer
Rte. #1, Mitchell, Nebr.
Shaffer, Leo E., Pfc
Anna Shaffer
Logan, Iowa
Shain, Lewis E., 2nd Lt
Marie H. Shain
3330 Blach Ave., Fresno, Calif.
Sheedy, James G., Pvt
Sarah N. Sheedy
Rte. #1, Seneca, Ill.
Shelton, Morris M., S/Sgt
Elizabeth B. Shelton
114 S. Spring St., Independence, Mo.
Sherrill, William T., Pvt
Mildred L. Sherrill
361 Kelly St., Statesville, N. C.
Shimer, Leland C., Pfc
Charles Shimer
Rte. #2, Binghamton, N. Y.
Shook, John E., Pvt
Eleanor S. Shook
2108 Freeport Rd., Natrona Hts., Penna.
Sievers, Edward L., Pfc
Georgia Sievers
1115 Superior Ave., Dayton, Ohio
Simko, Francis J., Pvt
Mary Simko
309 E. Elm St., Torrington, Conn.
Simmons, Carl M., Sgt
Nannie Simmons
521 Clark St., Bluefield, W. Va.
Simpson, Hubert S., Pvt
Hazel Simpson
Rte. #1, Box 509A Concord, N. C.
Six, Ralph, Pfc
Hattie Six
Rte. #1, Holbrook, Penna.

Small, James D., Sgt
Dorothy V. Small
1117 Tower Grove, St. Louis, Mo.
Smelcer, John R., Pvt
Georgie C. Smelcer
Rte. #4, Sevierville, Tenn.
Smith, Cecil L., Pvt
Betty Leona Smith
Rockport, Mo.
Smith, John D., Pvt
Gerothy P. Smith
Rte. #3, Toccoa, Ga.
Smith, Richard R., S/Sgt
Violet Smith
844 N. Washington, Bridgeport, Ill.
Smith, Robert M., Pfc
Edith H. Smith
4135 N. Juneau St., Portland, Ore.
Smyth, Harry A., Pvt
Hazel Marie Smyth
711 W. 9th St., Dixon, Ill.
Snead, Walter R., Pvt
Flora Snead
112 Atkins St., Shreveport, La.
Snider, Elmer, Pvt
Rose Snider
Qulin, Mo.
Solganick, Isadore, Pfc
Elizabeth R. Solganick
3735 W. 19th St., Chicago, Ill.
Solsman, Wm. R., Pvt
Esther Solsman
1208 N. Main St., Scranton, Penna.
Spears, Barney F., Pvt
Alma Ramey Spears
Rte. #1, Nauvoo, Ala.
Spiewak, Edward J., Pvt
Marie Spiewak
1987 E. Grand Blvd., Detroit, Mich.
Spirito, Nicholas D., Pfc
Rose Spirito
405 Pacific Ave., Jersey City, N. J.
Split, John E., Pfc
Minnie Split
1060 Sheridan S.W., Grand Rapids, Mich.
Staples, Leroy E., Pvt
Rose Staples Clark
2571 Lincoln Ave., Ogden, Utah
Stark, Junior G., Pfc
1208 Bluff St., Keokuk, Iowa
Starr, John C., Pvt
Josephine Starr
Rte. #1, Sharon, Penna.
Stephens, Charles P., Jr., T/Sgt
Charles P. Stephens
Rte. #2, Mendon, Mo.
Stephens, John A., Pvt
Lorena E. Stephens
Rte. #1, Arden, N. C.
Stern, Alvin J., Sgt
Emma Stern
Crandell, S. D.
Stoll, Theron E., Pfc
Myrtle Stoll
10 Hillside Ave., Neltong, N. J.
Strange, James E., Pfc
Louise Strange Moody
Rte. #2, Allendale, S. C.
Strobel, Charles D., Pfc
Helen M. Strobel
226 S. 1st St., Vandalia, Ill.
Strubbe, Clarence C., T/Sgt
Mrs. Clarence C. Strubbe
6605 S. Figueroa, Los Angeles Calif.
Sylvester, Ralph, Jr., Sgt
Blanche Sylvester
Rte. #4, Bloomington, Ind.
Szczombrowski, Henry J., Pvt
Mary Szczombrowski
Rte. #4, Hagaman, N. Y.
Tamker, John F., S/Sgt
Agatha Tamker
2021 Jane, Pittsburgh, Penna.
Taylor, Joseph L., Pvt
Louis T. Taylor
Box 284 Priest River, Idaho
Taylor, Wm. B., Pfc
Mrs. Wm. Taylor
3301 Broad Acres, Vanport City, Ore.
Tedford, Johnnie T., Pfc
Maudie Tedford
Farmersville, Calif.
Thomas, Cletus W., Pvt
Alvin Thomas
St. Boniface, Penna.
Thomas, George R., Pvt
Betty Thomas
Mentor Ave., Box #1, Mentor, Ohio
Thomas, Harry D., Pfc
Blanche Thomas
Rte. #2, Palisade, Minn.
Thomas, John E., Pvt
Roy S. Hunt
2020 S. Broad St., Philadelphia, Penna.
Thomas, Roy H., Jr., Pfc
Mrs. Roy H. Thomas, Jr.
Rte. #1, Snow Hill, N. C.
Thompson, John E., Pvt
Frances Thompson
Rear 128 2nd St., Conemaugh, Penna.
Thornburgh, Wilbur C., Pvt
Carrie B. Thornburgh
Rte. #1, Perry, Iowa
Thornhill, Theo T., Pfc
Pauline F. Thornhill
545 Alviso St., Santa Clara, Calif.
Thornton, Emory G., Pfc
Lauvina E. Thornton
Valleyview Apts., Pulaski, Va.
Thums, Daniel J., S/Sgt
Anna Thums
Rte. #1, Westboro, Wisc.
Tice, Gary P., Pfc
Mrs. Myrtle Rothijen
Rte. #1, Boring, Ore.
Tidwell, James H., Pvt
Flora K. Tidwell
Rte. #1, Delmar, Ala.
Trotta, Dominic P., Pfc
Filomena E. Trotta
3239 W. Flournoy St. Chicago, Ill.
Tucker, Jim, Pvt
Viley L. Tucker
Gen. Del., Ashland, Miss.
Turner, Arlyn R., S/Sgt
Gertrude Turner
1750 N. Humboldt Blvd., Chicago, Ill.
Turner, Lucius L., S/Sgt
Frances J. Turner
Box 2, Downieville, Calif.
Tyson, Lynn E., Pfc
Grace M. Sells
Rte. #1, Kearney, Nebr.
Urban, Arnold A., Pfc
Lillian Urban
825 Penn St., Sturgeon Bay, Wisc.
Vaca, Charles R., Cpl
Lucille Vaca
1437 Berkeley Way; Berkeley, Calif.
Vaca, David L., Pfc
Lupe Vaca
Box 36, Templeton, Calif.
Valdez, Albert C., Pfc
Mercedes Valdez
Box 56 Caliente, Calif.
Vana, Alex R., Pvt
Josephine Vana
Rte. #2, Halettsville, Texas
Vasques, Frank, Pfc
Ignacio Vasques
Elgin, Ohio

Vaughan, Henry S., Pvt
Marie Vaughan
2500 N. Richmond St., Chicago, Ill.
Vinovich, Daniel D., Pfc
Danica Vinonich
3659 S. Grand St., St. Louis, Mo.
Vitagliano, John, Pvt
Millie Deck
2 Aspen St., Auburn, N. Y.
Waggoner, Samuel H., T/Sgt
Mrs. Anna M. Waggoner
100 Luttrell St., Knoxville, Tenn.
Walker, William H., Pfc
Grace M. Walker
Rte. #2, Box 261 Brawley, Calif.
Walters, Simon A., Pfc
Anna G. Walters
4712 Greenhill Ave., Baltimore, Md.
Warner, Benjamin J., Pfc
John Warner
Rte. #1, Siren, Wisc.
Warren, Arthur, Pfc
Nancy Warren
Main St., Cold Springs, N. Y.
Warthen, Gerald L., Pfc
Mrs. Gerald Warthen
3046 Manhattan Ave., La Carscenta, Calif.
Waters, Charles W., Pfc
Mary Waters
Lake City, Iowa
Watson, George E., Pvt
Ida Anderson
3034 Hudson, Detroit, Mich.
Weaver, W. J., Pfc
Della Weaver
Rte. #1, Box 108, Chestnut, La.
Wesseler, Bernard H., Pvt
Della A. Wesseler
Rte. #8, Rushville, Ind.
Wesson, Clint A., S/Sgt
Vassie McClure Wesson
Box 222 Bessemer City, N. C.
West, Arbie J., Sgt
Eddie J. West
Rte. #5, Fayetteville, N. C.
Wheeler, Raymond R., Pvt
Roberta A. Wheeler
619 E. Princeton Ave., College Park, Ga.
White, William C., Pvt
Alice B. White
Rte. #2, Bufords, Tenn.
Whitefield, Ralph H., Pfc
Gladys A. Whitefield
328 Louisa, Peoria, Ill.
Wieland, Harold W., Pfc
Harold F. Wieland
7420 Reilly Ave., St. Louis, Mo.
Wilson, Leonce L., Pfc
Leon Wilson
212 E. Clinton St., Johnstown, N. Y.
Winfrey, Wiley J., Pfc
Marian D. Winfrey
Portage Des Sioux, Mo.
Wixon, James R., Pvt
Elizabeth Wixon
25 Madison Ave., Yonkers, N. Y.
Wolf, Albert, Pvt
Next of Kin Unknown
1940 E. 3rd St., Los Angeles, Calif.
Wolf, Francis N., Pvt
Pearl V. Wolf
938 Lancaster Ave., Reading, Penna.
Wolfe, Lloyd A., Pvt
Irene E. Wolfe
Cambridge, Nebr.
Wolkenfelt, Alexander J., Pfc
Clara Wolkenfelt
103-30 42nd Ave., Corona, N. Y.
Wood, Paul E., Pvt
Florence C. Wood
Rte. #1, Roseville, Ohio
Woodland, Ward C., Sgt
James M. Woodland
635 E. Halliday, Pocatello, Idaho
Woody, John C., Pfc
Kate S. Woody
Rte. #2, Hoschton, Ga.
Wright, Charles L., Pfc
John A. Wright
Rte. #2, Corryton, Tenn.
Wroten, Alfred I., Pvt
Iva M. Wroten
Andrews, Md.
Yanocha, Matthew J., Pvt
Stella Yanocha
Linfield, Penna.
Yeazel, Samuel L., Pfc
Annabel Yeazel
2723 Atlantic Ave., Cincinnati, Ohio
Yonkers, Russell W., Pvt
Sophie Yonkers
1247 Ballard St., Grand Rapids, Mich.
Yontz, Billy S., Sgt
Ollie Shutt Yontz
Rte. #6, Winston-Salem, N. C.
Young, Arch B., Jr., Cpl
Mavis Hicks Young
289 SE St., Globe, Ariz.
Zapf, Newell P., Cpl
Chas. F. Zapf
57 N. Laurel, Ventura, Calif.
Zivicki, Hubert D., Pfc
Ben Zivicki
311 Walker St., Stevens Point, Wisc.

91st SIGNAL COMPANY

Erickson, Lloyd R., Sgt
Mrs. Mary Erickson
Rte. 1, Arkdale, Wis.
Messinger, John W., T/5
Mr. John Messinger
806 Beauer Ave., York, Neb.
Schrader, Robert L., Pvt
Mrs. Elizabeth P. Schrader
437 South Hillside, Wichita, Kan.

91st QUARTERMASTER COMPANY

Farnell, Lloyd D., Pfc
Norman P. Farnell
Rte. 10, Box 13A, Tacoma, Wash.
Johnson, Harris, Jr., S/Sgt
Harris Johnson, Sr.
30 No. Front St., Wheeling, W. Va.
Lynch, William P., Pfc
Mrs. Dile H. Lynch
Rte. 3, Sanford, N. C.
McCoy, Raymond I., T/4
John E. Hayes
RFD 1, Clarkton, Mo.
Ward, Vance J., Sgt
Mrs. Frank J. Ward
587 Cass Ave. SE., Grand Rapids, Mich.

791ST ORDNANCE COMPANY

Friar, Howard W., T/3
Mr. Harvey Friar
Montpelier, Ind.
Gorius, Wilfred F., T/3
Mr. Joseph Gorius
Rte. 2, Anamosa, Ia.
Gray, Lynn R., Sgt
Mrs. Charles W. Gray
Viewfield, S. C.
Jagielski, Norbert B., T/5
Mrs. John Jagielski (Mother)
1241 S. 31st St., Milwaukee, 4, Wis.
Sorrells, Clyd H., Pvt
Mrs. Velora R. Sorrells
Box 652, Haskell, Tex.

DIVISION ARTILLERY

Clark, Paul F., Jr., Pfc
Paul F. Clark
485 Warren St., Brookline, Mass.
Howell, Madison L., Cpl
Mrs. Ruth Howell
5092 Elmonte, Fresno, Calif.

916TH FIELD ARTILLERY BATTALION

Adgeoff, Theodore, Pvt
Mrs. Alice Adgeoff
1217 W. Grand Blvd., Detroit, Mich.
Allen, Erastus, Pfc
Mrs. Josephine S. Allen
Cherrylog, Ga.
Burton, John W., Pvt
Mrs. Martha C. Burton
Gen. Del., Jacksonville, N. C.
Dalziel, William, Pvt
Mrs. Helen Dalziel
20207 Coventry, Detroit, Mich.
Davis, Robert P., T/5
Mrs. Delcie C. Davis
Goldston, N. C.
Hand, Delmar L., T/5
Mrs. Maude Hand
Rte. 1, Box 64, Damascus, Va.
Higgins, Roy, T/5
Mrs. Mary B. Higgins
Box 58, LaMesa, N. Mex.
Horton, Fred L., Cpl
Mrs. Lissie S. Horton
Rte. 2, Lutts, Tenn.
Hughes, Leonard, Pvt
Mrs. Bernice Hughes
Rte. 3, Paris, Tenn.
Hunter, Howard M., Jr., Cpl
Mrs. Clara Hunter
3437 Mockingbird Lane, Dallas, Tex.
Gailey, Newton C., S/Sgt
Mrs. Maudie T. Gailey
413 Copeland St., Gainesvile, Ga.
Gurr, Howard A., Pvt
Mr. Verne Gurr
321 S. Bell Ave., Chicago, Ill.
Koenig, Arthur M., Pvt
Mrs. Monica P. Koenig
842 N. Rockwell St., Chicago, Ill.
Lyons, Shelly R., Pvt
Mrs. Edna Lyons
RR 1, Clifty, Ky.
Moore, Carl F., Pvt
Sidney E. Moore
Rte. 3, Box 283, Benton, Ark.
Nichrasz, Anthony, Pvt
Mrs. Genevieve J. Nichrasz
347 Henderson St., Jersey City, N. J.
Nort, Elroy A., T/5
Mrs. Sophia Nort
Box 195, Ilwaco, Wash.
Nunn, Robert E., Pfc
Mrs. Grace Nunn
Rte. 1, Clear Lake, Minn.
Richardson, Connor W., Pvt
Mrs. Artha Roberts
Arapahoe, N. C.
Royce, Frederick O.
Mrs. Rose M. Royce
Star Rte. Gilman, Wis.
Rybicki, Stanley F., Pvt
Mrs. Margaret Rybicki
288 Potomac Ave., Buffalo, N. Y.
Spooner, William A., 1st Lt
Tekell, James L., 1st Sgt
Mrs. Elizabeth Tekell
601 N. 9th St., Waco, Tex.

346TH FIELD ARTILLERY BATTALION

Birk, Richard O., 1st Lt
Mr. Otto J. Birk
Rte. 2, Jackson, Mo.
Bonds, Clyde R., Cpl
Mrs. Irene A. Bonds
Rte. 1, Guntersville, Ala.
Kunesh, Emanuel J., T/5
Mrs. Joseph Schwiner
2501 No. 4th St., Milwaukee, Wis.
Walker, Hubert G., Pvt
Mrs. Bessie R. Walker
Rte. 1, Box 61, Franklinville, N. C.

347TH FIELD ARTILLERY BATTALION

Atwell, Louis E., Cpl
Mary Atwell
5640 Harding Ave., Detroit, Mich.
Bonham, Carl T., Pvt
Tom M. Bonham
Rte. 1, Arden, N. C.
Bush, Oscar J., T/4
T. M. Moore
Gen. Del., Jean, Tex.
Gillies, William J., Pvt
George H. Gillies
40 Marshall Field Estate, Hunnington, Suffolk, N. Y.
Jurgens, Lorance H., T/5
Edna M. Jurgens
1615 3d Ave., Kearney, Neb.
Murphy, Albert L., Cpl
Jane Murphy
5929 Lansdowne Ave., Philadelphia, Pa..
Neumann, Milton A., Pfc
Elsie Neumann
3379-A, No. 24th St., Milwaukee, Wis.
Oswald, Edgar H., T/5
Julia O. Bewig
5505 Finkman, St. Louis, Mo.
Smith, Ernest L., Cpl
Carniece L. Smith
Rte. 2, Box 275, Sikeston, Mo.

348TH FIELD ARTILLERY BATTALION

Dziedzic, Walter, T/5, **NBC**
Paul Dziedzic
2433 Homen Ave., Chicago, Ill.

Polinick, Edward., Pvt
Helen Polinick
519 Nipsic Road, Glastonbury, Conn.

316TH ENGINEER BATTALION

Barton, Francis C., 1st Lt, **NBC**
Mrs. Dorothy L. Barton
Box 144, Aruada, Colo.

Bostic, Paul W., T/5, NBC
Mrs. Gladys G. Bostic
2107 6th Ave., Parkersburg, W. Va.

Ellione, Felix E., Pfc
Mrs. Mary Hanson
213 S. Madison, Hinsdale, Ill.

Gardner, Francis E., Pvt
Mrs. Katherine C. Gardner
Como Sta., Apt. 3, St. Paul, Minn.

Golubic, John J., Pvt
Mrs. Anna McCreary
Harlan, Ky.

Gonzales, John H., Sgt
Mrs. Juanita Gonzales
Mt. Harris, Colo.

Holloway, Ollie J., T/4
Mrs. Maggie V. Holloway
Civ. Housing Project, AAB Clovis, N. Mex.

Iwers, Joseph H., T/5
Mrs. Minnie Struck
4530 Orchid Ave., Los Angeles, Calif.

Lane, Robert R., S/Sgt
Mrs. Edith Lane
Box 4, St. Francisville, Ill.

Marling, Ross P., S/Sgt
Mrs. Bessie L. Marling
3449 2d Ave., Los Angeles, Calif.

Motley, John J., 1st Lt
Mrs. Alice Motley
236 E. Mineola Ave., Valley Stream, N. Y.

Nevils, Alvin F., Pfc
Mrs. Blanche Nevils
Rte. 2, Box 41, Kinder, La.

Pollis, Joseph, Pfc
Mrs. Anna Pollis
Box 184, Stone, Ky.

Ray, Loyd C., S/Sgt
Mrs. Claud Ray
Rte. 2, Sulphur Springs, Tex.

Smiderle, Ralph, Cpl, **NBC**
Mrs. Magdalena Smiderle
Box 136, Guasti, Calif.

Smith, Earnest B., Pfc
Mrs. Bertha B. Smith
Rte. 2, Box 230, Marigold, Miss.

Southern, Dean H., Pvt
Mrs. Treva Southern
Mingo, Ia.

316TH MEDICAL BATTALION

Birchfield, James S.
Sallie S. Birchfield
311 Thornton Ave., Brookfield, Huntsville, Ala.

Cullen, Phillip B., Cpl
Mrs. Mary I. Cullen
2358 Gatewood St., Los Angeles, Calif.

Farber, Donald A., Sgt
Olive Farber
1171 Valencia St., San Francisco, Calif.

Garrels, William C., S/Sgt
Mrs. Ida B. Martin
Payson, Ariz.

Gil, Gilberto, Pvt
Candido Gil
Box 963, Raymondville, Tex.

Gonsowski, Thomas J., **NBC**
Mrs. Mary Gonsowski
Weyerhauser, Wis.

Nerbas, Arthur H., 1st Lt
Mrs. A. H. Nerbas
Medford, Oreg.

Roberts, Harold D., Sgt
Mrs. Mary McCart
Baker, Fla.

Stewart, Lloyd H., Pfc, **NBC**
Mrs. Faye Stewart
Corona, Calif.

91ST CAVALRY RECONNAISSANCE TROOP

Dessen, Donald I., Cpl
Mrs. Elinor R. Dessen
c/o Mrs. Hazel Mckeen, 1807 Franklin Ave. SE, Minneapolis, Minn.

Farmer, William C., Pfc
Mrs. Maggie L. Farmer
Rte. 1, Pikeville, Tenn.

Gaskins, Henry H., Pfc
Henry Luke
505 S. Blvd., Tallahassee, Fla.

Hall, Leo L., Sgt
Leo A. Hall
4529 Main St., Kansas City, Mo.

Riede, Edwin P., 1st Lt
Mr. A. W. Riede
100 Peasly St., Boise, Ida.

Seeddorff, Clifton L., Pfc
Mrs. Alice I. Rawlings
1117 W. White Oak St., Independence, Mo.

Wisz, Henry, Pfc
Mrs. Mary Wisz
10615 Bensley Ave., Chicago, Ill.

INDIVIDUAL AWARDS

MEDAL OF HONOR

Harmon, Roy W.,[1] *Sgt,*[2] *C* 362[3]
Pixley, California

Johnson, Oscar G., Pfc, B 363
Foster City, Michigan

DISTINGUISHED SERVICE CROSS

Ambrose, Francis G., S/Sgt, E 362
American Fork, Utah
Anderson, John W., 1stLt, F 361
Bend, Oregon
Bastron, LeRoy A., 1stLt, K 363
Walla Walla, Washington
Brown, Harry L., 1stLt, L 363
Henrietta, Texas
Carlson, Franklin, T/5, Med 362
Coeur d'Alene, Idaho
Conley, Edward J., Capt, G 363
Lorain, Ohio
Czinki, John, Pfc, B 362
Lincoln Park, Michigan
Darragh, Ralph F., Pfc, C 361
Fords, New Jersey
Douglass, Harry W., S/Sgt, I 361
Raleigh, North Carolina
Dullum, Roy H., T/Sgt, I 361
Alamo, North Dakota
Eggers, Carroll D., Sgt, H 362
Vilas, North Carolina
Fulton, William B., Capt, K 363
Berkeley, California
Gibson, Thomas, Pfc, L 361
Canton, North Carolina
Green, Jack E., Pfc, E 362
Seattle, Washington
Greig, Alexander M., S/Sgt, L 363
Stroughton, Wisconsin
Hamilton, Harry E., Pfc, E 362
Portland, Oregon
Hutson, Robert J., S/Sgt, F 363
San Diego, California
Kaich, Elmer W., T/Sgt, F 362
Tyler, Texas
Lake, Johnny D., Sgt, F 363
Santa Rosa, California
Lerma, Rosario V., T/5, Med 361
Los Angeles, California
Lloyd, Clyde W., T/Sgt, E 361
Houston, Texas
Loe, Joseph D., Pfc, I 361
Coffeyville, Kansas
Martin, Kenneth E., S/Sgt, B 757 Tk Bn[4]
Clarksburg, West Virginia
Montooth, William A., S/Sgt, L 363
Sweetwater, Tennessee
McKelvey, Russell M., 2dLt, F 363
Houston, Texas
Rexall, Teddie B., Pvt, Hq 362
Magnolia, Arkansas
Sheff, Bertram N., Maj, 2d Bn 361
Unknown
Thompson, Robert C., 1stLt, 755 Tk Bn[4]
Seattle, Washington
Turner, Bruce K., Pfc, Med 361
Huron, Michigan
Van Osdol, Joseph O., Pfc, E 362
Jefferson City, Missouri
Weaver, Howard E., Pvt, F 363
Harrisburg, Pennsylvania
Westfall, Welton C., S/Sgt, F 362
Pomona, California
Wilson, George G., 1stLt, I 361
Albany, New York

OAK LEAF CLUSTER TO SILVER STAR

Christopher, James L., T/5, Med 363
East Cleveland, Ohio
Cohea, Jesse L., Capt, E 362
Springfield, Tennessee
Doornhaag, Arnold, T/Sgt, C 361
West Branch, Michigan
Drazkowski, Frank A., 2dLt, B 363
Bessemer, Michigan
Fox, Raymond W., Cpl, Med 361
Portland, Oregon
Gnemi, Martin L., S/Sgt, E 362
St. Louis, Missouri
Jewkes, Rex W., Pfc, I 363
Orangeville, Utah
Krieger, Mervin, 1stLt, K 361
San Mateo, California
Little, Ralph V., 1stLt, I 362
Bellows Falls, Vermont
Meyer, Frank J., 1stSgt, I 361
St. Meinard, Indiana
McDermott, James L., T/Sgt, G 363
San Luis Obispo, California
Pope, David E., 1stLt, A 361
Forrest City, Arkansas
Turner, William H., Pvt, Med 363
Manson, Tennessee
Van Scoyk, Reed A., S/Sgt
Willamina, Oregon
Med 361

SILVER STAR

Abrego, Manuel B., Pfc, C 361
San Benito, Texas
Ace, Paul A.
Adams, Dewane J., Cpl, S 363
Washington, Utah
Adams, Samuel A., Pfc, E 361
Eugene, Oregon
Akers, Alfred, Pfc, E 363
Joplin, Missouri
Allen, Lee D., T/5, Med 363
Santa Rosa, California
Allen, Vaughn E., 2dLt, A 755 Tk Bn[4]
Layton, Utah
Anders, James D., 1stLt, L 363
Oil City, Louisiana
Anderson, Cletus O., Pfc, E 363
Williamsport, Maryland
Anderson, Robert C., Pfc, K 362
Appleton, Wisconsin
Anstaett, Robert L., Pfc, E 362
Osage City, Kansas
Arquette, William, Pfc, 361
Yakima, Washington
Aubrey, Sam B., 1stLt, C 361
Tulsa, Oklahoma
Bacciglieri, Louis T., Pfc, L 363
Martinez, California
Bacheliar, Clifford J., Sgt, K 361
Amelia, Ohio
Bahr, George R., Sgt, C 361
Esterville, Iowa
Bailey, Douglas R., Sgt, H 361
Cheney, Washington
Baird, Clare W., 1stLt, L 362
Flint, Michigan
Baker, Beryl H., S/Sgt, E 363
San Diego, California
Ball, Walter G., Maj, Hq 2d Bn 361
Scranton, Pennsylvania
Ballard, James O., Cpl, M 362
Lawrenceville, Illinois
Baptista, David, Pfc, A 361
Hughson, California
Bargsten, Klaus W., Pvt, G 363
Beall, Edward C., Pfc, A 362
Elk, California
Beck, James M., Capt, Hq 2d Bn 362
Valdosta, Georgia
Benckart, Robert G. Jr., 1stLt, E 363
Bloomington, Indiana
Berg, Frederick W., Pfc, B 361
Highland Park, New Jersey
Berry, Eugene A., Pfc, E 361
Banner Elk, North Carolina
Biye, Charles T., Pfc, K 361
Manuelito, New Mexico
Black, Eli, Capt, C 361
Lake Ariel, Pennsylvania
Blakeney, James B., Pfc., M 363
Pahokee, Florida

---

[1]Names in *italic* indicate posthumous award.
[2]At time deed was done, unless otherwise noted.
[3]At time deed was done.
[4]Attached unit.

Bleclic, Carl, Pfc, Hq 3d Bn 362
Tiffin, Ohio
Blood, Everett G., 2dLt, B 362
Manhattan, Kansas
Bloom, Eugene H., Capt, C 361
Emporia, Virginia
Blum, Carl L., 1stLt, G 363
York, Pennsylvania
Blunt, Robert W., Sgt, B 363
South Milwaukee, Wisconsin
Bodle, Dale, 2dLt, C 362
Salina, Kansas
Boecheler, Andrew J., 1stLt, F 361
Karlsruhe, North Dakota
Boling, Ferdinand A., 1stLt, B 361
Clyde, Kansas
Boone, Jay D., 1stLt, L 361
Philadelphia, Pensylvania
Booth, Frederick W., 1stLt, L 362
Clearlake, Wisconsin
Borba, William A., Pfc, A 361
Niles, California
Bothman, Stanley F., T/Sgt, K 363
Hazel, Minnesota
Bowen, Grover E., Pfc, A 361
Seaton, Missouri
Bowles, Edward L., Pfc, E 362
Kamiah, Idaho
Bramlett, Thomas D., Sgt, F 361
Emmett, Idaho
Branch, Aldert D., T/Sgt, Hq 2d Bn 363
Jacksonville, Florida
Brigham, Horace D., Pvt, 91 Recon
Washington, D.C.
Brockmiller, Leonard C., T/4, Med 362
Milton Center, Ohio
Broedlow, Rudolph W., Col, CO 361
Chippewa Falls, Wisconsin
Brooks, Billy M., 1stLt, E 361
Little Rock, Arkansas
Brooks, James C., Pfc, K 363
Fresno, California
Brooks, James F., S/Sgt, M 363
Warsaw, Missouri
Brott, Kenneth C., Pfc, Med 362
Mountain Grove, Missouri
Brown, Arthur W., S/Sgt, F 363
Bakersfield, California
Brown, David A., Pvt, A 347
Childress, Texas
Brown, L. D., T/Sgt, A 362
Englewood, Colorado
Brozenich, Edward S., Pfc, K 361
Tanentum, Pennsylvania
Bruehl, George J., 2dLt, F 361
Shiocton, Wisconsin
Bryant, Bert L., S/Sgt, B 361
Disney, Oklahoma
Buchanan, Bennie D., Pfc, A 361
Pleasant Hill, N. C.
(MIA when Award made; KIA as of 17 March on 15 Oct. 44)
Bryant, Miller J., Pfc, B 363
Blue Sulphur Springs, West Virginia
Buchanan, Jack E., Pfc, L 363
Elk Park, North Carolina
Buie, Herman F., T/Sgt, I 362
Akron, Ohio
Bunnell, George D. Jr., Capt, Hq 347
Blairstown, New Jersey
Bunte, Albert G., 2dLt, Med 363
Berkeley, California
Burgess, Leo J., S/Sgt, K 361
Tony, Wisconsin
Burnett, Marion R., Pfc, K 362
Cushing, Oklaholma
Byars, Hilma L., S/Sgt, E 361
Martinez, California
Byington, Vernon H., Pfc, A 362
Grace, Idaho
Callaghan, John E., 2dLt, I 363
Cohoes, N. Y.
Campbell, Hiram E., Pfc, G 361
Neodesha, Kansas
Carlin, Thomas A., Pfc, E 363
Cleveland, Ohio
Carr, Robert D., Pvt, Med 361
Gallatin, Tennessee
Carron, John J., 1stLt, Hq 1st Bn 361
Pocatello, Idaho
Carroum, Omer D., 1stLt, L 362
Snackover, Arkansas
Caruso, Carmel Pfc, M 363
Oakland, California
Castillo, Ines G., Pfc, Med 363
Tulare, California
Castro, Leland G., Pfc, E 362
Visalia, California
Chacon, Ezequiel Jr., Pfc, C 362
El Paso, Texas
Chaffin, Lon D., Pfc, I 361
Gainsboro, Tennessee
Chambers, George K., S/Sgt, A 361
Meriden, Connecticut
Chappelle, Fred J., Pfc, AT 361
St. Louis, Missouri
Chastain, William R., Pfc, C 362
Little Rock, Arkansas
Christopher, James L., Pvt, Med 363
Columbus, Ohio
Cioci, Richard A., Pfc, A 363
Alliance, Ohio
Clark, James W., 1stLt, I 361
New Orleans, Louisiana
Clayton, Arvil O., Pvt, K 363
Dorris, California
Clift, Frank W. Jr., Pfc, D 361
Pontiac, Michigan
Coady, Thomas A., S/Sgt, B 363
Pawnee, Illinois
Cobb, James C., Pvt, 916
Bath, New York
Cochran, John V., S/Sgt, A 363
Sterling, Iowa
Cofield, Rudolph, S/Sgt, K 362
Rt. 1, Philcampbell, Alabama
Cohea, Jesse L. Jr., Capt, E 362
Springfield, Tennessee
Cohen, Albert, Pfc, E 361
Chicago, Illinois
Collins, Lloyd B., S/Sgt, A 362
Richmond, California
Colquitt, Newell, Sgt, I 361
Houston, Texas
Conn, Earl R., Sgt, D 363
Beloit, Kansas
Connelly, Robert J., S/Sgt, L 362
Phoenix, Michigan
Cook, Roy H., 2dLt, Hq Co 3d Bn 362
Loudon, Tennessee
Cortez, Alfred L., S/Sgt, L 361
Houston, Texas
Couch, Malba H., Sgt, L 362
Senoia, Georgia
Cousino, Edsol J., Pvt, Med 363
Toledo, Ohio
Cox, Louis A., Pfc, C 362
Iowa City, Iowa
Coz, Eifren G., 1stLt, G 361
Yuma, Arizona
Cozine, Alfred H., S/Sgt, C 361
Centerville, Missouri
Crawford, Enos V., S/Sgt, C 362
DeSota, Missouri
Crawford, Raymond R., Sgt, K 363
Glendive, Montana
Crayton, Clinton N., S/Sgt, K 361
Tickfau, Louisiana
Crego, Russell F., 2dLt, C 361
Depew, New York
Crews, Robert F., S/Sgt, F 363
Lawrenceville, Illinois
Cross, Millard, 1stLt, K 361
Oneida, Tennessee
Cullar, Lewis B., Pvt., E 363
Steamboat Springs, Colorado
Cummings, Herbert D., Sgt, C 916 FA
Marysville, California
Curtiss, Stanley M., Pvt, L 363
Kirkland, Washington
Cuthrell, Amos O., Pfc, K 362
New Bern, North Carolina
Dano, Maxwell E., Pvt, D 362
Clayton, New York
Daughtry, John P., T/Sgt, L 362
Smithfield, North Carolina
Davenport, Perry F., Pfc, D 361
Grover, Missouri
Davis, Joe L., Sgt, C 361
Crestview, Florida
Dawson, Andrew J., Pfc, C 362
Highland Park, Michigan
Dean, Riter H., Pvt, C 362
Grand Junction, Colorado
Derouen, David P. Jr., S/Sgt, G 361
Houston, Texas
Dewart, Lloyd J., 1stLt, D 361
Los Angeles, California
Diamond, Seymour, 2dLt, E 361
New York, New York
Dickens, Loman R., Cpl, K 361
Old Hickory, Tennessee
Dill, Kenneth B., Pfc, B 362
Neosho Rapids, Kansas
Diner, Irving, S/Sgt, I 363
Brooklyn, New York
Donadio, Carl J., Pfc, Can 361
Little Falls, New York
Donahue, Thomas F., III, T/5, S 362
West Orange, New Jersey
Doornhaag, Arnold, S/Sgt, C 361
West Branch, Michigan
Dougherty, Charles E., S/Sgt, K 361
Terre Haute, Indiana
Draper, Charles P., Pfc, S 361
Van Nuys, California
Drazhkowski, Frank A. Jr., S/Sgt, B 363
Bessemer, Michigan
Duke, George E., Pfc, L 362
Summerville, Louisiana

[1]Names in *italic* indicate posthumous award.
[2]At time deed was done, unless otherwise noted.
[3]At time deed was done.
[4]Attached unit.

Dugger, Peter, Pfc, I 361
Watsonville, California
Dunblazier, Max M., 1stSgt, E 361
McComb, Illinois
Dunham, Gerald L., Pvt, B 363
Sanford, Maine
Duran, Johnny S., Pfc, Hq 3d Bn 362
Big Spring, Nebraska
Eckard, Chester R., 1stLt, E 363
Gilroy, California
Eckroat, Virgil O., S/Sgt, G 363
Pampa, Texas
Edwards, Delbert L., S/Sgt, F 361
Richmond, Indiana
Edwards, Theodore F., Capt, Med 363
Louisville, Kentucky
Eisfeldt, Henry A., Pfc, F 363
Taylor, Texas
Elder, Inue E., Pvt, K 362
Jefferson, Georgia
Elliott, Leslie H., 2dLt, K 362
Coronado, California
Ellis, Ernest M., Pfc, F 361
Porterville, California
Emmert, John W., Pfc, D 362
Marysville, Ohio
England, Marvin D., Pfc, F 363
Green City, Missouri
Enlow, Roy, S/Sgt, E 362
Shelton, Washington
Evans, Joe M. Jr., 1stLt, L 363
El Paso, Texas
Eyherabide, Stephen P., 1stLt, L 363
Bakersfield, California
Fayant, Adolphlis J., Pfc, G 361
Lewiston, Montana
Fellows, Albert E., Sgt, E 362
San Antonio, Texas
Fernandes, Joseph R., S/Sgt, K 361
Los Angeles, California
Fieberling, Ernest L., 1stLt, F 361
Oakland, California
Finnell, Louis C., S/Sgt, M 363
Carona, California
Fitzgerald, John J., 1stLt, I 362
Brooklyn, New York
Fitzsimmons, Clifford E., Pfc, I 361
Sallisaw, Oklahoma
Flesher, Charles W., 2dLt, G 363
Indianola, Iowa
Fletcher, James J., 1stLt, B 361
Alderson, West Virginia
Flores, Simon, Pvt, E 362
Laredo, Texas
Flory, Harold E., Pvt, C 362
Chicago, Illinois
Foley, Thomas P., S/Sgt, A 362
Germantown, Pennsylvania
Foote, Gordon H., Capt, K 362
Whitehall, New York
Foster, John P., Pfc, K 362
Troy, Pennsylvania
Foster, Marion B., Pfc, Hq Btry 916
Fontana, California
Fox, Raymond W., Pvt, Med 361
Portland, Oregon
Fraser, Bruce H., 1stLt, E 362
Aberdeen, Mississippi
Frederick, Alvin P., Pfc, E 363
St. Louis, Missouri
Freggiaro, Joseph R., 2dLt, Med 363
Jackson, California
Fries, Robert E., Sgt, A 362
Jackson Heights, Long Island, New York
Frye, Robert M., 1stLt, E 361
St. Joseph, Missouri
Gallopo, Salvatore P., Pvt, D 363
Lodi, New Jersey
Galyon, Charles A., Pvt, Med 362
Forest Hills, New York
Gann, Williard C., Pfc, E 362
Pima, Arizona
Garcia, Fernando C., Pfc, K 363
Salinas, California
Garner, Alonza J., Pfc, B 363
Goodsprings, Alabama
Garretson, Richard T., Sgt, E 363
Ocean City, New Jersey
Gaultney, John O., 2dLt, I 361
Memphis, Tennessee
Geisen, Fred H., 1stLt, G 362
Dresden, North Dakota
Gendusa, Joseph F., Pfc, F 362
Milwaukee, Wisconsin
Genschaw, Walter W., 1stLt, F 361
Madison, Wisconsin
Georgopulos, Nick, Pfc, K 361
Indianapolis, Indiana
Gibson, Matthew E., 2dLt, 362
Hamilton, Ontario, Canada
Giddings, Carl R., 2dLt, B 363
Jerome, Michigan
Gifford, Harry F., Pfc, C 363
Steele, Missouri
Gilbert, William J., T/3, Med 362
Los Angeles, California
Ginocchio, Joseph B., S/Sgt, G 362
Reno, Nevada
Glee, Paul W., Sgt, C 361
Rochelle, Illinois
Gloor, Alton V., 2dLt, C 361
Gonzales, Texas
Gnemi, Martin L., S/Sgt, E 362
St. Louis, Missouri
Goar, Charles L., Pfc, M 361
Athens, Alabama
Godfrey, Arthur L., S/Sgt, Med 363
Mount Clemens, Michigan
Goe, Ralph H., S/Sgt, Med 363
Prescott, Washington
Gonzales, Henry G., Sgt, E 362
Los Angeles, California
Goodwin, Ernest L., T/5, Med 363
Portland, Oregon
Goolsbee, Rolland G., Pfc, AT 361
St. Claire Shores, Michigan
Graham, Lawrence P., S/Sgt, B 361
Fayetteville, North Carolina
Greczek, Alex S., S/Sgt, I 362
Chicago, Illinois
Green, James M., Sgt, Hq Co 3d Bn 362
Clayton, North Carolina
Green, John T., Pvt, 362
Buckman, Tennessee
Greenberg, Morris Pvt, E 363
Chelsea, Massachusetts
Gregory, Joe H., Pfc, 1st Bn Hq Co 361
Perry, Florida
Grobe, Robert P., Cpl, Med 362
Lebanon, Oregon
Hahn, Lowell R., Pvt, 916
Mason City, Iowa
Hale, Warren C., 2dLt, F 362
Corbin, Kentucky
Hall, David A., S/Sgt, C 804 TD[4]
Timber, Oregon
Hall, Donnie B., T/5, Med 363
White Bluff, Tennessee
Halpern, Jerome B., Pvt, C 362
New York, New York
Hambrick, William R., 1stLt, I 363
Albany, Georgia
Hammons, John H., Pvt, Med 361
Marysville, Tennessee
Hanger, Everette W., Pfc, H 362
Middle Brook, Virginia
Hanley, Jean D., T/Sgt, L 362
McKenzie, Montana
Hanson, Sherman R., Pfc, L 362
Klamath Falls, Oregon
Harding, James J., Pfc, K 363
Kansas City, Missouri
Harlin, Clarence R., Capt, F 362
Rock Springs, Wyoming
Harmon, Paul B., Pvt, C 362
Charleston, West Virginia
Harnden, Donald W., 2dLt, 916 FA
St. Paul, Minnesota
Harriman, Merwin T., 1stLt, M 362
Kansas City, Missouri
Harrison, James D., Cpl, AT 361
Mineral Wells, Texas
Hartman, John H., Pfc, B 362
Exeter, California
Harvey, William R., Pvt, M 361
St. Paul, Minnesota
Harze, Charles H., Pvt, K 363
Philadelphia, Pennsylvania
Hatfield, Herman W., Pvt, A 361
Bonifay, Florida
Hawkins, Richard I., S/Sgt, D 362
Arimo, Idaho
Hayes, Preston, C., S/Sgt, B 361
Hobbs, New Mexico
Heisinger, Ralph W., 1stLt, A 362
Jefferson City, Missouri
Hernandez, Ralph, Pfc, Med 363
New York, New York
Hewitt, William R., Pfc, A 361
Coalport, Pennsylvania
Hickman, Ray E., S/Sgt, G 361
Erbie, Arkansas
Higdon, Joseph D. Jr., S/Sgt, A 363
Waco, Texas
Hines, Robert B., Pvt, A 362
Timonium, Maryland
Hixson, Walter G., S/Sgt, S 363
Lamar, Colorado
Hockreiter, Francis, Pvt, K 363
Ashley, Pennsylvania
Holland, George W., Pfc, B 361
Mahan, Mississippi

---

[1]Names in *italic* indicate posthumous award.
[2]At time deed was done, unless otherwise noted.
[3]At time deed was done.
[4]Attached unit.

Holscher, Paul F. Jr., Capt, AT 362
Rocky Mount, North Carolina
Honstedt, Jack K., 2dLt, C 361
Lincoln, Nebraska
Horne, Aaron R. Jr., 2dLt, A 362
Fort Worth, Texas
Horne, Richard C., T/Sgt, I 362
Iron River, Michigan
Hospital, Ralph, BrigGen, CO Div Arty
Ithaca, New York
Hrelec, William J., Pfc, A 361
Aliquippa, Pennsylvania
Hughes, Leon, Pfc, L 362
Collins, Mississippi
Hunter, Howard M. Jr., Cpl, 916
Dallas, Texas
Hurst, Berdon C., S/Sgt, B 757 Tk Bn[4]
Gilroy, California
Huter, Theodore F., T/5, Med 362
Raymond, California
Hyland, Frank T., Capt, I 363
Chicago, Illinois
Hyland, Kenneth L., 2dLt, E 362
Pocatello, Idaho
Inman, Lloyd J., Capt, C 363
Metropolis, Illinois
Irvin, C. B., Pfc, I 362
Mayfield, Kentucky
Jasper, George J., Pfc, D 361
La Salle, Illinois
Jenkins, Horace B., S/Sgt, D 363
Midland, North Carolina
Jenkins, Kelly R., Sgt, A 362
Gloucester County, Virginia
Jenkins, Newell N., 1stLt, B 362
Humbolt, Kansas
Jennings, Bert H., Pfc, H 361
Kansas City, Kansas
Jewkes, Rex W., Pfc, I 363
Orangeville, Utah
Johnson, Donald V., Pvt, B 362
Fort Wayne, Indiana
Johnson, Howard T., Pvt, C 363
Atlanta, Georgia
Johnson, John W., Pfc, AT 361
East Prairie, Missouri
Johnson, Kenneth A., S/Sgt, A 361
Los Angeles, California
Johnston, George, Pfc, E 361
Baltimore, Maryland
Jones, Clyde E., Pvt, D 361
Cumberland Furnace, Tennessee
Jones, Robert L., 1stLt, Hq Co 1st Bn 361
Dickens, Iowa
Jones, Vernon F., Pfc, F 362
Idabel, Oklahoma
Jopke, Henry C., Pfc, G 363
Madison, Wisconsin
Karl, Frank Jr., Pfc, C 361
Cleveland, Ohio
Kasney, James J., 1stLt, I 363
Cleveland, Ohio
Katz, Norman P., Pvt, B 363
Chelsea, Massachusetts
Kearney, Philip J., T/Sgt, E 361
Platteville, Wisconsin
Keeling, Cleo, Pfc, E 361
Rockford, Illinois
Kelch, Clarence J., T/5, Med 363
Overland, Missouri
Keller, Forrest E., Pfc, C 361
Crestline, Ohio
Kenny, James F., Sgt, C 362
Crockett, California
Kern, Benjamin, Pfc, C 363
Eaton, Colorado
Kettman, Julius C., Pfc, K 363
San Jose, California
Keys, William, Pvt, B 361
Wauna, Oregon
King, Martin Jr., Pfc, I 363
Waco, Texas
Kline, John H., T/4, Med 362
Jones, Michigan
Klippert, Edward, Pfc, A 361
Lincoln, Nebraska
Kloeb, William C., 1stLt, G 362
Cincinnati, Ohio
Kopelov, Jerome H., Pvt, G 361
Kokomo, Indiana
Kovas, March P., 1stLt, E 363
South Bend, Indiana
Kracht, Gerald W., 1stLt, F 361
Centerline, Michigan
Kreuser, Joseph H., Pvt, Med 361
Phoenix, Arizona
Krieger, Mervin, 2dLt, I 361
San Mateo, California
Kubasinski, Harry M., S/Sgt, K 362
Detroit, Michigan
Kunicki, Casmer B., Pfc, F 361
Trenton, New Jersey
Lancaster, Edwin H., 1st Sgt, K 362
Weston, West Virginia
Land, Larry L., S/Sgt, B 363
Abernathy, Texas
Lane, Earl L., Pfc, A 361
Saginaw, Michigan
Laninga, Clarence, Pfc, L 362
Grand Rapids, Michigan
Lassiter, Buford, Pfc, B 361
Pocahontas, Tennessee
Lausten, Raymond L., Sgt, A 361
Cooks, Michigan
Lay, Forest A., Capt, L 362
Aurora, Illinois
Leath, Don C., Sgt, I 363
Lafayette, Tennessee
Lentz, William A., Pfc, B 361
Royal Oak, Michigan
Lerma, Rosario V., Pfc, Med 361
Los Angeles, California
Lester, Robert G., Pfc, A 362
Pasadena, California
Levin, Malcolm H., Pfc, C 361
New York, New York
Lewis, Dilmus L., Pfc, L 361
Lakeland, Florida
Lewis, Kenneth M., T/Sgt, I 362
Francisville, Illinois
Lewis, Ray G., 2dLt, C 755 Tk Bn[4]
Los Angeles, California
Lien, Floyd L., 1stLt, L 362
Clayton, Wisconsin
Lierman, Herbert G., 1stLt, M 361
Scotts Bluff, Nebraska
Lile, Edward F., Pfc, C 363
Greensburg, Kentucky
Little, Ralph V., 1stLt, I 362
Bellows Falls, Vermont
Locklear, William H., Pfc, A 362
Harrisburg, Illinois
Long, Glen C., LtCol, Hq 3d Bn 363
Oklahoma City, Oklahoma
Ludwig, Robert M., Pvt, L 362
Scranton, Pennsylvania
Lusk, Stewart R., S/Sgt, A 361
Herdon, West Virginia
Lynn, Thomas L., Sgt, L 362
Durham, North Carolina
Malec, Frank J., Sgt, M 363
Leavenworth, Kansas
Mann, Ray H. Jr., Pfc, I 362
Canton, North Carolina
Marcum, Wade H., Pvt, Med 363
Dunlour, West Virginia
Martin, Eugene C., Pfc, M 363
Shelby, North Carolina
Martin, Warren H., Sgt, H 362
New Orleans, Louisiana
Martin, William R., T/Sgt, 363
Oakland, California
Martinez, Alfred G., Pvt, E 362
Houston, Texas
Martinez, Joe S., Sgt, B 361
Bakersfield, California
Mason, Jesse, T/Sgt, G 361
Depew, Oklahoma
Massey, Alan, S/Sgt, E 362
Kansas City, Missouri
Matthews, Claude E., T/Sgt, I 363
Fairmount, West Virginia
Matthews, Norman P., Pfc, C 361
East Prairie, Missouri
Maxson, Kenneth Richard, 2dLt, C 916
Buffalo, New York
Maxwell, Ralph A., Pvt, E 362
Baxter, Tennessee
May, John L., Pfc, C 361
Pollock, Louisiana
Mayfield, William R., S/Sgt, C 362
Waldrow, Arkansas
Meadows, Walter T., 1stLt, B 361
Oklahoma City, Oklahoma
Meek, William I., 1stLt, L 361
Redondo Beach, California
Melcher, Donald R., Pvt, Med 362
Joliet, Illinois
Menchaca, Rudy, Pvt, B 363
San Antonio, Texas
Mesa, Tony, Pfc, C 361
Fresno, California
Metheny, Ralph J., Pfc, B 363
Altoona, Pennsylvania
Meyer, Frank J., 1stSgt, I 361
St. Meinrad, Indiana
Meyners, Ferdinand H., T/5, Med 363
Boyden, Iowa
Michael, Francis M., T/3, Med 361
Bakersfield, California
Miller, Rex D., Pvt, A 363
Lansing, Michigan
Mink, Eugene M., 1stLt, 362
Los Angeles, California
Mitchell, Thomas J., Pfc, C 757 Tk Bn[4]
Liverpool, West Virginia
Mitrano, Daniel C., 2dLt, M 362
North Easton, Massachusetts

[1]Names in *italic* indicate posthumous award.
[2]At time deed was done, unless otherwise noted.
[3]At time deed was done.
[4]Attached unit.

Mizerski, Walter W., Pvt, C 363
Pittsburgh, Pennsylvania
Mohler, Charles L., Capt, A 361
Arkansas City, Kansas
Montalvo, George A., Pfc, K 361
Watsonville, California
Moore, Byrel A., 1stLt, C 804 TD Bn[4]
Cap Rock, New Mexico
Moore, Daniel M., 1stLt, I 361
Bradford, Pennsylvania
Moore, Thomas K., Pfc, B 363
Marshville, North Carolina
Morrison, James J., Pvt, Med 361
Chuckey, Tennessee
Morton, Thomas C., 1stLt, G 362
Rochester, New York
Mudd, Elbert S., Pfc, E 363
Carreyville, Kentucky
Muller, Robert F., Capt, 3d Bn 363
Whitestone, New York
Mungle, Juvin J., T/5, Med 361
Atoka, Oklahoma
Murdoch, Craig J., S/Sgt, C 348
Renton, Washington
Murphy, Charles J., 2dLt, B 363
Sioux Ctiy, Iowa
Murray, Warren D., Pfc, F 362
Bell Gardens, California
Myers, Huston O., Pfc, K 363
Knoxville, Tennessee
McCarthy, Daniel T., Pfc, B 363
Chicago, Illinois
McCormick, French H., Pfc, I 361
Woodville, West Virginia
McDermott, James L., T/Sgt, G 363
San Luis Obispo, California
McDonald, Stanley W., 1stSgt, E 362
Petaluma, California
McGowen, Jess Jr., Sgt, F 362
Fanshawe, Oklahoma
McIlhargey, James D., Cpl, Med 363
Priest River, Idaho
McLane, Walter L., Pfc, B 362
Sacramento, California
McLaren, Hugh D., Pfc, L 363
Enosburg Falls, Vermont
McLean, Richard C., Pfc, B 362
Duluth, Minnesota
McNabb, Dale M., Pfc, Med 363
Moshein, Tennessee
Nanneman, Aloysius A., S/Sgt, I 362
Bynumville, Missouri
Nash, William J., Pfc, M 362
Minneapolis, Minnesota
Nelson, Earl T., Pfc, K 363
Anson, Maine
Nethig, Victor A., Pfc, G 361
Barron, Wisconsin
Newell, Lester G., S/Sgt, K 361
Coeur d'Alene, Idaho
Nisson, Ivan G., T/5, Med 362
Swan Lake, Idaho
Noakes, Ray W., Pvt, M 363
Rolla, Missouri
Notaro, Ross A., 1stLt, A 363
Chicago, Illinois
Ocasio, Ferminy, Pfc, M 363
Bronx, New York
Odell, Fred H. Jr., Pfc, L 362
Union, South Carolina

Ogden, Jesse A., 2dLt, C 916
Corpus Christi, Texas
Olson, Lyle L., 2dLt, L 363
Myrtle Point, Oregon
Osenbaugh, Walter M., Pvt, L 363
Folson, West Virginia
Owens, John W., T/Sgt, A 362
Woodward, Oklahoma
Palesky, Daniel, 1stLt, K 361
Scranton, Pennsylvania
Palmer, Crafton A., S/Sgt, A 361
Kansas City, Missouri
Parcells, Frank M. Jr., 2dLt, F 361
Los Angeles, California
Parinella, Joseph J., T/3, Med 363
Oakland, California
Paris, Francis R., Pfc, I 363
Lawrence, Massachusetts
Pastore, Leonard W., Pfc, E 362
Hoboken, New Jersey
Pate, Alvin J., Sgt, B 361
Marin City, California
Paz, John, Pfc, K 361
Los Angeles, California
Pelletier, Ulysse R., 2dLt, C 757 Tk Bn[4]
West Warwick, Rhode Island
Pendergrass, Earl, S/Sgt, I 361
Grapeland, Texas
Perles, Albert J., Pfc, A 755 Tk Bn[4]
Shamokin, Pennsylvania
Peter, George L., S/Sgt, C 361
Chicago, Illinois
Peterson, Arthur D., Pfc, C 361
Angleton, Texas
Peterson, Harry, Pfc, K 362
Albert Lea, Minnesota
Petroff, Stoian V., Pfc, Med 363
Champaign-Urbana, Illinois
Pickle, Coy R., Capt, Med 361
Ennis, Texas
Pigg, H. S., Pvt, I 362
Hilham, Tennessee
Pioch, Richard G., Pfc, F 361
Grosse Point Park, Michigan
Plata, Natalie, Sgt, M 361
Chicago, Illinois
Platt, Hancel D., Pfc, Med 363
Wauchula, Florida
Platt, Matthew J., T/5, 916
Algona, Iowa
Plummer, Dennis M., T/3, Med 363
Shady Valley, Tennessee
Poland, Lawrence H., Pvt, E 362
Indianapolis, Indiana
Pope, David E., 1stLt, A 361
Forrest City, Arkansas
Pope, Ernest R., T/Sgt, I 361
Hope Mills, North Carolina
Popoff, Max J., Pfc, G 361
Sheridan, California
Porter, Clifford, Pfc, L 361
Forrest Hill, California
Porter, Joe S., Pvt, B 362
Milfay, Oklahoma
Prall, Robert W., Cpl, I 362
Phoenix, Arizona
Prieto, Ines G., Pfc, Can 361
Poth, Texas
Prior, Earl B., Pfc, A 361
Mechanicsburg, Illinois

Pyle, John E. Jr., Pvt, L 362
Lebanon, Virginia
Pyle, Woodrow W., S/Sgt, F 362
Wasco, California
Quale, Clayton J., Pvt, K 363
Quinn, James F., Pfc, G 361
Girardville, Pennsylvania
Radicella, Anthony J., Pvt, L 361
Brooklyn, New York
Randall, James T., T/Sgt F 362
Wichita, Kansas
Rawls, Niles E., Pvt
Corvallis, Oregon
Ray, William G., T/Sgt, L 363
Detroit, Michigan
Reavis, Isham, 1stLt, G 362
St. Louis, Missouri
Reels, Haskel M., Pfc, Med 363
Brotherton, Tennessee
Regner, William J., Capt A 363
Portland, Oregon
Respass, Heber, Pfc, K 362
Plymouth, North Carolina
Ressiguie, Leo H., Sgt, K 361
Binghamton, New York
Restifo, Frank P., Pfc, L 361
Saginaw, Michigan
Reynolds, Fred R., T/4, Med 363
Jefferson, Texas
Rice, Perry H., S/Sgt, K 361
Marshall, North Carolina
Richard, Willie J., Pfc, Med 361
Thibodaux, Louisiana
Richardson, George C., LtCol Hq 362
Hilo, Hawaii
Richardson, William H., Pfc I 361
Trenton, Michigan
Riede, Edwin P., 1stLt, 91 Recon
Boise, Idaho
Riesgo, Joe B., Pfc, C 361
Corcoran, California
Ritter, Crum M., Pvt, E 363
Sheperdstown, West Virginia
Robbins, Robert A., Capt, E 361
Los Angeles, Calif.
Roberts, Roy R., S/Sgt, I 362
Dearborn, Michigan
Robertson, Oscar, Cpl, Med 361
Savannah, Tennessee
Robinson, Howard N., Pvt K 362
Brooklyn, New York
Rodgers, George T., Capt, A 757
Allentown, Pennsylvania
Romagnoli, Albert
Roquet, Corwin E., S/Sgt C 362
Anaheim, California
Rosellini, Bruno R., 1stLt, B 363
San Francisco, California
Rothracker, Robert E., Sgt, C 362
San Francisco, California
Rowell, Fred G., LtCol, 804 TD Bn[4]
Roswell, New Mexico
Rowzee, Oscar B., Jr., Pvt C 362
Baltimore, Maryland
Roy, Ferris G., Pfc, K 362
Hedley, Texas
Ruggles, Robert S., Pfc 3d Bn 362
Kalispell, Montana
Russell, Wade F., Sgt, K 362
Denton, North Carolina

---

[1]Names in *italic* indicate posthumous award.
[2]At time deed was done, unless otherwise noted.
[3]At time deed was done.
[4]Attached unit.

Samaniego, Augustine M., Pvt
G 361
Bronte, Texas
Sanchez, David L., Cpl, G 363
Austin, Texas
Sanders, Howard F., S/Sgt,
A 361
St. Bernard, Ohio
Sansom, William T., S/Sgt,
G 361
El Monte, California
Santos, Jacinto, Pfc, B 362
Yuma, Arizona
Sarle, John L., 1stLt, D 362
Midland, Michigan
Sawyer, Charlie R., S/Sgt
K 361
Roper, North Carolina
Saylor, Daniel D., Pfc, K 363
Bruset, Montana
Senn, Charles A., Jr., LtCol
1stBn 362
Laurens, South Carolina
Schreiner, Elmer, Sgt, I 362
Gibbon, Minnesota
Schuck, Joseph A., Pfc, A 361
Carmel, Pennsylvania
Schwartz, William H., T/5
Med 362
Bronx, New York
Scott, Claude J., Pfc, B 362
Seattle, Washington
Scott, Robert E., Jr., Capt
346 FA
Torrance, California
Scott, Russell E., Pvt, I 361
Vanceburg, Kentucky
Scudder, James H., 1stLt, K 361
Durham, N. H.
Sears, David M., S/Sgt, E 362
Tallahasse, Florida
Sears, John R., 1stLt, L 362
Durham, N. C.
Sheppard, William K., Sgt
K 361
Oakland, California
Sheridan, Thomas E., Pvt, C 361
Bronx, New York
Shipman, D. C., Cpl, 2d Bn 362
Grishom, Oregon
Shishkowsky, John J., Pvt
Med 362
Baker, Montana
Silva, Raymond A., Pfc, C 363
San Jose, California
Simmons, Chester B., Pvt, I 362
Waterloo, Iowa
Simon, Dalferes J., Pvt, A 362
Lafayette, Louisiana
Skelton, Oscar, Jr., Pfc, G 361
London, Kentucky
Skipper, Everette H., Sgt, I 362
Loris, South Carolina
Sladeski, Joseph C., Capt, D 361
Newburgh, New York
Smart, Relphard E., Pfc, B 363
Benton, Arkansas
Smith, Cecil L., Pvt, G 363
Rock Port, Missouri
Smith, Marcellus, Pfc, A 361
Dudley, North Carolina
Smith, Stanley L., 1stSgt, K 361
Los Angeles, California
Smith, Zeb H., S/Sgt, F 362
Canton, North Carolina
Smolin, Gustave, Pvt, I 362
Philadelphia, Pennsylvania

Snyder, Hope, S/Sgt, C 361
Portersville, Alabama
Solberg, Granville O., S/Sgt
C 361
Crookston, Minnesota
Solomon, Max R., Pvt, L 362
Hudson, Michigan
Sorensen, Vernon S., Pfc, E 361
Pocatello, Idaho
Sowards, Windel A., 2dLt, G 361
Columbus, Ohio
Spatz, Ralph R., Pvt, L 362
Shartlesville, Pennsylvania
Specht, Eugene B., Pvt, E 362
Webster, New York
Spencer, Donald F., Sgt, I 361
Yale, Michigan
Spencer, Elmer G., T/3, Med 362
Grace, Idaho
Spooner, William A., 1stLt, 916
Lodi, California
Sprik, Ysbrand, S/Sgt, E 363
Grand Rapids, Michigan
Stanchowicz, Stanley, T/5,
Med 361
Menasha, Wisconsin
Stanislav, Herman R., Pfc
Hq Co L Bn 361
Linwood, Nebraska
Starr, John C., Pvt, B 363
Erie, Pennsylvania
Steele, Charles R., Pvt, AT 362
Texarkana, Texas
Stein, Nick, Pfc, E 362
Laramie, Wyoming
Steinman, David C., T/5
Med 361
Gresham, Oregon
Steinberg, Julius D., Pvt, G 361
Macon, Georgia
Stewart, Thomas W., Sgt, L 363
Berea, Kentucky
Stockton, Lloyd D., Pfc, I 361
Gardena, California
Stolp, Merrill G., Pfc, K 361
Rochester, Minnesota
Stone, Thomas W., 2dLt, AT 362
Fort Pierce, Florida
Storosko, Theodore, Pvt, A 361
Nanticoke, Pennsylvania
Swanton, George E., Jr., Pfc
E 363
Springfield, Ohio
Sweeten, Kenneth W., S/Sgt
A 348
Mt. Juliet, Tennessee
Sympson, Decil E., T/Sgt, F 361
Woodland, California
Szaflarski, Stanley A., Pfc
F 361
Rib Lake, Wisconsin
Tanksley, Jeptha C., 1stLt
B 363
Mt. Airy, Georgia
Taylor, Jack L., 1stLt, G 361
Waelder, Texas
Taylor, Jesse L., Pfc, K 363
Roganville, Texas
Teter, Herman F., T/3, Med 362
Moberly, Missouri
Tharp, Charles D., 2dLt, A 363
Sacramento, California
Thompson, Archie M., Pfc
A 361
Seattle, Washington
Thompson, Donald S., T/5
Med 361
Corvallis, Oregon

Thompson, Theodore R., Sgt
K 363
Tomblin, Clifford C., Pfc, B 361
Dingess, West Virginia
Trusheim, James C., Pfc, L 362
Wellsville, Kansas
Turkowitch, Jacob J., Pvt,
G 362
Milwaukee, Wisc.
Turner, William H., Pvt
Med 363
Manson, Tennessee
Valine, Eugene L., Sgt, A 361
Sacramento, California
Van Matre, John R., T/Sgt
I 361
McAllen, Texas
Van Oss, Orval R., Sgt, B 363
Channing, Michigan
Van Scoyk, Reed A., S/Sgt
Med 361
Willamina, Oregon
Verkovich, Joe J., Sgt
A 804 TD Bn[4]
Fort Bragg, North Carolina
Vittolino, George, Pfc, G 362
Philadelphia, Pennsylvania
Voran, Bruce H., 1stLt
A 757 Tk Bn[4]
Kingman, Kansas
Waesche, Thomas L., Pvt, B 362
Pikesville, Maryland
Waggoner, Raymond L., Pfc
H 361
Roseburg, Oregon
Wahl, George H., Pvt, E 362
Winchester, Ohio
Waite, Samuel A., S/Sgt, I 362
El Paso, Texas
Walker, Walter C., S/Sgt, C 361
Wanilla, Mississippi
Wall, William A., Pfc, I 361
Lodgegrass, Montana
Walraven, Johnnie E., Sgt
Hq Co 361
Dallas, Texas
Walrond, Darrel E., S/Sgt
AT 362
Lafayette, Tennessee
Walsh, George D., Cpl
Hq Bat 916
Boston, Massachusetts
Walsh, Robert J., T/Sgt, E 362
Los Angeles, California
Watson, George B., Pfc, L 361
Alton, Illinois
Weitzel, William C., Pfc, K 362
Portland, Oregon
Wendt, Henry H. F., S/Sgt
K 362
Blair, Nebraska
Wessendorff, Joseph C., Capt
Hq Co 3d Bn 363
Richmond, Texas
Wheetley, Earl B., Pfc, G 362
Success, Arkansas
White, Cecil G., Jr., Capt
Med 362
Atlanta, Georgia
Whitman, Willie L., Pfc, K 361
Eclectic, Alabama
Whittenton, Ralph M., T/Sgt
K 361
Forrest City, Arkansas
Wiechel, Robert K., 2dLt, C 361
Port Clinton, Ohio
Wilkinson, Robert R., Pvt, F 362
Macy, Nebraska

---

[1]Names in *italic* indicate posthumous award.
[2]At time deed was done, unless otherwise noted.
[3]At time deed was done.
[4]Attached unit.

Williams, Tom E., S/Sgt, I 361
Clarksville, Arkansas
Williamson, Edgar S., 1stLt
C 804 TD Bn[4]
Little Rock, Arkansas
Williamson, Raymond E. S.
BrigGen, 91st Div
New York, New York
Willis, Albert J., Jr., Pfc, F 363
Cohoes Albany, New York
Willis, James H., 2dLt, G 362
Hammond, Indiana
Wilson, John L., Pvt, A 362
Navy Yard, South Carolina
Wilson, Leoance E., Pfc, I 363
Albany, New York
Wireman, Charles E., Pfc
L 362
Levee, Kentucky
Wisner, Albert F., T/4
C 757 Tk Bn[4]
Sacramento, California
Wondrak, Henry J., Pvt, A 362
Aurora, Illinois
Wood, Emmet L., Pvt, E 362
Dobson, North Carolina
Woods, R. N., LtCol, 1st Bn 363
Athens, Ohio
Wright, William V., 1stLt I 363
Murray, Utah
Wynn, Clair C., S/Sgt, C 362
Moccasin, California
Yepez, Gildo R., Pfc, I 363
Chicago, Illinois
Yoquelet, Ray A., 2dLt, C 361
Macon, Georgia
Young, Fred D., Pfc, E 363
Hibing, Minnesota
Zuzga, Walter L., Pvt, Med 361
Detroit, Michigan

LEGION OF MERIT

Alexander, Thomas L., Jr., Capt
Hq 347
Atlanta, Georgia
Ambo, Walter J., 1stSgt, H 361
Binghamton, New York
Anderson, Tyndall R., WOJG
Sv 361
Shickley, Nebraska
Antinone, Rinaldo J., 1stSgt
Hq 3d Bn 363
Philadelphia, Pennsylvania
Bair, Gordon L., S/Sgt
Hq 91st Div
Indianapolis, Indiana
Barry, Calvin E., LtCol, CO
346
Topeka, Kansas
Blass, Carel S., M/Sgt
Hq Div Art
Midian, Kansas
Broedlow, Rudolph W., Col
CO 361
Chippewa Falls, Wisconsin
Burtch, Anthel D., S/Sgt, A 361
Baraboon, Virginia
Caster, Orville O., M/Sgt
Sv 348
Central Point, Oregon
Coleman, Charlie C., Maj
Hq 3d Bn 362
McCormick, South Carolina
Collier, Robert B., LtCol,
CO 348
Houston, Texas
Cotton, John W., Col,
CO 362
Santa Rosa, California
Cowin, Earl A., Sgt, 2d Bn 361
Van Buren, Missouri
Donnovin, Joseph P., Col,
Hq 91st Div
Gainesville, Florida
Driessen, Harvey T., T/Sgt
Hq 347
Chilton, Wisconsin
Faries, Robert B., 1stSgt, B 362
Itasca, Texas
Ferguson, Raymond A., Sgt
1st Bn 363
Baker, Montana
Fletcher, George E., LtCol
Hq 91st Div
Newark, Delaware
Gele, Louis N., Jr., 1stLt, H 363
New Orleans, Louisiana
Gibson, Gerald F., T/Sgt
Hq Det 316 Med
Dubuque, Iowa
Griffith, Richard, Cpl, S Bat 916
Newhall, California
Grover, Lionel F., T/4
Hq 91st Div
Los Angeles, California
Hospital, Ralph, BrigGen,
CG Div Arty
Ithaca, New York
Hawkinson, David F., Maj
Hq 361
Highmore, South Dakota
Heinze, Leon W., T/4
Hq Bat 916
Portage, Wisconsin
Holley, William C., LtCol
CO 316
Klamath Falls, Oregon
Jackson, Paul E., T/Sgt
Hq Bat 916
Seattle, Washington
Jernigan, Curtis D., Pfc
Hq Co 3d Bn 363
Gainesville, Florida
Ketchum, Harry T., LtCol,
173 FA
Abbotsford, Wisconsin
Kirschenman, Aaron, S/Sgt
I 361
Bakersfield, California
Kleinert, Harry R., Sgt
A 316 Eng
Virgilla, California
Land, Ernest H., Capt, Can 363
Sylacauga, Alabama
Lawson, Willie R., 1stSgt A 346
Goldwaithe, Texas
Loesch, Edward, T/Sgt, S 363
Goodrich, New Jersey
Lynch, David, 1stSgt, 91 Recon
El Paso, Texas
Lynn, Woodrow L., LtCol
CO 347
Tampa, Florida
Magill, W. Fulton, Jr., Col
CO 363
Portland, Oregon
Millard, Duane F., Sgt, B 363
Jamesville, Wisconsin
Montgomery, Marvin D., T/Sgt
Hq 361
Winfield, Missouri
McCleskey, Earnest C., S/Sgt
B 316
Wheeler, Texas
McHaney, Flake, Maj, Hq 347
White Oak, Missouri
Neeley, William G., LtCol
Hq 91st Div
Rock Hill, South Carolina
Newman, Campbell W., LtCol
Hq 91st Div
Dallas, Texas
Oldfield, Ados M., S/Sgt
Hq Co 361
Portersville, California
Oshlo, Richard C., Lt Col, 361
Council Bluffs, Iowa
Patsko, John, 1stSgt, C 346
West Homestead, Pennsylvania
Patterson, James H., Maj
Hq 3d Bn 361
Kearny, New Jersey
Pinnick, Floyd V., Maj, 363
Ulysses, Kansas
Pittman, Beuford A., LtCol
CO 91 QM
Alexandria, Louisiana
Reid, George, 1stSgt, E 363
Flint, Michigan
Reibe, Francis A., S/Sgt, Hq 363
Kalispell, Montana
Stapleton, Herbert T., T/Sgt
8 Ord Bn[4]
Overland, Michigan
Sanden, James, Maj, Hq 348 FA
Ault, Colorado
Sargent, Wilburn F., S/Sgt
Hq 362
Dallas, Texas
Scarborough, Richard, LtCol
Hq 91st Div
Macon, Georgia
Shaw, James E., Jr., Lt Col
916 FA
Granada, Mississippi
Smith, Stanley L., 1stSgt, K 361
Los Angeles, California
Thomson, Edwin K., LtCol
Hq 2d Bn
Laramie, Wyoming
Warren, Victor C., LtCol
Co 91 Sig
Rochelle, New York
Weckstein, Leon, S/Sgt
Hq Co 1st Bn 363
Los Angeles, California
White, Vander, M/Sgt, S 347
Cheraw, South Carolina
Wilder, Alvin D., Jr., Capt, 316
Oakland, California
Wood, Hollis O., 1stSgt, H 362
Woods, Ralph N., LtCol
1st Bn 363
Athens, Ohio

---

[1]Names in *italic* indicate posthumous award.
[2]At time deed was done, unless otherwise noted.
[3]At time deed was done.
[4]Attached unit.

DISTINGUISHED UNIT CITATIONS

WAR DEPARTMENT
Washington 25, D.C., 19 July 1945

GENERAL ORDERS
NO. 58

BATTLE HONORS.—As authorized by Executive Order 9396 (sec. I, WD Bul. 22, 1943), superseding Executive Order 9075 (sec. III, WD Bul. 11, 1942), citations of the following units in the general orders indicated are confirmed under the provisions of section IV, WD Circular 333, 1943, in the name of the President of the United States as public evidence of deserved honor and distinction:

* * *

The *3d Battalion, 361st Infantry Regiment,* is cited for outstanding performance of duty in action, from 7 to 14 October 1944, near Livergnano, Italy. Committed to attack along Highway 65 in the drive beyond the enemy Gothic Line, the *3d Battalion* in 7 days of continuous fighting over rough mountainous terrain, ideally suited for defense, decisively defeated elements of three German divisions and captured the town of Livergnano, the key position in the enemy's prepared line of defense. Unable to use supporting armor because of the terrain and enemy demolitions on the highway, the *3d Battalion* successfully repelled several strong enemy counterattacks, inflicting heavy losses on the enemy, and advanced continuously through the heaviest type of enemy mortar and artillery fire. In a skillful maneuver, one company knifed 1,700 yards through German lines, cutting the enemy's main line of communication and forcing the withdrawal of 300 enemy and six tanks, which had been counterattacking other elements of this battalion. This small force of only 80 men gallantly repelled fanatical, tank-supported enemy counterattacks for 8 hours, even after every machine gun had been destroyed by the terrific enemy fire and ammunition had been all but exhausted. Members of another company, attempted to take Livergnano, fought into the outskirts of the town and resisted enemy counterattacks until the buildings they were defending crumbled as a result of terrific incessant fire from enemy self-propelled guns, tanks, and mortars. Advancing under direct enemy observation against numerically superior enemy forces over rough mountainous terrain, which made it necessary to hand-carry supplies forward and made evacuation of the wounded a difficult and exhausting task, the men of the *3d Battalion* overcame every natural obstacle and each fiercely defended enemy strongpoint. Thus, the courageous infantrymen of the *3d Battalion* succeeded in seizing Livergnano, the keypoint in the enemy's "Caesar Line." By spearheading the Fifth Army's drive, the battalion penetrated the German line at one of its strongest points, thereby rendering the entire line untenable. The indomitable courage and fighting spirit displayed by the men of the *3d Battalion, 361st Infantry Regiment*, in the face of great odds and extreme personal danger, are a credit and inspiration to the armed forces of the United States. (General Orders 70, Headquarters Fifth Army, 10 June 1945, as approved by Commanding General, United States Army Forces, Mediterranean Theater of Operations.)

* * *

BY ORDER OF THE SECRETARY OF WAR:

G. C. MARSHALL
Chief of Staff

OFFICIAL:

EDWARD F. WITSELL
Major General
Acting The Adjutant General

WAR DEPARTMENT
Washington 25, D.C., 20 October 1945

GENERAL ORDERS
NO. 90

BATTLE HONORS.—As authorized by Executive Order 9396 (sec. I, WD Bul. 22, 1943), superseding Executive Order 9075 (sec. III, WD Bul. 11, 1942), citations of the following units in the general orders indicated are confirmed under the provisions of section IV, WD Circular 333, 1943, in the name of the President of the United States as public evidence of deserved honor and distinction. The citations read as follows:

* * *

The *2d Battalion, 361st Infantry Regiment*, is cited for outstanding performance of duty in action from 16 to 18 April 1945, in the vicinity of Mount Adone, Italy. During the 3-day battle to drive the Germans from their heavily fortified positions forming the defense line before Bologna, the *2d Battalion* decisively defeated an advantageously entrenched foe and successfully accomplished its mission. At first all attacks were frustrated by intense enemy resistance as the *2d Battalion* sought to advance through numerous mine fields and over confining routes which led to the objective. The enemy, enjoying clear and full observation, opened automatic, rifle, and mortar fire upon the troops attempting to advance through the mountainous terrain. The assaults were repulsed, and the *2d Battalion*, entrenched under fire at the foot of their objective, evacuated casualties, and hand-carried supplies over the difficult terrain. The next day, having reorganized, the *2d Battalion* attacked frontally and from the flanks, but so heavy was the German resistance that all assaults were stopped and severe casualties suffered. Attacking again and again, the Americans sought to ascend the mountain and reach the enemy, but the Germans repulsed each effort, using their well dug-in positions to the fullest advantage. On the third day, a coordinated assault again was launched, and the *2d Battalion* moved forward, determined to wrest its objective from the foe. Onward and upward, the men advanced into the face of enemy machine-gun fire, skillfully maneuvering through rugged terrain and mine fields, and closed with the Germans. As a result of the ensuing victory, the defense belt before Northern Italy was broken. The intrepid courage and aggressive determination of the men of the *2d Battalion, 361st Infantry Regiment,* in the face of great odds, are a notable tribute and inspiration to the armed forces of the United States. (General Orders 95, Headquarters Fifth Army, 4 August 1945, as approved by the Commanding General, United States Army Forces, European Theater.)

* * *

BY ORDER OF THE SECRETARY OF WAR:

G. C. MARSHALL
Chief of Staff

OFFICIAL:

EDWARD F. WITSELL
Major General
Acting The Adjutant General

WAR DEPARTMENT
Washington 25, D.C., 22 December 1945

GENERAL ORDERS
NO. 123

BATTLE HONORS.—As authorized by Executive Order 9396 (sec. I, WD Bul. 22, 1943), superseding Executive Order 9075 (sec. III, WD Bul. 11, 1942), citations of the following units in the general orders indicated are confirmed under the provisions of section IV, WD Circular 333, 1943, in the name of the President of the United States as public evidence of deserved honor and distinction. The citations read as follows:

* * *

The *3d Battalion, 363d Infantry Regiment,* is cited for outstanding performance of duty in action from 12 to 19 September 1944, near Monticelli, Italy. Attacking the strongpoints of Il Giogo Pass in order to pierce the formidable Gothic Line, the *3d Battalion, 363d Infantry Regiment,* successfully flanked and occupied Pgio Di Castro, capturing a large number of enemy outposts. Resisting several intensive counterattacks, the battalion effected plans to attack M. Altuzzo and Monticelli on the following day. Struggling up the steep slope of M. Altuzzo, one company stubbornly advanced against heavy enemy mortar, artillery, and rifle fire, and through minefields, pillboxes, and wire entanglements to the crest of the hill. An intensive concentration of mortar and artillery fire forced the company to withdraw to cover on the reverse slope. Meanwhile, the remaining two companies, engaged in a coordinated attack on Monticelli, met bitter resistance for 2 days. On 17 September, the *3d Battalion* again was ordered to attack the formidable defenses of Monticelli and to take it at any cost. One platoon, grimly pushing forward through the hail of fire, managed to reach the summit with 11 men remaining. The battalion commander and the artillery liaison officer, realizing the precarious situation, gathered 30 men together and stormed the mountain to augment the small force tenaciously holding the summit. The Germans, retreating down the reverse slope, were caught in a terrific artillery and mortar barrage. Monticelli, the dominating peak in this key defense point, had fallen before the courageous and determined men of the *3d Battalion.* The aggressiveness and intrepid spirit displayed by the men of the *3d Battalion* against overwhelming odds forced the initial break-through of the Gothic Line and led to a general enemy withdrawal. Their performance is a notable tribute to the Army of the United States. (General Orders 89, Headquarters Fifth Army, 10 July, 1945, as approved by the Commanding General, United States Army Forces, Mediterranean Theater.)

* * *

BY ORDER OF THE SECRETARY OF WAR:

DWIGHT D. EISENHOWER
Chief of Staff

OFFICIAL:

EDWARD F. WITSELL
Major General
Acting The Adjutant General

DIVISION CITATIONS

HEADQUARTERS 91ST INFANTRY DIVISION
UNITED STATES ARMY
APO 91

2 July 1945

GENERAL ORDERS
NO. 84

COMMENDATION OF UNIT.—Under the provisions of Army Regulations 600-55, the following named organization is commended:

Company B, 361st Infantry Regiment, 91st Infantry Division, is hereby commended for outstanding accomplishment in combat. During the period 11 October 1944 to 14 October 1944, near Livergnano, Italy, *Company B, 361st Infantry,* was given the mission of securing from the determined enemy the steep, bare hill that was the key terrain feature in the enemy's prepared line of defense. The enemy positions, deeply entrenched and blasted into solid rock, withstood days of continuous bombing and heavy concentrations of artillery and mortar fire. On 11 October, *Company B* attacked the heavily defended hill. Repulsed after several attempts to advance by withering enemy mortar and machine-gun fire, the Company withdrew to positions at the base of the hill. Early in the morning of 12 October, a second attack was launched. Advancing slowly in the face of intense mortar and machine-gun fire, the men of *Company B,* by skillful and daring maneuvering, succeeded in assaulting and capturing the main enemy cave positions, taking nineteen prisoners and three machine guns. That night, supported by machine guns, rifles, grenades and flame throwers, the enemy counterattacked in force. Although some of the enemy reached positions on the cliff directly above and threw grenades on them, *Company B* gallantly withstood the attack, driving off the enemy and inflicting heavy casualties with accurate concentrated fire. On the morning of 13 October, after a night of ceaseless enemy mortar and machine-gun fire, the Company attacked again. After a fierce fire fight, the men advanced in the face of severe enemy machine-gun cross-fire, courageously assaulting and reducing each enemy strongpoint, driving the routed enemy from the crest of the hill. In the engagement, thirty enemy were killed or wounded, fifty captured, and six machine guns taken. With the loss of this key position, the enemy was forced to abandon the entire line of prepared defenses hinged on Livergnano. The aggressive fighting spirit and unwavering devotion to duty displayed by members of *Company B* reflect the finest traditions of the armed forces of the United States.

BY COMMAND OF MAJOR GENERAL LIVESAY:

RALPH N. WOODS
Lt. Col., Inf.
Actg. Chief of Staff

OFFICIAL:

FAUSTIN F. JEHLE
Major, A.G.D.
Actg. Adj. Gen.

HEADQUARTERS 91ST INFANTRY DIVISION
UNITED STATES ARMY
APO 91

2 July 1945

GENERAL ORDERS
NO. 83

COMMENDATION OF UNIT.—Under the provisions of Army Regulations 600-55, the following named organization is commended:

Company I, 362d Infantry Regiment, 91st Infantry Division, is hereby commended for outstanding accomplishment in combat. On 19 September 1944, near Poggio, Italy, *Company I, 362d Infantry,* was given the mission of taking the town of Poggio, situated on the crest of a knoll, guarding the approaches to Mt. Gazzaro. The objective, Mount Gazzaro, was guarded on three sides by steep impassable mountains, making it necessary to occupy Poggio in order to secure it, the highest ground in the sector. The enemy was well entrenched behind barbed wire entanglements, supported by machine-gun cross-fire covering every approach to the town, and by mines tactically laid to prevent any infiltration of their defenses. Intense artillery and air bombardments could not penetrate the individual positions that the enemy had selected. It was evident to the men of the company that in order to occupy the town they would have to engage the enemy in hand-to-hand combat. The first attack was launched but resulted in failure although many individual acts of heroism were displayed by the men of *Company I.* In the action many casualties had been inflicted on the fanatic enemy and the company suffered the loss of several officers and men. New plans for a second assault were made immediately. Minute details were arranged with the artillery for the company to attack under cover of a rolling barrage. During the second attack, as the barrage moved forward, the infantry followed closely and was able to over-run completely the enemy's defensive positions. Poggio was entered and fierce hand-to-hand fighting took place within the town. From emplacements on the reverse slope of the hill, machine guns harassed the troops. These emplacements were so well entrenched that they had to be blown open by bazooka fire. Although the company suffered heavy casualties, Poggio was captured. Sixty prisoners were taken along with a large amount of material. This action, and the display of determination and courage by *Company I,* made possible the advance of the entire 362d Infantry which would otherwise have been impossible. The action reflects the highest traditions of the Division and the United States Army.

BY COMMAND OF MAJOR GENERAL LIVESAY:

RALPH N. WOODS
Lt. Col., Inf.
Actg. Chief of Staff

OFFICIAL:

FAUSTIN F. JEHLE
Major, A.G.D.
Actg. Adj. Gen.

DIVISION HEADQUARTERS, THE STAFF, SPECIAL UNITS, AND DIVISION TROOPS

This work is essentially a record of the engagements of the infantry regiments for they were the units in close contact with the enemy. It is justly and properly so, however, it must be remembered that the Division as a whole was a well knit team with all elements functioning together in harmony and that the Division as a whole was many times greater in combat efficiency than the sum of its subordinate units. The accomplishments of the close-combat elements were made possible by good staff work and the efficiency of the Special Units and Division Troops.

The Division with its normal battle attachments numbered some 20,000 individuals that had to be trained, equipped, clothed and fed under all kinds of conditions. Most of the requirements for living that are found in a small city were carried on within the Division but under infinitely more difficult circumstances for the Division was never static like a city, everything had to be constantly mobile. While in training, in movement by rail, water, motor or marching, during and after battle in all kinds of weather and terrain throughout every day of the year the staff, administrative and supply elements of the Division were continuously caring for the needs of its personnel so that they might live and fight. How well these elements accomplished their mission is a matter of official reports that show good spirit and high morale, good discipline, high sanitary standards and excellent care of all equipment. To those elements that made these things possible this additional chapter is dedicated. It will not describe the operations of the Division Headquarters staff and special troops in detail for this would be a volume within itself. Rather it merely highlights the personalities whose leadership and efficient work made the Division a cohesive combat unit. If this chapter were not included in the Division History, there would be a failure to give due credit to the elements of the Division that helped materially to make the 91st Infantry Division known as one of the smoothest operating and toughest fighting units the United States Army has ever produced.

The Division was fortunate in having few changes of command. Maj. Gen. William G. Livesay was assigned command on July 14, 1943 and continued actively in command with only short periods of absence as interim commander of the II Corps and the Fifth Army until the Division was inactivated.

General Livesay brought to the Division a broad and sympathetic understanding of human nature, a kindly spirit, great energy and an almost inspired tactical knowledge. Subordinate only to the tactical situation was his consideration for the welfare of his men and officers. He spared himself in no way, but he knew how to delegate authority and to back those to whom he delegated it. Habitually present wherever the going was rough, his quiet stability and his kindly counsel comforted and supported his subordinates. He operated on the theory that his commanders were the best in the world and this spirit and knowledge made his subordinate commanders produce results beyond all normal expectations. General Livesay rarely sent for subordinate commanders, but he and his staff went to them instead. Slow, deliberate and positive in censure, he was quick in praise. Throughout the training period there had been a quiet sifting of the unqualified thus making it unnecessary to relieve commanders

General Livesay and the Division General Staff. Left to right: Lt.Colonel Neely, Lt.Colonel Newman, General Williamson, Lt.Colonel Foster, General Livesay, Colonel Fletcher, Colonel Donnovin

in combat. It is remarkable to note that except for replacements made necessary by wounds and sickness, all but two key officers who accompanied the Division overseas remained with it throughout its many days of battle. No finer tribute to the quality of a commander could be written.

Demands of other organizations made necessary frequent changes in office of Assistant Division Commander during its initial training phase. Brig. Gen. Percy W. Clarkson, Brig. Gen. Charles L. Bolté and Brig. Gen. William Crist served in turn for short periods and each contributed his share to the training of the Division. The Division was fortunate in its final and permanent Assistant Division Commander in that Brig. Gen. Raymond E. S. Williamson joined in January 1943 from long and distinguished service in the Pacific Theater and remained with the Division until its inactivation.

The Division Commander used General Williamson as an assistant in the truest sense of the word. His principal use was as a direct contact with the infantry regiments, assisting, advising, counseling wherever his services were most needed. He supervised or commanded units widely separated from the bulk of the Division. This role was well illustrated by the famous "lateral pass" of Task Force Williamson which captured the long-beleaguered city of Leghorn, and then immediately wheeled north to surprise the Germans and invest the south half of the city of Pisa. It was again marked in the Po Valley where a succession of rivers presented formidable obstacles. Here the Assistant Division Commander initially accompanied the unit assigned the most difficult task in forcing river crossings and he remained at the site to insure that the crossings were made expeditiously. The welcome with which General Williamson was received by the unit commanders in their area was proof of the value of his wise counsel and assistance.

Colonel Fenton S. Jacobs, the original Chief of Staff left the Division in July 1943 for duty in England, where he was promoted to the grade of Brigadier General. Upon the departure of Colonel Jacobs, Colonel Joseph P. Donnovin, Executive of the Division Artillery, was designated Chief of Staff. He ably filled that position until 30 June 1945 when he returned to the United States on emergency leave due to illness in his family. Because of administrative rulings, his assignment to the division was terminated and he was replaced by Lieutenant Colonel Ralph N. Woods, a battalion commander of the 363d Infantry.

Colonel Donnovin will be long remembered for his meticulous attention to detail and his acceptance of nothing short of the best that could be achieved under the circumstances. Upon his shoulders were placed the multitudinous details of operation thereby leaving the Division Commander free for the longer aspects of command. Colonel Donnovin by his splendid leadership and devotion to duty ably organized the staff into a cheerful and efficient team that surmounted all difficulties of terrain and weather in the forward progress of the Division. Endowed with mature judgment and a remarkable capacity for sustained work, he effectively directed the efforts of the less experienced members of the staff along the road to efficient operation.

There was a succession of G-1s in the first year of the Division's history. The position was permanently filled when Lt. Col. William G. Neely took over the duties upon the hospitalization of Lt. Col. J. G. Faes. Lt. Colonel Neely functioned with great enthusiasm constantly keeping in mind the needs of the troops. He was responsible to a great extent for the supply of filler replacements and the organization and operation of rest centers and recreation facilities for the personnel of the Division. He will be best remembered for his cheerful smile and helpful attitude towards all ranks.

Upon activation of the Division the G-2 section was organized by Lt. Col. O. M. Barton. When duty called him elsewhere, Lt. Col. E. O. Foster was designated G-2, and remained in this capacity throughout the war. It was his responsibility to accumulate information of the terrain and enemy forces and the collation of this information into the knowledge of the enemy's dispositions, capabilities and intentions. One important contribution of the G-2 section was a scale relief map of the Mt. Adone area which permitted detailed study of the terrain by the staff and commanders.

Lt. Col. Harry M. Bartlett was G-3 of the Division until he was transferred elsewhere in February 1943. Upon his departure, Col. John W. Cotton took over the duties of G-3 until his assignment as commanding officer of the 362d Infantry in August 1943, whereupon Lt. Col. George E. Fletcher was transferred from the 70th Division to become G-3. He continued in this capacity until June 1945 when he joined the G-3 Section of II Corps. In the meantime a change in tables of organization had permitted his well deserved promotion to colonel.

Under Colonel Fletcher's guidance, the G-3 section worked continuously keeping up to date information of the location and progress of combat, formulating the details of plans of operation and passing to the subordinate units the will of the Division Commander in the form of orders, messages, and other directives. Colonel Fletcher's tactical judgment and direct method of operation contributed much to the success of the Division.

The G-4 Section of the Division Staff was organized by Lt. Col. Don Riley who was succeeded early in the training phase by Lt. Col. Campbell W. Newman. Colonel Newman, his able assistant, Major Mose Wander, and his entire force were indefatigable in their constant attention to the supply needs of the Division. He operated on the principle that there are twenty-four hours in each day and to the best of his ability, he used each one in the interest of the officers and men of the Division. There was no detail too small for his personal attention. From socks to parkas, from K rations to prime beef and from .30 caliber ammunition to the largest caliber, whatever the troops needed, he would find some way of getting it no matter what difficulties lay in the way. During the long cold winter in the Apennines when extra socks were needed, he procured in one month 123,000 pairs. When it became apparent that even this supply was too small, considering the long time it took for laundry to be returned to the front, he installed a sock laundry capable of turning out five thousand pairs of clean dry socks daily. The washing machines, placed well forward under cover of a hill, were improvised from oil drums with wringers made by the Ordnance Company. Italian *paesanos* supplied the labor. This is only one of the many examples of ingenuity and untiring effort expended by the G-4 section in the interests of the Division.

To give full credit by name to each officer and enlisted man of the staff deserving of recognition would take far more space than can be allocated in this history. Each section performed their duty in an exceptionally efficient manner always maintaining the relations of a happy family and functioning without friction or hestitation. Suffice to record that the superior condition of the records of the Division was a result of the high quality of the work performed by the Adjutant General's Section headed by Lt. Col. C. F. Primm. Much assistance was given in this work by Lt. Col. Donald L. Roberts and his Inspector General's Section. Lt. Col. Samuel J. Taggart, the Finance Officer and his section, could always find a way to pay the men even under almost impossible situations. His work will long be remembered by all the personnel of the Division. The spiritual affairs of the Division were well looked after by Chaplain Vernon P. Jaeger and his corps of fifteen chaplains, while Lt. Col. K. V. Larsen, the Division Surgeon, and his assistants kept a close watch on the physical welfare of the personnel of the Division. The desires of the men for post-exchange supplies and athletic facilities, when they could be used, were well met by the Division Special Service Section under Major Robert Greenberger.

The volume of mail that passed through the division post office had to be viewed before it could be realized just how big this job was. This section, under Capt. Robert Hunter, always managed some way to get the mail to its destination.

Happily, the Chemical Warfare Section, headed by Lt. Colonel Boyles was not used for chemical war except for smoke to cover tactical operations and occasional use of the mortar companies for white phosphorus shelling. This left Lt. Colonel Boyles free to accomplish many and varied tasks not specifically assigned to other sections.

The Judge Advocate's Section under Lt. Col. Richard F. Scarborough contributed much to the well-being of the Division. Lt. Colonel Scarborough brought to his work a kindly nature, great dignity, superior judgment, kindness and a knowledge of the law. All court-martial cases were handled with dispatch but with keen understanding of human nature to which can be attributed in a great part the extremely low delinquency rate of the Division. The policy of the Division Commander was that there should be no delay when court-martial was necessary, and the Division maintained an average of fourteen days from the commission of an offense to the forwarding of the completed record of general court-martial cases; an average of nine days in special courts-martial cases and an average of three days in summary court-martial cases. The Division Commander took a personal interest in all courts-martial cases and it is doubted that any officer or man in the Division ever felt that he was unjustly or unfairly treated.

Shortly before the Division went overseas, a change in tables of organization brought all the Division company-size special troops under one commander. Major Paul R. Jeffrey was promoted to lieutenant colonel and given this assignment. He functioned as administrative commander of Headquarters Company, the Signal Company, the Ordnance Company and the Quartermaster Company. In addition to his duties as Commanding Officer, Special Troops, Lt. Colonel Jeffries was also Headquarters Commandant. He brought to both of his busy jobs a great enthusiasm and a vital personality.

The Division Signal Company under the operational supervision of the Division Signal Officer, Lt. Col. Victor Warren, maintained communications despite the most adverse conditions of weather and destruction of lines by enemy shelling and our own combat vehicles. Division Headquarters was never out of communication with the units of the Division for more than a few minutes at any time. One could not fully appreciate this effort until he saw linemen working waist-deep in snow repairing lines under shell fire. The company was thoroughly imbued with the spirit that "communications must remain open."

The officers and men of the Division saw little of the Reconnaissance Troop as a whole, but they were constantly conscious of its presence. Under the able command of Capt. Clifford E. Lippincott the Reconnaissance Troop was constantly engaged in obtaining enemy information for the Division by manning observation posts, reconnoitering for roads and approaches, and enemy locations by combat patrols into enemy territory. Many times it operated by platoon protecting open flanks or gaps between units, and in this role it saw its full share of close combat. The troop developed in a special pride that made its members unafraid to tackle any mission.

The work of the Ordnance Company was not spectacular but the freedom of movement of the Division was entirely dependent upon the work of the motor maintenance section. Weapons of all caliber had to be repaired and replaced and they all consumed tremendous quantities of ammunition. How well the company operated, under the guidance of the Division Ordnance Officer, initially Lt. Colonel Van Rue and later Lt. Colonel John Ford, is demonstrated by the fact that the Division was never short ammunition or ordnance equipment and the motors were kept running under the most trying conditions.

The Quartermaster Company under the operational control of the Division Quartermaster, Lt. Col. Beuford A. Pittman, operated around the clock procuring and distributing clothing, rations and other quartermaster supplies. From the supply point to the Division distributing point was often many miles and much credit must be given the men who drove their trucks over tortuous roads through the dark nights to deliver the needed supplies. The Quartermaster Company well knew the maxim that "An army moves on its stomach."

The Headquarters Company, under the command of Capt. Vincent A. Conk, supplied the enlisted personnel for the Division Headquarters. Much of the credit for the efficient operation of the staff is due to these men. They had been carefully selected during the training periods and their smooth and efficient functioning throughout made possible the superior efficiency of the staff.

Despite the crowded conditions and the limited road net at the front, the Division Commander kept his command post well forward. This resulted in shorter lines of communication and closer and more efficient contact by the Division staff. The normal procedure was to leapfrog the elements of the command post, closing at the old location and opening at a new location simultaneously. Under the Headquarters Commandant, reconnaissance for command posts was constantly pushed to the front and predicted as far in advance as was possible. Prior to the move of the command post, a "jump team" of the Signal Company would establish communications, an advance element of the command post would move out and establish themselves. The balance of the forward command post would close to the front as soon as it was established that communications and control had been taken over by the advance party. Except during rest periods, the Division operated a forward and a rear command post. In the forward command post were included the General Staff and such elements of the command post as were needed at the front. Somewhere in the rear, at a place best suited for their functioning, the Adjutant General and the rear command post was maintained. The rear command post moved less frequently than did the forward command post to the end that its work could be less interrupted. The elements which normally functioned in the forward command post were the Division Commander, Assistant Division Commander, Chief of Staff, the General Staff, the Inspector General, the Chemical Officer, the Chaplain, the Surgeon, Headquarters Commandant, Signal, Engineer and a small increment of the Adjutant General's Section. Those which functioned in the rear command post were normally the Adjutant General (less the forward increment) the Judge Advocate, the Finance Officer and the Personnel Sections of the regiments and separate battalions during the combat phases. Distances between the forward command post and the rear command post varied with conditions. Normally the rear command post was set up about twenty miles back. However, in the wild race across the Po Valley before VE-day, there was one occasion when the forward command post got over 150 miles ahead of the rear command post before the latter could be moved.

The 91st Infantry Division enjoyed an enviable record of high morale and low disciplinary rate. The Division also occupied a high place in the regard of the Army Commander for its efficiency in combat as evidenced by the fact that the Division spearheaded the operations of the Fifth Army continually. This enviable condition was brought about by many factors, among which were the superb leadership of the Division and of all its subordinate commanders. Ranking high among the factors that contributed to the outstanding success was the superb functioning of the elements of the staff. The staff as a whole provided one of the outstanding examples of the war of smooth, integrated, coordinated functioning. Each element knew its job and was keen to prosecute it. There was no clashing of gears. The staff worked like a well oiled machine, so smoothly performing its function that it in itself was not in evidence.

In the early stages of its training, Division Artillery was exceptionally fortunate in having the services of Brig. Gen. Edward S. Ott, whose splendid leadership and profound knowledge of field artillery training made it possible for Division Artillery to pass its ground force tests the first time and laid the foundation for the magnificent support by Division Artillery and its attachments in the later actions of the Division. His departure shortly before completion of the training phase to become Artillery Officer of XV Corps was deeply regretted by the Division. General Ott was ably succeeded by Brig. Gen. Ralph Hospital, who continued in command of Division Artillery until its inactivation. Under his able leadership Division Artillery maintained its high standard and built an enviable combat record that was unsurpassed.

Col. Joseph P. Donnovin was the original Executive and commanded Division Artillery for two months during the absence of Brigadier General Ott. When Colonel Donnovin was appointed Chief of Staff, he was succeeded as Artillery Executive by Col. Frank Dodd, who in turn was succeeded by Col. John Reynolds. Colonel Reynolds was evacuated from Italy by reason of illness and was in turn succeeded by Col. Harry T. Ketcham, who remained with Division Artillery until its inactivation.

Division Artillery alone, not counting the battalions of Corps artillery which were attached for the major engagements, delivered an average of 170 tons per day of death-dealing projectiles to clear the way for the infantry elements of the Division during the three months preceding the Futa Pass operation. Its fire power reached a crescendo on the 17th, 18th and 19th of September. On these three successive days, over 300 tons per day of high-explosive and time-fuzed projectiles crashed into the German positions denying them food and replacements and leaving them shocked and shattered for the infantry's *coup de grâce.* The confidence expressed by the infantry in its own artillery and the close and affectionate relations of the combat teams with their associated artillery battalions is the highest compliment that could possibly be paid to the standards and ideals instilled in Division Artillery.

Col. Harry T. Ketcham

Lt. Col. William C. Holley took command of the 316th Engineer Battalion and became Division Engineer early in the training period of the Division. This young officer brought to his engineers a boundless enthusiasm and a "can do" spirit that won the respect of each member of the Division. In the mountainous terrain of Italy, the roads were the key to each operation and the Germans well knew how to use mines and demolitions. On one occasion, the road north of Vaglia had been blown from the side of the mountain into the ravine below in eleven places in one kilometer. It was nothing short of miraculous how by-passes were carved from the mountains and how few times the advance of the Division was held up by the German demolitions. In the Po Valley there were many formidable rivers to cross and the German demolitions and the Allied air forces left little behind in the way of passages. Bridges and ferries appeared overnight under the Division Engineers. It was normal for the 316th Engineers to be working in front of the infantry and under the intense fire of the enemy, removing mines and opening roads. A near impossible task was a challenge that the 316th Engineer Battalion always met.

There were occasions when even armored bulldozers found it impossible to work, but the unarmored members of the 316th Engineer Battalion managed miraculously to get the job done. Reconnaissance for river crossings was carried on under the machine guns of the enemy and there were numerous occasions where a lone swimmer from the 316th Engineer Battalion, crawling on his stomach in the bitter cold, brought back invaluable information as to bridgehead sites. The work of the Engineer Battalion under the leadership of Lt. Colonel Holley deserves a history of its own.

The history of the Medical Corps in this war is one of outstanding accomplishment and the work of the 316th Medical Battalion, under Lt. Col. Paul W. Breecher, was no exception to this meritorious service. The men of the Division knew that should they become a casualty they would be quickly and efficiently cared for. Not counting the great numbers that were treated and returned to their units, there were 6,659 battle casualties and 15,790 non-battle casualties that passed through the clearing station during the period of combat.

Lt. Col. K. V. Larson, the Division Surgeon, assisted by Major A. L. Kauffman, Division Psychiatrist, and the personnel of the Medical Battalion maintained a rehabilitation center well forward for combat exhaustion cases. It is a great tribute to the Medical Section that sixty per cent of the battle-exhaustion cases were returned to their units in fit condition to resume their combat duties.

Perhaps the most widely known unit in the Division was the Military Police Platoon commanded by the Provost Marshal, Major Alvin Laird. Contrary to the popular concept of police, the main function of the 91st Military Police Platoon was to be helpful to all members of the Division. The highest compliment that can possibly be paid them was the fact that the officers and men of the Division liked them, respected them and had confidence in them. Day and night, in all types of weather, terrain and battle conditions, they were present to point the way, direct traffic and to assist the elements of the Division.

They performed their duties thoroughly, but always with consideration, kindness and good judgment.

SOURCE MATERIALS USED

I. *Narrative Histories*

First World War History
1942-43 History
History, 20 January to 1 June 1944
Monthly History, 1 June through July 1944
History of 316th Engineer Battalion
Fifth Army History, Volumes I through III
Attached Long Histories

II. *G-2 Records*

91st Division G-2 Periodic Reports, July 1944 through April 1945
91st Division G-2 Journal, 1 April to 16 July 1945
91st Division G-2 Intelligence Summaries, September 1944 through June 1945
91st Division Interrogation of Prisoners of War Reports
Prisoner of War Tabulation
Prisoner of War Breakdown File
Miscellaneous Interrogation of Prisoners of War (34th and 88th Divisions)
Shell Reports, II Corps and 91st Division, August 1944 to April 1945
Report on Patrols in 91st Division, 1 March 1945
G-2 Intelligence Annexes (Summaries), July 1944 to May 1945
Miscellaneous G-2 File, May 1945
91st Division Artillery S-2 Reports
91st Division Artillery S-2 Catalog of Enemy Installations
Miscellaneous Artillery S-2 File
Arno River Reconnaissance File

III. *Other Units' G-2 Reports*

II Corps G-2 Reports, 30 July 1944 to May 1945
Fifth Army G-2 Reports, July 1944 to July 1945
Miscellaneous G-2 File
88th Division G-2 Reports
10th Mountain Division G-2 Reports

Allied Force Headquarters Intelligence Summaries, 1945 (Complete)
Allied Force Headquarters Intelligence Notes, Nos. 85 to 106
II Corps Order of Battle Overlays, September 1944 to May 1945
Fifth Army Order of Battle Overlays, September 1944 to May 1945
Office of Strategic Services Reports
10th Indian Division Situation Report File
II Corps Artillery Intelligence Reports
II Corps Artillery Information File
Miscellaneous Intelligence Summaries

IV. G-2 Miscellaneous File

91st Division Photo Interpretation File
Fifth Army PIC File
Futa Pass Defenses (PI)
Technical Intelligence Bulletins (Incomplete)
Technical Intelligence Summaries (Incomplete)
Material on Germans File
Material on Jugoslavs File
Gothic Line File
Po Valley Defenses
Austria Pamphlet (Eighth Army)
Maps, Overlays, and Minefields Files
PWB Interrogations
Tactical Reconnaissance File
G-2 Miscellaneous File
Captured Documents File
Preliminary Terrain Study, Northeast Italy

V. 91st Division G-2 Journal and Field Orders

91st Division G-3 Journal File, July 1944 to July 1945
91st Division G-3 Journal Message File, April to July 1945
91st Division Field Orders, Nos. 1, 12, 32-52
91st Division General Orders, August 1944 to July 1945

VI. 91st Division Periodics and Other Unit File

91st Division Periodic Files, July 1944 to April 1945
II Corps and Fifth Army Periodics, February to June 1945
XIII Corps Miscellanous Files, May, June, July, 1945
XIII Corps Situation Reports, May to August 1945
XIII Corps Orders File, May to June 1945
XIII Corps G-3 Information Log, May to August 1945
II Corps Operations Instructions File

VII. Shipment and Miscellaneous Files

G-3 Shipment File
G-3 Planning File
G-3 TWX File
G-3 Information (Infantry)
G-3 Overlays and Sketches
G-3 Charts and Graphs

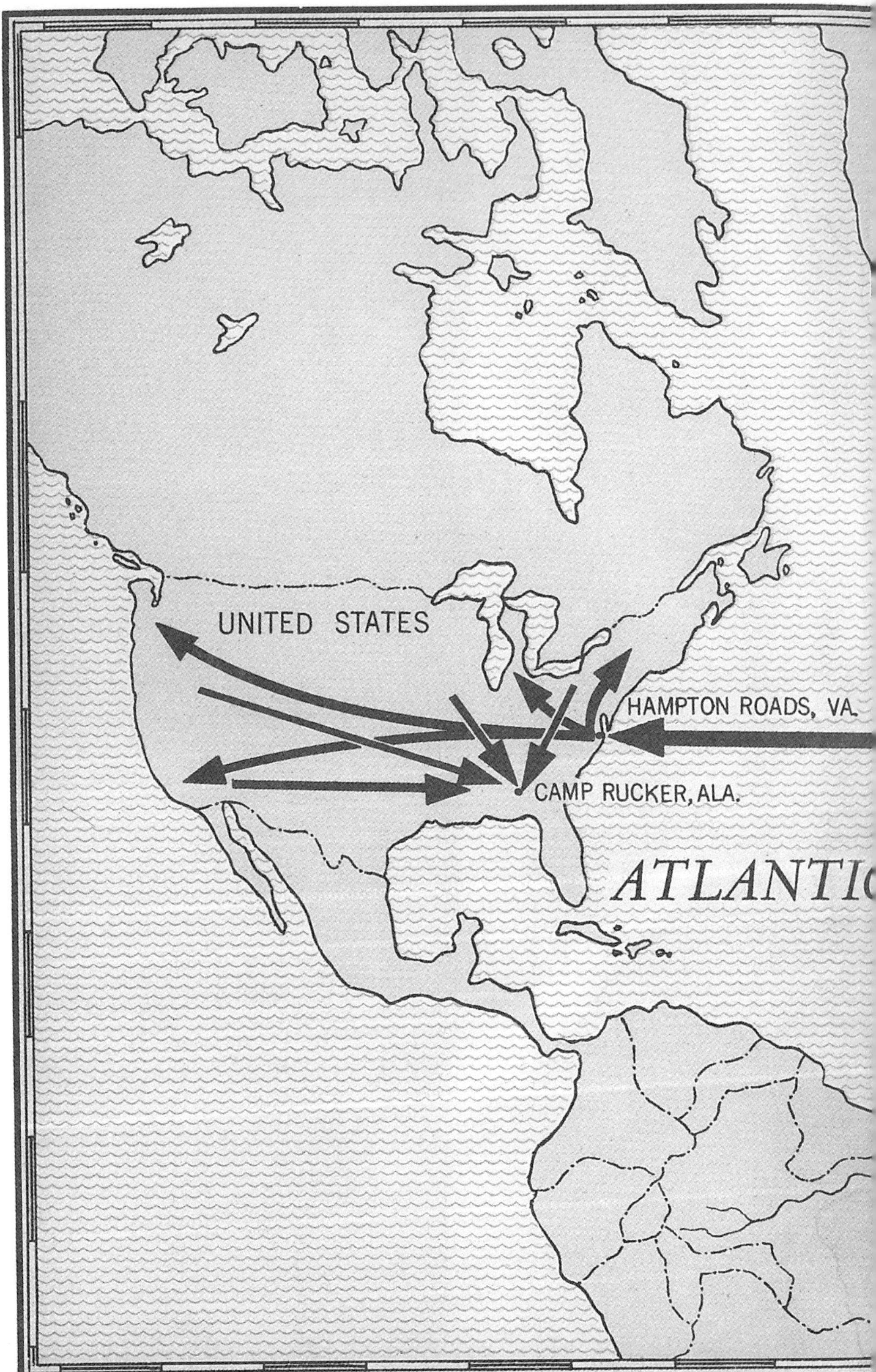

UNITED STATES
HAMPTON ROADS, VA.
CAMP RUCKER, ALA.
ATLANTIC